PERGAMON INTERNATIONAL LIBRARY
of Science, Technology, Engineering and Social Studies

*The 1000-volume original paperback library in aid of education,
industrial training and the enjoyment of leisure*

Publisher: Robert Maxwell, M.C.

Plate Tectonics
&
Crustal Evolution

Pergamon Titles of Related Interest

Anderson THE STRUCTURE OF WESTERN EUROPE
Anderson/Owen THE STRUCTURE OF THE BRITISH ISLES
Bowen QUATERNARY GEOLOGY
Lowe THE LATEGLACIAL ENVIRONMENT OF NORTH-WEST EUROPE
Lunar & Planetary Institute BASALTIC VOLCANISM ON THE TERRESTRIAL
  PLANETS
Roberts GEOTECHNOLOGY
Roberts INTRODUCTION TO GEOLOGICAL STRUCTURES

Related Journals*

COMPUTERS & GEOSCIENCES
CONTINENTAL SHELF-RESEARCH
GEOTHERMICS
INTERNATIONAL JOURNAL OF ROCK MECHANICS AND MINING SCIENCES
JOURNAL OF STRUCTURAL GEOLOGY
OCEAN ENGINEERING

*Free specimen copies available upon request.

Plate Tectonics & Crustal Evolution

Second Edition

Kent C. Condie

New Mexico Institute of Mining and Technology
Socorro, New Mexico

Pergamon Press

New York Toronto Oxford Sydney Paris Frankfurt

Pergamon Press Offices:

U.S.A.	Pergamon Press Inc., Maxwell House, Fairview Park, Elmsford, New York 10523, U.S.A.
U.K.	Pergamon Press Ltd., Headington Hill Hall, Oxford OX3 0BW, England
CANADA	Pergamon Press Canada Ltd., Suite 104, 150 Consumers Road, Willowdale, Ontario M2J 1P9, Canada
AUSTRALIA	Pergamon Press (Aust.) Pty. Ltd., P.O. Box 544, Potts Point, NSW 2011, Australia
FRANCE	Pergamon Press SARL, 24 rue des Ecoles, 75240 Paris, Cedex 05, France
FEDERAL REPUBLIC OF GERMANY	Pergamon Press GmbH, Hammerweg 6 6242 Kronberg/Taunus, Federal Republic of Germany

Copyright © 1982 Pergamon Press Inc.

Second printing, 1983.

Library of Congress Cataloging in Publication Data

Condie, Kent C.
　　Plate tectonics and crustal evolution.

　　Bibliography: p.
　　Includes index.
　　1. Earth--Crust. 2. Plate tectonics. I. Title.
QE511.C66　1982　　　551.1'3　　　81-10710
ISBN 0-08-028076-5　　　　　　AACR2
ISBN 0-08-028075-7 (pbk.)

Printed in the United States of America

*To Carolyn,
Tamara, Linda, and
Nathan*

Contents

Plate I. Tectonic Map of the World (in pocket)

Preface

This book has grown out of a course I teach with the same title. The rapid accumulation of data related to sea-floor spreading and plate tectonics in the last decade has necessitated continued updating of the course. Although new data are still coming in, the rate of increase of fundamentally important data has decreased somewhat and the time seems right for a textbook to accompany the course. The book is written for an advanced undergraduate or graduate-level student. It assumes a basic knowledge of geology, chemistry, and physics that most students in the earth sciences acquire during their undergraduate education. It also may serve as a reference book for various specialists in the geological sciences. I have attempted to synthesize and digest data from the fields of oceanography, geophysics, geology, and geochemistry and to present this information in a systematic manner addressing problems related to the evolution of the Earth's crust over the last 3.8 b.y. The role of plate tectonics in the geologic past is examined in light of existing geologic evidence, and examples of plate reconstructions are discussed.

Since the first edition of this book was published, a great wealth of information related to plate tectonics has appeared in scientific journals. The second edition is updated in light of these new contributions. Included are examples from the results of seismic profiling in both continental and oceanic areas, which have enhanced our understanding of crustal structure, especially along continental margins. Another new development is the application of ultra-high-pressure experimental studies to possible mineral assemblages in the Earth's mantle. Geochemical and isotopic tracer studies have also been important in defining mantle heterogeneities and in studying mantle and crustal evolution. The book also includes new information on arc systems, continental rifts, and cratonic basins, as well as a new section on mineral and energy deposits and their relation to plate-tectonic settings. Chapter 9 contains more examples of Phanerozoic orogenic systems, and a new chapter has been added to accommodate information on the rapidly growing field of Precambrian geologic history. Sections have been added on comparative planetary evolution and paleoclimates, and on the origin and evolution of life. The origin and development of the atmosphere and oceans is also considered in greater detail. Also, the tectonic map of the world (Plate I) has been completely redone, and considerably greater detail has been added.

In order to keep the book to a reasonable length and avoid duplicating extensively information that is widely available in other books, some subjects are covered in only a cursory manner and others not at all. For instance, the methods by which geologic, geochemical, and geophysical data are gathered are only briefly mentioned, as books on these subjects are already available. Extensive mathematical treatments are omitted for the same reason. Because the book is designed primarily as a textbook, references are kept to a minimum. I have attempted, however, to reference the major papers and some of the minor ones that have strongly influenced me in regard to many of the interpretations set forth in the text. More extensive bibliographies can be found in these papers and in the references listed under "Suggestions for Further Reading" at the end of each chapter.

It is not possible to acknowledge everyone who has influenced my opinions in writing the book. The following individuals, however, gave freely of their time to review and criticize portions of the text and I gratefully acknowledge their help in arriving at a final version: A. T. Anderson, Richard L. Armstrong, Gale K. Billings, James E. Case, Peter J. Coney, S. R. Hart, Peter W. Lipman, John R. MacMillan, Paul Mueller, Michael A. Payne, Marshall Reiter, John W. Schlue, Denis M. Shaw, and S. R. Taylor. It should be pointed out that the interpretations I present in the book do not necessarily reflect those of any of the above. I am also grateful to those publishers and authors who have allowed reproduction of many of the figures used in the book. Dennis Umshler, Charles O'Melveny, Michael Graham, George Ross, and Susan Williams are acknowledged for their assistance in compiling data and references and Carolyn Condie for her help in editing. I am also grateful to the many secretaries who typed and retyped chapters, beginning with my almost illegible handwriting and evolving to the finished product.

Chapter 1

Introduction

A PERSPECTIVE

The origin and evolution of the Earth's crust is a tantalizing question that has stimulated much speculation and debate dating from the early part of the 19th century. Some of the first problems recognized—such as how and when did the oceanic and continental crust form—remain a matter of considerable controversy even today. Results from the lunar landings and satellite data collected from other planets indicate that the Earth's crust may be a unique feature of bodies in the solar system. The rapid accumulation of data in the fields of geophysics, geochemistry, and geology in the last 25 years has added much to our understanding of the physical and chemical nature of the Earth's crust and of the processes by which it has evolved. Most evidence favors a source for the materials composing the crust from within the earth. Partial melting of the Earth's interior appears to have produced magmas that moved to the surface and produced the first crust. The continental crust, being less dense than the underlying mantle, has risen isostatically and has been subjected to weathering and erosion. Eroded materials appear to have been added partially to continental margins, causing the continents to grow laterally, and then partially returned to the mantle (the region between the crust and the Earth's core), to be recycled and perhaps again to become part of the crust at a later time. Specific processes by which the crust was created and has subsequently grown are not well known, but the large amount of data currently available allows some important boundary conditions to be invoked. In this book, important physical and chemical properties of the crust and upper mantle are presented and discussed in terms of models for crustal origin and evolution.

The theories of sea-floor spreading and plate tectonics that have so profoundly influenced geologic thinking in the last decade have also provided valuable insight into the mechanisms by which the crust has evolved. One of the major problems regarding crustal evolution is that of discovering when in geologic time plate-tectonic and sea-floor spreading processes began. Some scientists consider the widespread acceptance of sea-floor spreading and continental drift as a "revolution" in the earth sciences (Wilson, 1968). Scientific disciplines appear to evolve from a stage primarily of data gathering, characterized by transient hypotheses, to a stage where some new unifying theory or theories are proposed that explain a great deal of the accumulated data. Physics and chemistry underwent such revolutions around the beginning of the 20th century, whereas the earth sciences may be just entering such a revolution. As with scientific revolutions in other fields, new ideas and interpretations in the earth sciences do not invalidate earlier observations. On the contrary, the theories of sea-floor spreading and plate tectonics offer for the first time a unified explanation for heretofore seemingly unrelated observations in the fields of geology, paleontology, and geophysics.

At the outset, it is necessary to briefly discuss sea-floor spreading and plate tectonics and to introduce a few terms that will be used throughout the book. Most geophysical data suggest that the surface of the earth is composed of rigid plates 50–150 km thick, known collectively as the *lithosphere*. These plates rest on a hotter, more dense layer that deforms plastically and is known as the *asthenosphere* (fig.

1

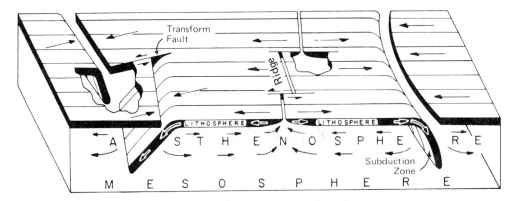

1.1. Schematic three-dimensional diagram showing the major features of sea-floor spreading
(after Isacks *et al.*, 1968).

1.1). The upper part of the lithosphere (6–40 km deep) is composed of the *crust*. An important part of the sea-floor spreading theory is that new lithosphere and crust are continually being created at oceanic ridges by injection and eruption of magma derived from the mantle. This lithosphere spreads laterally away from ridges and is finally consumed by the asthenosphere in subduction zones. The lithosphere can be considered, to a first approximation, as a mosaic of plates bounded by oceanic ridges, subduction zones, and transform faults (boundaries along which plates slide by each other) (fig. 1.1). Intracontinental compressive zones (a less common type of plate boundary) are not depicted in fig. 1.1. The study of the interactions of lithospheric plates is known as plate tectonics.

THE APPROACH

The general format of the book will evolve from one primarily of presentation of data critical to models for the origin and evolution of the crust to one of interpretation and speculation. Chapter 2 is concerned chiefly with the gross physical and chemical features of the Earth-moon system; it serves as a basic framework for later discussions. Hypotheses regarding the origin of the Earth and moon are also briefly reviewed in this chapter.

An account of the structure and composition of the mantle and core is given in Chapter 3. This provides important information bearing both on the source of the crust and on sea-floor spreading, which appears to result from dynamic processes in the mantle. The seismic, heat flow, gravity, magnetic, and electrical properties of the crust, together with a discussion of crustal composition, are presented in Chapter 4. In Chapter 5 the chief methods of radiometric dating used to define crustal provinces are discussed, as are the types of events that can be dated. A survey of both Phanerozoic and Precambrian crustal provinces is then presented, in which overall structure, rock distributions, and orogeny (mountain building) are considered.

In Chapter 6 the theory of sea-floor spreading is presented and the evidence that led to its formulation is reviewed. Lithospheric plates are described and hypotheses for the causes of sea-floor spreading are discussed. Magma associations on the Earth are described in Chapter 7 with reference to a plate-tectonic framework. The origin of magmas, which is of considerable importance in models for crustal origin and growth, is also considered in light of existing field, experimental, geochemical, and geophysical data.

Chapter 8 is concerned with the principles of plate tectonics and continental drift and presents a discussion of the methods available for reconstruction of plate positions in the geologic past. The relationship of orogeny to plate tectonics is also reviewed in this chapter. In Chapter 9 specific examples of plate reconstructions are given to illustrate methods set forth in Chapter 8. Specific Phanerozoic orogenic belts are also discussed.

Precambrian crustal evolution is discussed in Chapter 10 with emphasis on the possible role of plate tectonics. The origin, composition, and growth of the crust, together with a summary of the Earth's thermal history, are discussed in Chapter 11. Also included in this chapter is a discussion of comparative planetary evolution, secular compositional changes in the crust, and a brief summary of the origin and history of the atmosphere, oceans, and life.

Although methods and techniques of acquiring data will not be extensively discussed in the book, it is perhaps appropriate to briefly review some of the more important methods and at the same time introduce some basic terms.

METHODS AND DEFINITIONS

Seismic Methods

When an earthquake or an explosion occurs in the Earth, two types of elastic waves are produced—*body waves* and *surface waves*. Body waves travel through the Earth and are reflected and refracted at interfaces. They are of two types: *P waves* (or compressional waves), which are characterized by alternate compression and expansion in the direction of propagation, and *S waves* (or shear waves), with particle motion normal to the direction of propagation. P waves are always faster than S waves, and S waves cannot be transmitted through a liquid. Surface waves are propagated along or near the surface of the Earth and also are of two types: *Rayleigh* and *Love* waves. Rayleigh waves exhibit elliptical particle motion confined to a vertical plane containing the direction of propagation, while Love waves are characterized by horizontal motion normal to the propagation direction. The region in the Earth at which elastic waves are produced by an earthquake (or explosion) is defined as the *hypocenter* or *focus*, and the point on the Earth's surface vertically above as the *epicenter*. These various types of seismic wave motion are illustrated in fig. 1.2.

Elastic waves are detected by seismometers, which respond to ground movements. Computerized arrays of seismometer stations have recently made it easier to separate interfering signals, to improve signal-to-noise ratio, and to measure wave velocities directly.

Several seismic methods are used in investigating the interior of the Earth (Cleary and Anderssen, 1979). The gross features of the Earth's interior are determined from travel-time distance studies of body waves traveling through the Earth. Detailed structure of the crust and uppermost mantle is determined by seismic refraction and reflection methods. Large underground explosions are particularly useful in these studies because the time and location of such explosions can be known more accurately than are earthquake times and hypocenter locations. The refraction

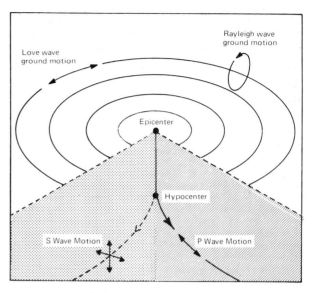

1.2. Basic types of body- and surface-wave motion relative to the hypocenter and epicenter of an earthquake (after Davies, 1968).

method, which is used both on land and at sea, is based on measuring the travel times of P waves between shot points and seismic recorders located various distances apart, usually along straight-line profiles. The method is limited, in that very detailed crustal structure cannot be determined. Evidence for low-velocity layers is obtained from modeling of surface-wave data and from amplitude studies of refraction data. The use of super-critical reflections (i.e., reflected waves that have incident angles greater than the critical angle) can enhance the interpretation of refraction data. Vertical incidence reflections occur only at sharp discontinuities and often allow the seismologist to distinguish between sharp and gradational discontinuities.

Travel-time anomaly studies have proved valuable in evaluating upper-mantle structure. A *travel-time anomaly* (or residual) is the difference between observed and calculated body wave arrival times at a given seismograph station. Calculated arrival times are azimuthally corrected and based on idealized models. Maps constructed by contouring travel-time anomalies are useful in relating such anomalies to geological and other geophysical features.

Earthquakes produce natural vibrations in the Earth known as *free oscillations*. Two types of oscillations occur: torsional oscillations, involving particle displacements normal to the Earth's radius, and spheroidal oscillations, which are radial or tangential displacements. Long-period free oscillations are detected with strain seismometers and Earth-tide gravimeters. Free-oscillation studies have resulted in improved resolution and detection of interfaces within the Earth, as well as determination of density and seismic anelasticity of parts of the mantle. Rayleigh- and Love-wave dispersion (i.e., the variation of velocity and wavelength) provides a basis for detailed studies of crustal and upper-mantle structure.

Free oscillations produced by major earthquakes do not last indefinitely; the vibrational energy is gradually converted to heat. The oscillations are said to be attenuated, and the process is known as *anelasticity*. Body waves passing through the Earth are also attenuated. Anelastic attenuation is measured with a unitless factor Q, the *specific attenuation factor*. Low values of Q mean high seismic-wave attenuation. Measured Q values in the Earth range from about 10 to greater than 1,000. Anelasticity in the Earth appears to result from some combination of grain boundary damping, stress-induced ordering of crystal defects, and damping caused by vibration of dislocations (Gordon and Nelson, 1966). Q decreases rapidly as temperature and degree of melting increase in the Earth.

Seismic reflection profiling is used extensively in oceanic areas and, more recently, in continental areas as well (Brewer and Oliver, 1980; Stommel and Graul, 1978). In oceanic areas, powerful acoustic sources on ships are used to produce energy to study the stratigraphy and structure of sediments around continental margins and in ocean basins. The major reflection profiling studies on the continents are those of the COCORP research group in the United States (Brewer and Oliver, 1980). Vibrating trucks transmit energy into the Earth, and return echos are recorded by arrays of geophones. Data are collected and computer processed so as to produce a section through the crust as a function of seismic-wave travel times.

Magnetic Methods

The Earth's magnetic field is defined by its strength and direction. The direction is expressed in terms of the horizontal angle between true north and magnetic north—the *declination*—and the angle of dip with the horizontal—the *inclination*. The inclination becomes vertical at the two magnetic poles. The total magnetic field strength is strongest near the magnetic poles (70 μtesla at the South Pole) and weakest at the equator (about 30 μtesla). Both short- and long-term variations occur in the direction and strength of the magnetic field. Short-term variations (with periods of hours to years) are related chiefly to interactions of the magnetic field with the strongly conducting upper

layers of the atmosphere. Variations with periods of hundreds of years or more are known as *secular variations* and are interpreted to support an origin for the magnetic field in terms of fluid motions in the outer part of the Earth's core. Approximately 90 percent of the present field can be described by a magnetic dipole at the Earth's center, which makes an angle of about 11.5 degrees with the rotational axis. A general westward drift of the field is noted at a rate of about 0.18 deg/yr.

Local and regional variations in the magnetic field reflect, for the most part, rocks beneath the surface with varying degrees of magnetization. Such variations are measured with fluxgate or proton magnetometers on land or sea or in the air. Significant deviations from a magnetic background either on a local or regional scale are known as *magnetic anomalies*, the intensities of which are expressed in gammas (γ) or teslas (T) ($1\gamma = 100\mu$T). Small-scale anomalies extending over thousands of square kilometers reflect variations in the lower crust or upper mantle.

Rocks may become magnetized in the Earth's magnetic field by several mechanisms, which are described in Chapter 6. Such magnetization is known as remanent magnetization and is measured in the laboratory with spinner, astatic, or cryogenic magnetometers. The maximum temperature at which a mineral can possess remanent magnetization is known as the *Curie point* temperature. *Paleomagnetism* is the study of remanent magnetism in rocks of various geologic ages. If rock samples can be accurately oriented and the date of magnetization determined, it is often possible to determine the locations of earlier magnetic pole positions (see Chapter 8). Paleomagnetic studies have shown that the magnetic poles have reversed themselves many times in the geologic past. Such *reversals* are thought to be produced by instability in the outer core.

Gravity Methods

Gravity is the force of attraction between the Earth and a body on or in the Earth divided by the mass of the body. The average gravitational force of the Earth is 980 gals (1 gal = 1 cm/sec^2). Gravity is measured with a gravimeter and can be determined both on land and at sea. Accuracies are typically of about 1 mgal on land and 5–10 mgal at sea. The standard reference for gravity on the Earth is the gravitational field of a spheroid, and is dependent only on latitude. The gravity field on the Earth can be described using data derived from the directions and rates of the orbital shift of artificial satellites. From such data, it is possible to determine how much the Earth's average surface, or *geoid*, which is roughly equal to sea level, actually deviates from a spheroid. Existing data indicate that the Earth is pear-shaped, with an average equitorial radius of 6378 km and an average polar radius of 6357 km. Gravity distribution on the Earth can be calculated from spherical harmonic coefficients of the satellite gravitational data; this is discussed further in Chapter 3.

Local and regional gravity data must be corrected for latitude and elevation before interpretation. On land, gravity measurements are usually above the geoid surface, and hence an increase in gravity must be added to the observed value to account for the difference in elevation. This is known as the *free-air correction*. If the standard gravity value of the spheroid is now subtracted (i.e., the latitude correction), the *free-air anomaly* remains. Next, if the attraction of the rock between the geoid and the gravity station is subtracted (the Bouguer correction) and a correction is made for nearby topographic variations, we obtain the *Bouguer anomaly*. Measurements at sea require no free-air correction, since they are made at sea level, and the Bouguer correction, where used, is added to account for the change in gravity that would result if the oceans were filled with rock instead of water.

Early gravity measurements by Bouguer in the mid-1700s indicated that large mountain ranges exhibit smaller-than-expected gravitational attractions. Such data led to the principle of *isostasy*, introduced about 1900 by Dutton. This principle suggests that an equilibrium condition exists in the Earth whereby the load pressure due to overlying columns of rock is equal at some *depth of compensation*. Two main theories have been proposed to explain isostasy. Pratt's theory assumes that the density of rock columns in the outer shell of the Earth varies laterally above a constant depth of compensation and is expressed as a function of elevation on the Earth's surface. Airy's theory proposes that the outer shell is composed of low, rather constant density columns and that the depth of compensation varies as a function of the thickness of the columns. Both mechanisms probably contribute to isostatic compensation. Models suggest compensation depths of the order of 50 to 100 km for both the Airy and Pratt theories. *Isostatic anomalies* may be calculated by subtracting from Bouguer anomalies the mass distribution within some segment of the upper part of the Earth as determined from some combination of the Airy and Pratt compensation mechanisms.

Electrical Methods

The Earth's magnetic field induces electrical currents, known as *telluric currents*, which flow in the crust and mantle. Most short-period variations in the magnetic field are produced by interactions with the strongly conducting ionosphere (upper atmosphere). A *magnetic storm* produces large magnetic variations lasting for a few days and is caused by strong currents of high-energy particles emitted by solar flares that are trapped in the ionosphere. Magnetic variations can be used to estimate conductivity in the Earth, since the strength of induced currents depends on electrical conductivity distribution. Short-period variations of such currents penetrate only to shallow depths, while longer periods penetrate to greater depths.

Four methods have been used to estimate conductivity distribution in the crust and mantle (Creer, 1980; Keller, 1971): (a) direct-current sounding; (b) magnetotelluric sounding; (c) electromagnetic sounding; and (d) geomagnetic deep-sounding. Direct-current sounding involves driving a current into the ground between widely spaced electrodes and measuring voltage drops between electrodes. The depth of penetration of this method is limited to only several tens of kilometers. In the magnetotelluric method, both electric and magnetic variations in the Earth's field are measured simultaneously. An artificial electromagnetic field is generated, driven into the Earth, and measured in the electromagnetic method. The geomagnetic deep-sounding method involves measuring variations in naturally induced currents caused by magnetic storms. This currently provides the best method for estimating mantle conductivity distributions. Results of these methods may be presented as conductivities (Ω^{-1} m^{-1} $\equiv$ S/m) or resistivities (Ωm).

Geothermal Methods

Heat-flow measurements on the Earth involve two separate measurements, one of thermal gradient (dT/dx) and one of thermal conductivity (K). From these measurements, heat flow (q) is calculated as follows:

$$q = -K\frac{dT}{dx}. \qquad (1.1)$$

Heat flow may be expressed as μcal/cm^2 sec or as mW/m^2 where 1 μcal/cm^2 sec is defined as one heat flow unit (1 HFU) and 1 HFU = 0.0239 mW/m^2. Thermal gradient is measured with thermistors, which on land are attached to a cable and lowered down a borehole and at sea are attached to core barrels or mounted in a long, thin probe that is inserted into deep-sea sediments. In both cases time is allowed for thermal equilibration before measurements are taken. Thermal conductivity of water-saturated rocks is usually measured with a divided-bar apparatus in which a known heat flow is passed through a sandwich of copper discs, two standards, and a rock sample; thermal conductivity is calculated from the temperature difference across the sample and its thickness. The thermal conductivity of unconsolidated sediments is usually measured with a needle probe, which consists of a thermistor, an electrical heating element, and a hypodermic needle inserted into the sediment. Thermal conductivity is obtained from the rate at which the needle temperature rises for a given energy input to the heater.

In continental areas, significant ground water movement can produce anomalously low heat flow. Also, measured heat flow in areas that were covered by Pleistocene glaciers may be lower than actual heat flow. Although glacial corrections up to 30 percent have been proposed by some investigators, evidence is conflicting regarding the general importance of this effect.

The radiogenic heat production of a rock or of a geologic terrane may be calculated from the concentrations of U, Th, and K and the heat productivities of ^{235}U, ^{238}U, ^{232}Th, and ^{40}K. The concentrations of these isotopes can be determined by counting the natural radioisotopes with a gamma-ray spectrometer in the laboratory or in the field. Airborne gamma-ray spectrometers have been used to estimate concentrations of U, Th, and K over large areas of the crust. Radiogenic heat generation (A) is expressed in 10^{-13} cal/cm^3 sec or as $\mu W/m^3$. One heat generation unit (1 HGU) is defined as 10^{-13} cal/cm^3 sec and is equivalent to 0.0239 $\mu W/m^3$.

High-Pressure Studies

For many years it has been possible to reconstruct in the laboratory static pressures up to about 300 kbar, which is equivalent to about 1000 km burial depth in the Earth. A new era of high-pressure research began in 1972 with the development of the double-stage split-sphere apparatus and the diamond-anvil pressure cell (Bell, 1979; Liu, 1979). With these systems it is possible to study phase relations at pressures up to 1.7 Mbar and temperatures up to 3500°C, which allows direct investigation of lower mantle and core compositions. High-pressure experiments can be per-

formed with solid or liquid media at a large range of temperatures. It is also possible to measure a considerable number of properties of rocks at high pressures and temperatures: phase equilibria boundaries, elastic properties including P- and S-wave velocities, electrical and thermal properties, and fracture and flow characteristics are but a few. From such measurements, in conjunction with available geophysical data, it is possible to place limitations on the composition, mineralogy, and melting behavior of the crust and upper mantle and to evaluate quantitatively the origin of magmas. Using high-pressure and high-temperature rock-deformation studies, it is possible to understand more fully earthquake mechanisms and flow characteristics within the Earth.

Possible mineral assemblages and compositions of deeper parts of the Earth, including the core, can also be studied using the results of shock-pressure experiments. The method involves generating a strong shock (up to several megabars) in a material with explosives, producing a wave front that moves through the material at a velocity greater than sound and greater than the particle velocity of the shocked material (Ahrens and Petersen, 1969). The pressure and density within the wave can be deduced by measuring the shock and particle velocities. Results are generally expressed in terms of the hydrodynamic sound velocity plotted against density. Various elements, minerals, and rocks are examined, and the results are compared with hydrodynamic velocity data deduced from body-wave studies of the Earth. Such comparisons provide major limitations on the composition of the mantle and core.

Geochemistry, Geochronology, and the Geologic Time Scale

Geochemical data from a variety of rocks and minerals provide important information bearing on the composition of the upper mantle and evolutionary changes in the crust and mantle. Geochemical and isotopic research have advanced rapidly in the last decade in response to the development of new analytical methods and geochemical modeling. Trace element geochemistry, in particular, has been useful in studying the evolution of the crust and mantle (Frey, 1979). Important advances in our understanding of planetary evolution have also come from geochemical and isotopic studies of lunar and meteorite samples and of ultramafic inclusions derived from the Earth's mantle. Isotopic studies are important not only in terms of *geochronology* but also for *tracer studies*. Pb,

Table 1.1 The Geologic Time Scale

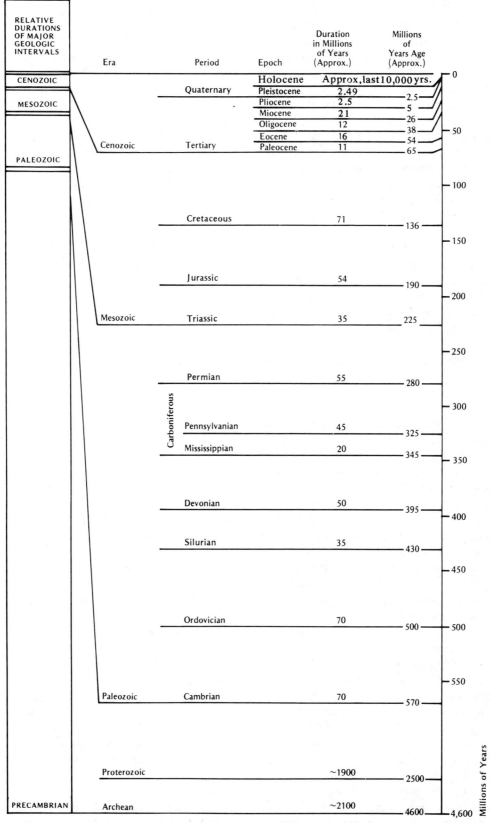

RELATIVE DURATIONS OF MAJOR GEOLOGIC INTERVALS	Era	Period	Epoch	Duration in Millions of Years (Approx.)	Millions of Years Age (Approx.)
CENOZOIC			Holocene	Approx. last 10,000 yrs.	0
		Quaternary	Pleistocene	2.49	2.5
MESOZOIC			Pliocene	2.5	5
			Miocene	21	26
			Oligocene	12	38
PALEOZOIC	Cenozoic	Tertiary	Eocene	16	54
			Paleocene	11	65
		Cretaceous		71	136
		Jurassic		54	190
	Mesozoic	Triassic		35	225
		Permian		55	280
		Pennsylvanian (Carboniferous)		45	325
		Mississippian (Carboniferous)		20	345
		Devonian		50	395
		Silurian		35	430
		Ordovician		70	500
	Paleozoic	Cambrian		70	570
	Proterozoic			~1900	2500
PRECAMBRIAN	Archean			~2100	4600

Millions of Years

Modified from Don L. Eicher, *Geologic Time*, 1968, p. 150 by permission of Prentice-Hall, Inc., Englewood Cliffs, N.J.

Nd, and Sr isotopes are especially important in both cases. Geochronology involves the study of time relationships in orogenic belts and in the evolution of continents and ocean basins. Tracer studies make use of daughter isotopes as "fingerprints" to study the origin of igneous rocks and to trace the evolution of the mantle and crust through geologic time (see Chapter 7).

Refinements in radiometric dating methods in recent years (see Chapter 5) have improved estimates of the beginning and duration of the various subdivisions of geologic time. A current version of the geologic time scale is shown in table 1.1. Geologic time is divided into eras, periods, and epochs. Five major eras are recognized; from youngest to oldest, they are the *Cenozoic, Mesozoic, Paleozoic, Proterozoic*, and *Archean*. The three younger eras are collectively known as the *Phanerozoic*, the two older ones as the *Precambrian*. The Precambrian comprises almost 90 percent of geologic time, and the oldest dated crust is about 3.8 b.y. in age.

Other Methods and Sources of Information

The *viscosity* of the mantle has been estimated by studies of isostatic recovery rates of large segments of the crust after removal of a surface load such as icecaps or large lakes, and from estimates of the seismic anelasticity Q (Anderson, 1966). The *mass* of the Earth can be estimated from surface gravity data after a rotational correction. The Earth's two principal *moments of inertia*—one about the polar axis and the other about an equatorial axis—can be estimated from rotational axis precessional data and the observed flattening of the Earth. Other physical properties as a function of depth within the Earth are estimated from measurements made on the Earth's surface and models of the Earth's interior.

Information from oceanic and continental drill cores allows a reliable projection of compositional data to shallow depths in the crust. The Deep Sea Drilling Project (DSDP), which began in 1968, has now recovered many cores from the sediment layer on the ocean floors, some up to several hundred meters in length. A specially designed drilling ship, the *Glomar Challenger*, is used as a floating drilling platform. A great deal of information pertaining to sediment ages and lithologies and to the history of the sea floor over the last 150 m.y. should become available from these and future cores as they are studied. Deep holes on the continents, other than oil wells, are rare. However, deep drilling into the conti-

nents in a variety of geologic environments is currently in the planning stages.

Last but not least are the conventional and well-established geological methods. Perhaps the most commonly overlooked yet extremely important source of data is *field geology*. The results of widespread geological mapping on the continents are of critical importance to the evaluation of the roles of sea-floor spreading and plate tectonics in the geologic past. *Stratigraphy, tectonics, volcanology, experimental petrology, sedimentation*, and *paleontology* are other important fields of investigation.

VARIATION OF PHYSICAL PROPERTIES IN THE EARTH

Internal Structure

The internal structure of the Earth is revealed primarily by body-wave studies and by free-oscillation data. The variation of P- and S-wave velocities and density in the Earth reflect changes in pressure, temperature, mineralogy, composition, and degree of partial melting. Although the gross features of seismic-wave velocity distributions have been known for some time, considerable refinement of data has been possible in the last ten years. A current model consistent with the Earth's mass, moment of inertia, periods of free oscillation, and travel times of body waves is given in fig. 1.3. Three major first-order discontinuities occur: the largest, at 2900 km, is the core-mantle interface; at 10–12 km beneath the oceans and 30–50 km beneath the continents is the *Mohorovičić discontinuity*, or *Moho*; and at about 5200 km is the inner-core–outer-core interface. The Moho separates the Earth's crust from the mantle. The core composes about 16 percent of the Earth by volume and 32 percent by mass. The discontinuities appear to reflect changes in composition or phase, or in both. Smaller velocity changes provide a basis for further subdivision of the mantle, as is discussed in Chapter 3.

The major regions of the Earth can be summarized as follows with reference to fig. 1.3.

1. The *crust* consists of the region above the Moho, and ranges in thickness from about 3 km in some oceanic regions to about 80 km in some continental areas.

2. The *upper mantle* extends from the Moho to 400 km, and includes the lower part of the lithosphere and the upper part of the asthenosphere. The

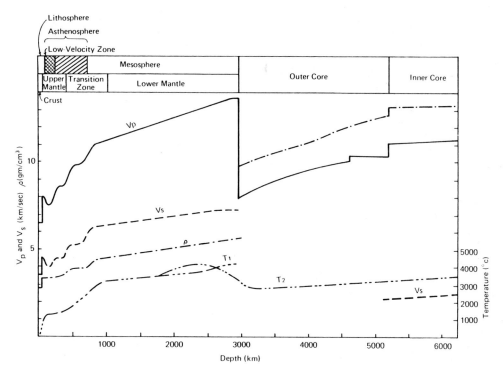

1.3. P-wave velocity (V_p), S-wave (V_s), density (ρ), and temperature (T_1, T_2) distribution in the Earth. (V_p, V_s, and ρ from Anderson *et al.*, 1971; T_1 from Tozer, 1959; and T_2 modified after Lubimova, 1969.)

lithosphere (50–150 km thick) is the strong outer layer of the Earth, including the crust, that reacts to stresses as a brittle solid. The *asthenosphere*, extending from the base of the lithosphere to about 700 km, is by comparison a weak layer that readily deforms by creep. A region of low seismic-wave velocities and high attenuation (low Q), the *low velocity zone* (*LVZ*), generally occurs at the top of the asthenosphere, and is from 50 to 100 km thick. Significant lateral variations in density and in seismic-wave velocities are common at depths of less than 400 km.

3. The *transition zone*, which extends from 400 to 1000 km, is characterized by several rather rapid increases in velocity that are considered in more detail in Chapter 3. Significant lateral variations in S-wave velocities appear to extend to about 1000 km.

4. The *lower mantle* extends from about 1000 km to the 2900 km discontinuity. For the most part, it is characterized by rather constant increases in velocity and density until just above the core-mantle interface, where a slight flattening of velocity and density gradients occurs. The lower part of the transition zone and the lower mantle are collectively referred to as the *mesosphere*, a region that may have strength and is relatively passive in terms of deformational processes.

5. The *outer core* will not transmit S waves and hence is interpreted to be liquid. It extends from 2900 km to the discontinuity at 5200 km.

6. The *inner core*, which extends from 5200 km to the center of the Earth, transmits S waves, although at very low velocities, suggesting that it is near the melting point or partly molten.

Gravity distribution can be calculated from the density profile and the pressure variation with depth can be determined from density and gravity distributions. Gravity changes very little to the core-mantle discontinuity and falls off rather rapidly thereafter. Pressure increases slowly to about 1.5 Mbar at this discontinuity and rapidly thereafter, to about 3 Mbar at the 5200 km discontinuity; the pressure in the inner core is about 3.5 Mbar.

Temperature Distribution

Considerable uncertainty exists regarding the temperature distribution in the Earth. It is dependent upon such features of the Earth's history as (a) the initial temperature distribution, (b) the amount of heat generated as a function of both depth and time, and (c) the process of core formation. Most estimates of the temperature distribution in the Earth are based on one or a combination of three approaches:

1. Models of the Earth's thermal history involving various mechanisms for core formation;
2. Variations of seismic-wave velocities and probable variations of electrical conductivity, thermal conductivity, and other physical and chemical properties with depth;
3. Models involving redistribution of radioactive heat sources in the Earth by melting and convection processes.

Current estimates using various models seem to converge on a temperature at the core-mantle interface of $3000 \pm 500°C$.

Two examples of calculated temperature distributions in the Earth are shown in fig. 1.3. Tozer (1959) calculated the temperature distribution in the mantle using electrical conductivity data. The results are shown as curve T_1 and indicate a rapid temperature increase in the lithosphere, a flattening of gradient in the asthenosphere, and then a moderate increase to the base of the transition zone. The lower mantle is characterized by a steady but gradual increase in temperature. Other models indicate that the process of core formation and fractionation of radioactive heat sources in the Earth by convection and melting are critical in estimating the Earth's temperature distribution. If the core formed during or soon after the accretion of the Earth, it should absorb gravitational energy; hence, a maximum in the temperature distribution should occur in the lower mantle, as shown by curve T_2. Movement of radioactive heat sources into the upper regions of the mantle early in the Earth's history would have the effect of producing a temperature maximum at even shallower depths in the mantle.

SUMMARY STATEMENTS

1. Three first-order seismic discontinuities occur in the Earth: the Moho (10–50 km), the core-mantle interface (2900 km), and the 5200 km discontinuity. These discontinuities define the crust, mantle, outer core, and inner core, in order of increasing depth.

2. In terms of strength and mode of deformation, the Earth is divided into the lithosphere (extending from the surface to 50–150 km depth), a strong outer layer that behaves as a brittle solid; the asthenosphere (extending from the base of the lithosphere to about 700 km), a region of low strength with a low-velocity zone in the upper part; and the mesosphere, a probably strong and homogeneous region extending to the base of the mantle.

3. A transition zone characterized by several rapid velocity changes occurs from 400 to 1000 km in depth and separates the upper from the lower mantle.

4. Temperature increases in the Earth most rapidly in the transition zone and reaches a possible maximum in the lowermost mantle. The probable temperature at the core-mantle interface is $3000 \pm 500°C$.

SUGGESTIONS FOR FURTHER READING

Cohee, G.V., Glaessner, M.F., and Hedberg, H.D., editors (1978) The Geologic Time Scale. *Am. Assoc. Petrol. Geologists, Studies in Geology* **6**, 388 pp.

Garland, G.D. (1979) *Introduction to Geophysics*, Second ed. Philadelphia: Saunders. 494 pp.

Mason, B., and Moore, C.B. (1981) *Principles of Geochemistry*, Fourth ed. New York: Wiley. 352 pp.

Pilant, W.L. (1979) *Elastic Waves in the Earth.* Amsterdam: Elsevier. 494 pp.

Volcanology, Geochemistry, and Petrology (1979) *Revs. Geophys. Space Phys.*, 17, **4**, 744–925.

Chapter 2

Origin of the Earth-Moon System

CHEMICAL COMPOSITION OF THE EARTH AND MOON

Since it is not possible to sample the interior of the Earth and moon directly, indirect methods must be used to estimate their composition. It is generally agreed that the Earth and other bodies in the solar system formed by condensation and accretion from a solar nebula, and that the composition of the sun roughly reflects the composition of this nebula. Nucleosynthesis models for the origin of the elements also provide limiting conditions on the composition of the planets. It is possible to enhance our knowledge of the composition of the Earth's crust and upper mantle with geochemical data obtained from crustal igneous rocks and from ultramafic rocks, some of which may represent samples of upper mantle material (see Chapter 3). Meteorite compositions and shock-wave experimental data, however, provide the most definitive information regarding the overall composition of the Earth.

Shock-wave data indicate a mean atomic weight for the Earth of about 27 (mantle = 22.4 and core = 47.0) and show that it is composed chiefly of iron, silicon, magnesium, and oxygen. As evidenced by the following mean atomic weights of meteorites, however, no single meteorite class has the appropriate composition to match the shock-wave data for the entire Earth: carbonaceous chondrites (H_2O-free), 23.4 to 24.0; ordinary chondrites, 24.4; high-iron chondrites, 25.1; enstatite chondrites, 25.6; and iron meteorites, 55 (Anderson et al., 1971). However, meteorite classes can be mixed in such a way as to give the correct core/mantle mass ratio (32/68) and mean atomic weight, three examples of which are

given in table 2.1. All three estimates indicate that iron and oxygen are the most abundant elements, followed by silicon and magnesium. Ninety percent or more of the Earth is composed of these four elements. Calcium, aluminum, nickel, sodium, and perhaps sulfur account for most of the remainder.

From lunar heat flow results, correlations among refractory elements, density, and moment of inertia, it is possible to estimate bulk lunar composition (table 2.1). Compared to the Earth, the moon is depleted in Fe, Ni, Na, and S, and enriched in other major elements. The bulk composition of the moon is commonly likened to the composition of the Earth's mantle because of similar densities. Data indicate, however, that the moon is enriched in refractory elements ($TiO_2 = 0.4\%$; $Al_2O_3 = 8\%$; $Ca = 6\%$) compared to the Earth's mantle (see table 3.2).

Trace-element concentrations in the Earth and moon are difficult to estimate and may depend on ad hoc assumptions. One approach involves estimates of upper and lower limits of potassium in the Earth (Hurley, 1968a, b). A lower limit is set by assuming that all of the Earth's ^{40}Ar is in the atmosphere and was produced from the radioactive decay of ^{40}K over geologic time, while an upper limit is set by establishing a lower limit for the fractional release of Ar from the Earth. By comparison with probable fractional release factors of H_2O, a lower limit for Ar release appears to be about 50 percent. The results suggest that the concentration of potassium in the Earth is about 200 ppm. By considering the lowest permissible K/Rb ratio for the Earth as deduced from meteorite data and the maximum permissible Rb/Sr ratio as determined from $^{87}Sr/^{86}Sr$ ratios of crustal and upper mantle rocks, Rb and Sr contents are then

Table 2.1. Estimates of the Bulk Composition of the Earth and Moon (in weight percent)

	Earth			Moon
	1	2	3	4
Fe	34.6	29.3	29.9	9.3
O	29.5	30.7	30.9	42.0
Si	15.2	14.7	17.4	19.6
Mg	12.7	15.8	15.9	18.7
Ca	1.1	1.5	1.9	4.3
Al	1.1	1.3	1.4	4.2
Ni	2.4	1.7	1.7	0.6
Na	0.6	0.3	0.9	0.07
S	1.9	4.7	–	0.3

1. 32.4% iron meteorite (with 5.3% FeS) and 67.6% oxide portion of bronzite chondrites (Mason, 1966).
2. 40% Type I carbonaceous chondrite, 50% ordinary chondrite, and 10% iron meteorite (containing 15% sulfur) (Murthy and Hall, 1970).
3. Nonvolatile portion of Type I carbonaceous chondrites with $FeO/FeO + MgO$ of 0.12 and sufficient SiO_2 reduced to Si to yield a metal/silicate ratio of 32/68 (Ringwood, 1966).
4. Based on Ca, Al, Ti = 5 × Type I carbonaceous chondrites; FeO = 12% to accommodate lunar density; and Si/Mg = chondritic ratio (after Taylor, 1975, 1979b).

calculated. From element ratios in meteorites and igneous rocks and from a knowledge of the general geochemical behavior of various trace elements, the concentrations of many trace elements in the Earth and moon can be estimated.

It is necessary to discuss volatile and refractory elements and to establish a basis for defining element enrichments and depletions in planetary bodies. A rather arbitrary division exists between refractory and volatile elements. In general, *volatile elements* are considered to be those elements that can be volatilized from silicate melts under moderately reducing conditions at temperatures of 1300–1500°C, while *refractory* (or *nonvolatile*) *elements* are not volatilized under the same conditions. Refractory elements can be further subdivided into *oxyphile* and *siderophile elements*, depending on whether they follow oxygen or iron, respectively, under moderately reducing conditions. Selected elements together with their probable chemical form under these conditions are given in table 2.2. The Earth and moon do not resemble either ordinary or Type I carbonaceous chondrites (C1) in terms of the trace elements presented in table 2.3. Compared to C1, the Earth is depleted in volatile elements and enriched in refractory elements. The moon is depleted in volatile elements and siderophile refractory elements and enriched in oxyphile refractory elements compared to both C1 and the Earth.

It appears that major vertical fractionation of many elements, and especially of those elements with large ionic radii, charge, or both (such as K, Rb, Cs, Th, U), has occurred in the Earth and moon. Such elements have been strongly enriched in the Earth's upper mantle and even more so in the crust in relation to the lower mantle and core. The crustal enrichment factors, expressed as percentages and calculated by dividing estimated element concentrations in the crust by estimated concentrations in the Earth, are shown in fig. 2.1. Those elements with

Table 2.2. Element Classification according to Relative Volatilities from Silicate Melts under High-Temperature Moderately Reducing Conditions.

Refractory Group		Volatile Group	
A. Oxyphile elements Mg, Al, Si, P, Ca, Sc, Ti, Sr, Y, Zr, Nb, Ba, Rare earths, Hf, Ta, Th, U		H, C, N	Probable volatile species H_2O, CO, N_2
		F, Cl, Br, I	Silicon and metal halides
		S, Se, Te, Na, K, Rb, Cs	Hydrides; elements
B. Siderophile elements		Zn, Cd, Hg, Tl, Pb, As, Sb, Bi, Se, Te	Elements
Fe, Co, Ni, V, Cr, Mn, Ca, Ag, Au, Mo, W, Rh, Pd, Re, Ir, Pt		Ga, Ge, Sn, In	Suboxides; sulfides

After Ringwood (1966).

Table 2.3. Trace-element abundances in Meteorites, Earth, and Moon

Element	Ordinary Chondrites	Type I Carbonaceous Chondrites (C1)	Earth	Moon
Volatile				
Na (%)	0.68	0.50	0.5	0.08
K (ppm)	850.	430.	200.	100.
Rb (ppm)	2.8	2.0	0.5	0.3
Cs (ppb)	4 to 619	190.	9.	0.01
Oxyphile Refractory				
Ca (%)	1.21	1.0	2.0	4.3
Sr (ppm)	11.	1.0	20.	40.
Ba (ppm)	4.1	4.0	5.5	18.
Sc (ppm)	8.0	4.0	17.	20.
Y (ppm)	2.0	1.6	4.	8.
La (ppm)	0.24	0.2	1.	2.
Th (ppb)	43.	65.	65.	150.
U (ppb)	12.	24.	20.	40.
Siderophile Refractory				
Cr (%)	0.28	0.25	0.26	0.12
Fe (%)	25.	25.	3.0	8.5
Co (ppm)	600.	400.	1000.	220.
Ni (%)	1.2	1.2	2.0	0.5

After Larimer (1971), Shaw (1972), and Taylor (1975).

enrichment factors less than 2 percent (Mn, Fe, Co, Ni, etc.) are concentrated chiefly in the mantle or core. Vertical fractionation of elements with large ionic size or charge into the crust probably reflects the difficulty with which these elements enter minerals in the mantle and core which are in a state of extreme close packing.

AGE OF THE EARTH AND MOON

The oldest dated rocks on the Earth, which now have been recognized on several continents, are about 3.8 b.y. in age. It is possible that older crustal rocks exist but have not as yet been dated. Gabbroic anorthosite samples from the moon exhibit isotopic dates of 4.4–4.5 b.y. and model lead ages of lunar soils give similar results. Rb–Sr, U–Th–Pb, and Sm–Nd dates of meteorites suggest an age for the solar system of about 4.6 b.y. Model lead ages of volcanic rocks and ores on the Earth suggest an age for the Earth of 4.55–4.6 b.y. Hence, it would appear that the major melting events following accretion of the Earth, moon, and other planetary bodies in the solar system occurred at about 4.6 b.y.

CONDENSATION AND ACCRETION OF THE PLANETS

The origin of the Earth-moon system is part of the general problem of the origin of the solar system,

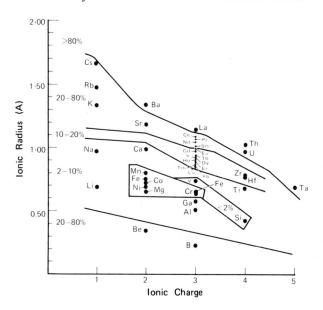

2.1. Estimated elemental crustal enrichment factors (in percent) as a function of ionic radius and charge (after Taylor, 1964a).

which can be considered in terms of three subproblems:

1. How the sun acquired the gaseous material from which the planets formed;
2. The history of condensation of the gaseous material;
3. The processes and history of planetary accretion.

Regarding the first question, one viewpoint is that the sun, already in existence, attracted material into a gaseous nebula about itself. Another proposes that the pre-existing sun captured a solar nebula of appropriate mass and angular momentum to form the solar system. Most theories, however, call upon condensation and accretion of the sun and planets from the same cloud at approximately the same time. All models have in common a gaseous nebula from which the planets form. The minimum mass of such a nebula is about 1 percent of a solar mass. The mechanisms by which the nebula is distributed into a disc with the sun at the center are not well understood. One mechanism is the transferring of angular momentum from the sun to the nebula, caused either by hydromagnetic coupling during rotational instability of the sun or by turbulent convection in the cloud. Condensing matter rapidly collapses into a disc about the sun (Safronov, 1972) or into a series of Saturn-like rings which condense into the planets (fig. 2.2). Small

planetesimals form within the cloud and spiral toward the ecliptic plane. Actual paths taken by planetesimals are complex. Regardless of the specific mechanism, the collapse of the nebula results in a cold cloud ($<0°C$) except in the region near the protosun within the orbit of Mercury.

Condensation (i.e., the formation of solid particles from gas) of the nebula begins during collapse and continues thereafter. The nebula is composed of silicate and oxide particles composed principally of Na, Mg, Al, Si, Ca, Fe, and Ni; "ices" of C, N, O, Ne, S, Ar, and Cl (as hydrides except for Ne and Ar); and a gaseous mixture composed chiefly of H and He (Ringwood, 1975). The ices are especially abundant in the outer part of the disc where the giant planets form; the silicate-oxide particles are concentrated in the inner part and give rise to the terrestrial planets. Thermodynamic considerations indicate that iron should be present in the cloud initially only in an oxidized state. It has been suggested the Type I carbonaceous chondrites (C1) may represent a sample of this primitive nebula. These meteorites contain only oxidized iron and large amounts of volatile components of the "ices" listed above (including organic compounds), and they have not been heated to more than 100°C to preserve such volatile constituents.

Although it is commonly assumed that the solar nebula was well mixed, differences in isotopic abundances and age of various meteorites suggest that it

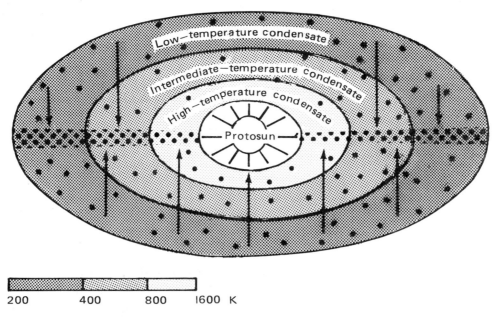

2.2. Schematic cross-section of the primitive solar nebula (after Ringwood, 1979). Temperatures in degrees K. Arrows indicate direction of gravitational sinking of solid particles.

was not (Lee, 1979). Variations in the abundance of rare gases and isotopic variations of such elements as Mg, O, Si, Ca, and Ba indicate large mass-dependent fractionation effects. The isotopic anomalies appear to require a sudden injection of neutrons into the early solar nebula. The source of such a neutron burst may have been a nearby supernova. It has also been suggested that such an injection event actually triggered the collapse of the solar nebula.

Accretion is the process by which solid particles collide to form planetary bodies in the disc around the sun. Accretion models fall into two general categories. *Homogeneous accretion models* call upon formation of the planets in a generally cold ($\lesssim 0°C$), well-mixed cloud, perhaps similar to C1 in composition. *Inhomogeneous accretion models*, on the other hand, involve progressive condensation and accretion of compounds from an initially hot cloud as it cools. Each of these models are considered in more detail in the following sections.

The specific mechanisms by which particulate matter evolves into small meter-sized bodies and these bodies, in turn, grow into planetesimals and planets are not well understood. Theoretical studies indicate that a large fraction of the gaseous cloud will condense and accrete into several large bodies rather than into a swarm of small particles. The model of Goldreich and Ward (1973) shows that the first generation of planetesimals will increase in radius to the order of 100 m in times of about one year after condensation. Clusters of these planetesimals contract to form second-generation planetesimals with radii of the order of 5 km in a few thousand years, and these planetesimals accumulate into the planets by collisions. Wetherill (1977) has recently simulated planetary growth rates by computer modeling. The results indicate that the accretion rate of the Earth and Venus increases rapidly reaching a value within 50 km of the final radius in the first 25 m.y. Accretion is 98 percent complete for both planets in times of the order of 10^8 y and large impact velocities (10–20 km/sec) during the final stages of accretion are capable of producing extensive melting in both planets.

There are important geochemical constraints which must be accounted for by any acceptable accretionary model. One constraint, as described in the previous section, is the depletion in volatile elements in the Earth and moon relative to C1 meteorites. Another is the enrichment of the refractory siderophile elements in the Earth compared to C1. Such enrichment strongly indicates that the Earth's mantle has not equilibrated with the core,

because if it had the siderophile elements would be carried into the core and depleted in the mantle. Although the mantle appears to be well mixed in terms of major elements, it is not well mixed in terms of Sr and Pb isotope distributions.

Inhomogeneous Accretion Models

Inhomogeneous accretion models call upon planetary growth by progressive condensation and accretion of various compounds as the temperature falls in an originally hot solar nebula. The zoned structure of the Earth, with an iron core surrounded by a silicate mantle, may have been produced by sequential condensation and accretion of the Earth from a cooling solar nebula. Thermodynamic calculations of condensation (Grossman, 1972) have provided a quantitative basis for the model. Cameron (1973) has suggested a model whereby the planets develop in a rotating, disc-shaped nebula amounting to two solar masses. Contraction of such a nebula produces high temperatures (1000–1700°C) in the region where the terrestrial planets form. Rapid dissipation of the nebula results in planetary accretion on a very short time scale, on the order of 10^3–10^4 years. Cooling of the gas as dissipation proceeds results in condensation over a wide range of temperature. The sequence of compounds condensed from the nebula at a pressure of 10^{-4} atm are summarized in table 2.4. Beginning with a temperature of about 1700°C, Clark, Turekian, and Grossman (1972) propose that the Earth accreted simultaneously with condensation as temperature fell over a period of 10^4 y resulting in a zoned Earth with an iron core surrounded by a mantle composed dominantly of Fe–Mg silicates. The upper mantle and crust accreted over a longer time scale (10^5–10^7 y), involving the addition of compounds with oxidized iron, iron sulfides, and hydrated silicates mixed with earlier, high-temperature condensates that failed to accrete (i.e., similar to the mineralogy of C1). The relative enrichment in the Earth in siderophile elements can be explained in the inhomogeneous accretion models by the fact that the Earth's core is never in contact with most of the mantle.

As discussed by Ringwood (1975), however, the inhomogeneous accretion models are faced with the following problems, which render them less appealing than homogeneous accretion models:

1. Existing thermodynamic data suggest a large degree of overlap in the condensation of iron and Mg-silicates (table 2.4), and hence the accretion of an early iron core is not clearly predicted.

Table 2.4. Condensation Sequence from Nebula of Solar Composition, at 10^{-4} atm

Phases	Temperature (°C)
Ca, Al, Ti, Zr, U, Th, REE oxides and some silicates	1250–1600
Metallic iron-nickel	1030–1200
Forsterite (Mg_2SiO_4) and enstatite ($MgSiO_3$)	1030–1170
Ca-plagioclase	900–1100
Na–K feldspar	~730
Troilite (FeS)	430
FeO in Mg-pyroxenes and Mg-olivine	300
Magnetite	130
Carbonaceous compounds	100–200
Hydrated Mg-silicates	0–100
Ice (H_2O)	< 0

After Grossman (1972).

2. The model predicts an Earth with a small core of Ca–Al–Ti oxides, which is not consistent with geophysical data today. This problem may be avoided if U and Th, also concentrated in the small core, produce enough heat to melt the surrounding iron and thus allow the Ca–Al–Ti-rich material to migrate upwards into a zone at the base of the lower mantle. This model also predicts that the upper mantle and crust should be depleted in Ca, Al, and Ti, a feature which is not observed.

3. Iron does not enter condensing Mg-silicates except at temperatures below 400°C, which is far less than the temperature at which the lower mantle accreted (1000–1200°C). Yet available geophysical and geochemical data indicate that the FeO/FeO + MgO ratio of the lower mantle must lie in the range of 0.1–0.2.

4. Sulfur should be a major component in the upper mantle of an inhomogeneously accreted earth, yet existing data indicate that it is extremely rare.

5. Because of the large mass of the Earth, late volatile condensates should collide with the Earth at such high velocities that they are evaporated and lost to the atmosphere. Hence, there is no way to account for the volatile constituents found in the upper mantle today.

6. It is necessary to have 10–20 percent of a low-atomic-weight element in the core to explain seismic wave velocity distribution (see Chapter 3), but the elements considered as possible diluents (Si, S, C, K, O) condense only after the iron core forms.

Homogeneous Accretion Models

The major difference between the homogeneous and inhomogeneous accretion models is the timing of accretion. In the inhomogeneous models, condensa-tion and accretion occur simultaneously, while in the homogeneous models condensation is essentially complete before accretion begins. In most homogeneous accretion models, the planets and the moon accrete from a cool ($\lesssim 100°C$), well-mixed, oxidizing solar nebula, similar perhaps in average composition to C1 meteorites. Time scales for accretion are 10^3–10^8 years. In these models, the Earth and other planets become zoned by later melting, which results in core formation and partial melting of the mantle. Most results suggest that more than enough heat was available to completely melt the Earth soon after or during the late stages of accretion (see Chapter 11).

One of the more popular and long-standing homogeneous accretion models for the Earth is the autoreduction model of Ringwood (1979). In this model, reduction of oxidized iron to metal, loss of volatiles, and fractionation resulting from core for-mation occur essentially simultaneously during the late stages of accretion. It is assumed that most of the Earth was accreted in $\lesssim 10^6$ years and before the sun evolved through a T-Tauri stage. During accretion, most of the gravitational energy is converted to heat and radiated away. The general features of the model are summarized in fig. 2.3 in terms of five evolution-ary stages.

Stage I. During the first stage, the accretional energy is small and the temperature is less than 700°C, maintained, in part, by evaporation of volatile components. A small nucleus (5–10 percent of the Earth mass) of cool, oxidized, and volatile-rich material is formed during this stage. Silicates, hy-drated silicates, and Fe-Ni sulfides coexist in the nucleus.

Stage II. As accretion continues, the impact energy of accreting bodies becomes high enough to cause near-surface heating, leading to reduction of

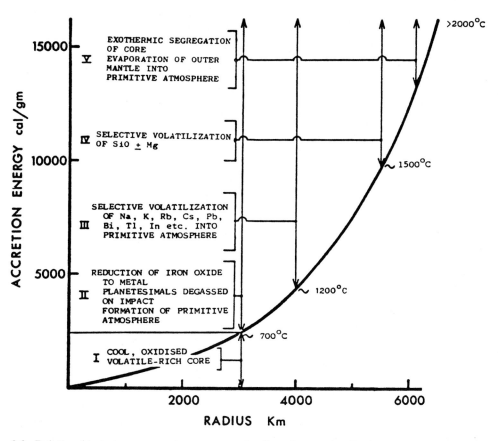

2.3. Relationship between accretion energy and radius of a growing Earth. Stages of accretion are shown in relation to accretion energy and approximate surface temperatures (after Ringwood, 1979).

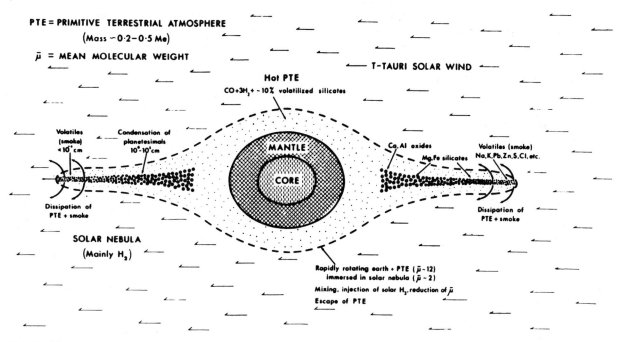

2.4. Structure of the Earth immediately after core formation, with its primitive atmosphere and sediment ring. (From Ringwood, 1972, with permission.)

oxidized iron to metal by carbonaceous matter. Incoming bodies are also completely degassed on impact, resulting in the formation of a primitive reducing atmosphere.

Stage III. By the time the mass is equal to 0.2 of an Earth mass, the temperature has risen to about 1200°C. Reduction of iron oxides to metal occurs in the atmosphere, and as temperature increases to 1500°C, volatile elements (table 2.3) are baked out of the Earth and enter the atmosphere. The accreting material is a mixture of Fe–Ni metal and iron-free silicates depleted in the volatile elements which remain in the atmosphere.

Stage IV. Increasing temperatures (> 1500°C) from increasing accretional energy reduce and volatilize silicate phases in accreting planetesimals. Silicon (as SiO) and to a lesser extent MgO are partially evaporated at such high temperatures, and enter the primitive atmosphere. Less volatile constituents, such as Ca and Al silicates, continue to accrete with olivine and smaller amounts of pyroxenes.

Stage V. Melting begins in the Earth near the surface, and iron, mixed with FeO or FeS, which reduce its melting point (Chapter 3), segregates into bodies which move toward the center of the Earth, forming the core. This process is highly exothermic, as well as probably catastrophic, liberating 600 cal/gm over the entire Earth, resulting in the evaporation of the outer part of the Earth into the atmosphere. This material collects in a sediment ring about the Earth, and the moon accretes from this ring (fig. 2.4). The early reducing atmosphere is lost either by being "blown off" as it accumulates or blown away by a solar wind coming from the sun as it passes through a T-Tauri stage of evolution.

ORIGIN OF THE MOON

Scientific results from the Apollo missions to the moon have provided a voluminous amount of data on the structure, composition, and history of the moon (Taylor, 1975, 1979b). Seismic data are generally interpreted in terms of five zones within the moon, which are, from the surface inwards: a plagioclase-rich crust (dominantly gabbroic anorthosite), 60–100 km thick; an upper mantle, composed chiefly of pyroxenes and olivine, 300–400 km thick; a lower mantle of the same composition (with olivine dominating), about 500 km thick; 400 ± 100 km of partly melted mantle; and 200–400 km of core.

Among the more important constraints that any model for lunar origin must satisfy are the following:

1. The moon does not revolve in the equatorial plane of the Earth, nor in the ecliptic.

2. Tidal dissipation calculations indicate that the moon is retreating from the Earth, resulting in an increase in the length of the day of 15 μsec/year (see next section).

3. The moon is enriched in refractory oxyphile elements and depleted in refractory siderophile and volatile elements relative to the Earth.

4. Geophysical and geochemical results suggest that pyroxenes are the major minerals in the lunar mantle compared to olivine in the Earth's mantle (Ringwood, 1975).

5. The lunar interior is much more reduced than the Earth's interior (Ringwood, 1975). Lunar rocks contain small quantities of metallic iron, while ferric iron is almost absent. The oxygen fugacity at 1200°C is six orders of magnitude smaller in the moon than in the Earth.

6. Radiometric dates from igneous rocks on the lunar surface range from about 3.1 to 4.46 b.y.

7. Model ages indicate that most lunar rocks formed during a major magmatic event at about 4.46 b.y.

Hypotheses for the origin of the moon generally fall into one of three categories: (1) fission from the Earth; (2) precipitation from a sediment ring around the Earth; and (3) capture by the Earth. All three hypotheses were with us prior to the Apollo landings, and although we now have a great deal of geochemical and geophysical data from the moon, each of the hypotheses still has a group of supporters. Any acceptable model must, however, account for the constraints listed above, and thus far none of these hypotheses is completely acceptable in this respect. A particularly difficult problem for any model is to deplete both siderophile and volatile elements in the moon relative to Cl. Each of the major hypotheses will now be briefly discussed.

Fission Models

Fission models involve separation of the moon from the Earth during an early stage of rapid spinning when tidal forces overcome gravitational forces. A modified version of the fission hypothesis (Wise, 1963) suggests that formation of the Earth's core reduced the amount of inertia, increasing the rotational rate and spinning off material to form the moon. Such a model is attractive in that it accounts for the similarity in density between the Earth's mantle and the moon and for the absence of a large metallic lunar core. The hypothesis is also consistent with the fact that the moon revolves in the same direction as the Earth rotates, the circular shape of the lunar orbit, and the existence of a lunar bulge facing the Earth.

Fission models, however, face several major obstacles. They do not explain the inclination of the lunar orbit, and they appear to require more total angular momentum than is available in the present Earth-moon system. Also, lunar igneous rocks are more depleted in siderophile elements than terrestrial igneous rocks, indicating that the lunar interior is not similar in composition to the Earth's mantle (Taylor, 1975). Sr and Pb isotopic data also indicate differences in composition between the moon and the Earth's mantle.

The Precipitation Hypothesis

The precipitation hypothesis was proposed by Ringwood (1972a, 1975). During Stage V of Ringwood's homogeneous accretion model for the Earth (fig. 2.3), silicates precipitate as the hot atmosphere expands and cools, forming a sediment ring of planetesimals around the Earth (fig. 2.4). This sediment ring becomes unstable and coagulates to form the moon. Because the volatile elements are largely lost with the removal of the early atmosphere by intense solar radiation from the T-Tauri wind, the material from which the moon accretes is enriched in oxyphile refractory elements and depleted in volatile elements, thus accounting for the major geochemical differences between the Earth and moon. Equilibrium between silicates and metal in the lunar interior accounts both for the low oxygen fugacity in the moon and for the depletion of siderophile elements in lunar igneous rocks. The greater amount of pyroxene in the lunar interior compared to the Earth's mantle is interpreted as reflecting evaporation of SiO from the Earth and addition to the sediment ring during the late stages of accretion.

The major difficulty with the precipitation hypothesis is that it does not readily account for the inclination of the lunar orbit. One way around this difficulty is to position an initial Earth with low viscosity such that the inclination of lunar orbit increases with time.

Capture Models

Capture models propose that the moon and Earth formed in different parts of the solar nebula and that early in the history of the solar system ($\lesssim 4$ b.y.) the moon approached the Earth and was captured (Kaula

and Harris, 1975). Both catastrophic and non-catastrophic models of lunar capture have been described, involving retrograde and prograde orbits for the moon prior to capture. Capture models explain the inclination of the lunar orbit and the retreat of the moon from the Earth. Chemical differences in the two bodies are accounted for by hypothesizing that each body was formed in a different part of the solar nebula, which is not well mixed. Six satellites in the solar system revolve in retrograde orbits, and hence capture may have been a common process in the early history of the solar system. One version of the capture model (Wood and Mitler, 1974) involves capture of a group of planetesimals which disintegrate as they come within the Earth's Roche limit, forming a sediment ring about the Earth. As in the precipitation model, this ring coagulates to form the moon.

Capture models do not readily account for the observed near-circular orbit of the moon, because most capture mechanisms predict highly elliptical orbits (Wise, 1963). Also, considering the orderly spacing of the planets (as deduced from Bodes' law), it is difficult to determine just where the moon may have formed in the solar system.

THE EARTH'S ROTATIONAL HISTORY

Integration of equations of motion of the moon indicate there has been a minimum in the Earth-moon distance sometime in the geologic past; that the inclination of the lunar orbit has decreased with time; and that the eccentricity of the lunar orbit has increased with time as the Earth-moon distance has increased (Lambeck, 1980). It has long been known that angular momentum is being transferred from the Earth's spin to the lunar orbital motion, which results in the moon moving away from the Earth. The current rate of this retreat is about 5 cm/yr. The corresponding rotational rate of the Earth has been decreasing at the rate of about 5×10^{-22} rad/sec. Tidal torques cause the inclination of the moon's orbit to vary slowly with time as a function of the Earth-moon distance. When this distance is greater than 10 Earth radii, the lunar orbit plane moves towards the ecliptic; when it is less than 10 Earth radii it moves towards the Earth's equational plane. For an initially eccentric orbit, the transfer of angular momentum is greater at perigee than apogee, and hence the degree of eccentricity increases with time. These evolutionary changes in angular momentum of

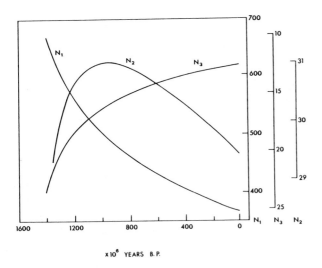

x 10⁶ YEARS B.P.

2.5. Number of solar days per year (N_1) and per synodic month (N_2), and number of synodic months per year (N_3) as a function of geologic time (after Lambeck, 1980). Model assumptions: phase-lag of semi-diurnal tides = 6 degrees; length of the year and mass of the Earth assumed constant; and tidal evolution of Earth's orbit is assumed small.

the Earth-moon system result in changes in the length of the terrestrial day and month. Estimates of the number of days per year N_1, the number of days per synodic month N_2, and the number of synodic months per year N_3, as a function of time, are given in fig. 2.5.

Many groups of organisms secrete sequential layers that are known to be related to cyclical astronomical phenomena. Such organisms, which are known as *paleontological clocks* (Pannella, 1972), provide a means of independently evaluating the curves given in fig. 2.5. The most important groups are corals and bivalves. Daily increments in these organisms are controlled by successive alternation of daylight and darkness. Seasonal increments reflect changes in the length of sunlight per day, seasonal changes in food supply, and, in some instances, tidal changes. Actual growth patterns are complex, and local environmental factors may cause difficulties in identifying periodic growth patterns. Results from Phanerozoic bivalves and corals, however, are consistent and suggest a decreasing rotational rate of the Earth of about 5.3×10^{-22} rad/sec, which is in good agreement with astronomical values. This corresponds to 420-430 days/year at 500-600 m.y. ago. Stromatolites (Chapter 11) also deposit regular bands,

although a quantitative relationship of banding to astronomical rhythm has not yet been established. The interpretation of stromatolite bands from the Biwabik Iron Formation in Minnesota is that at 2000 m.y. ago there were 800-900 days/year (Mohr, 1975). This is in good agreement with the extrapolation of N_1 in fig. 2.5.

If the moon came within the Roche limit of Earth (~2.9 Earth radii) after 3.8 b.y., a record of such a close encounter should be preserved. Even if the moon survived this encounter without disintegrating, the energy dissipated in the two bodies would largely melt both bodies and completely disrupt and recycle any earlier crust. The preservation of crust 4.46 b.y. in age on the moon strongly indicates that if such a close encounter did occur, it was prior to this time.

SUMMARY STATEMENTS

1. Radiometric dates from igneous rocks on the lunar surface range from about 3.1 to 4.46 b.y. Model ages indicate that most lunar rocks formed during a major magmatic event at about 4.46 b.y.

2. Isotopic dating indicates that planetary accretion was largely or entirely completed by 4.6 b.y., the first major period of melting in the terrestrial planets.

3. The Earth, moon, and other terrestrial planets appear to have condensed from a gaseous nebula and to have accreted from relatively high-temperature condensates in the inner part of the nebula. Homogeneous accretion models for the Earth are faced with fewer problems than inhomogeneous models.

4. Isotopic anomalies in meteorites suggest that the early solar nebula was injected with neutrons, perhaps from a nearby supernova.

5. Models for lunar origin must account for the fact that the moon does not revolve in the Earth's equatorial plane, for the retreat of the moon from the Earth, for differences in refractory and volatile element contents between the Earth and the moon, and for the fact that the lunar interior is depleted in oxygen relative to the Earth's interior.

6. Of the models for lunar origin, only the fission model encounters apparently insurmountable problems. Both the precipitation and capture models can be accommodated by existing data in one way or another.

7. Astronomical and paleontological data indicate that the moon has been retreating from the Earth at a rate of about 5×10^{-22} rad/sec for at least 1000 m.y. If the moon came within the Earth's Roche limit, it must have occurred within the first 100 m.y. after planetary accretion.

SUGGESTIONS FOR FURTHER READING

Lambeck, K. (1980) *The Earth's Rotation: Geophysical Causes and Consequences.* Cambridge: Cambridge Univ. Press. 449 pp.

Ringwood, A.E. (1979) *Origin of the Earth and Moon.* New York: Springer. 295 pp.

Smith, J.V. (1979) Mineralogy of the planets: A voyage in space and time. *Mineral. Mag.,* **43**, 1–89.

Taylor, S.R. (1975) *Lunar Science: A Post-Apollo View.* New York: Pergamon Press. 372 pp.

Chapter 3

The Mantle and Core

DETAILED STRUCTURE OF THE MANTLE

The Moho

A great deal has been learned about the Moho since it was first reported by Andrija Mohorovičić in 1910. It is generally well defined by an abrupt increase in P-wave velocity from about 6.6 km/sec to about 8.0 km/sec. It occurs from 10 to 12 km beneath the oceans to 30 to 50 km beneath the continents (ranging up to 80 km). Compressional waves that travel along the top of the mantle at the Moho are designated P_n *waves*. P_n velocities commonly range from 7.9 to 8.1 km/sec, although some areas exhibit velocities as low as 6.5 km/sec and other areas as high as 9.0 km/sec. The Moho appears to be caused by a phase or composition change, or both, as will be discussed later in this chapter. The sharpness of the Moho is not well defined, but appears to range from as little as 0.1 km in some oceanic areas and 0.5 km in stable continental shield areas to over 1 km in some tectonically active regions. Estimates of the Moho temperature beneath the continents range from 500°C to 700°C and beneath oceans from 150°C to 200°C.

Seismic anisotropy of P_n velocities has been recognized in several oceanic areas. Beneath the Northwest Pacific, for example, an anisotropy (ΔV_P) of 0.2–0.3 km/sec is observed (Raitt *et al.*, 1969). In general, most rapid P velocities occur at approximately right angles to oceanic ridges.

In some areas, one or both of the following problems have been encountered in defining the Moho (Knopoff, 1969): (a) two major discontinuities are detected at a depth less than 60 km, and (b) anoma-

lously low P_n velocities are found in some areas. In parts of the Basin and Range Province discontinuities occur at 25 km ($V_{P_n} = 7.7$ km/sec) and about 50 km ($V_{P_n} = 8.1$ km/sec), and in western Italy discontinuities are reported at 10 km ($V_{P_n} = 7.4$ km/sec) and 55 km ($V_{P_n} = 8.1$ km/sec). The question of which interface defines the crust has significance in terms of crustal evolution and the origin of both discontinuities. In regions where two discontinuities occur, the shallower one is usually accepted as the base of the crust. Low P_n velocities commonly ranging from 7.0 to 7.8 km/sec characterize the upper mantle beneath oceanic ridges and beneath continental rifts. Such low-P_n regions ($V_{P_n} \leq 7.8$ km/sec) are referred to as *anomalous upper mantle*. Low P_n velocities may also occur beneath island arcs (for instance, the Lesser Antilles exhibit V_{P_n} ranging from 6.5 to 7.5 km/sec), marginal-sea basins, and young orogenic areas. Anomalously low P_n velocities correlate well with high surface heat flow, thin lithosphere, and high electrical conductivity and seismic wave attenuation (low Q) immediately beneath the Moho. Unusually high P_n velocities of 8.6 km/sec or greater have been reported from some oceanic basins and continental shield areas. A correlation of P_n velocity distribution with crustal type suggests that the crust and upper mantle are not decoupled and that they evolved together as part of the same system.

Lateral variations in P_n velocity in the midwestern and western United States are shown in fig 3.1. Existing data for this area suggest that variations in P_n velocity correlate well with surface topography, average crustal velocities, and, in some cases, geologic provinces. In general, mountainous areas have lower P_n velocities than regions with little surface topogra-

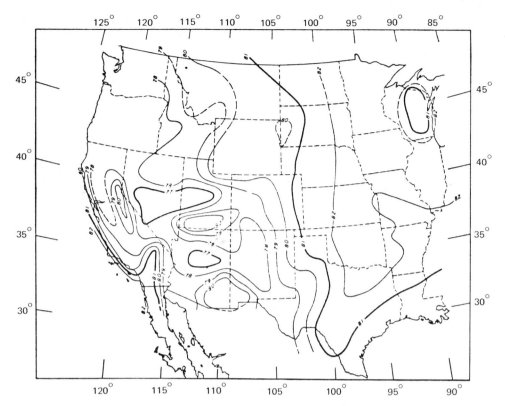

3.1. P_n velocity distribution in the western and central United States (from Archambeau *et al.*, 1969).

phy. Also, the Cordilleran orogenic belt in the western United States exhibits variable and, on the whole, lower P_n velocities than does the stable interior. P_n velocities of 7.7 to 7.8 km/sec characterize the Basin and Range Province, as does thin crust. The Great Plains, Colorado Plateau, and stable continental areas of the eastern United States generally exhibit P_n velocities of 8.0 to 8.2 km/sec. Velocity contours outline the Sierra Nevada Range in California, where P_n ranges from 7.8 to 8.0 km/sec. The E-W trending contours in Utah and Nevada, however, do not seem to reflect young surficial tectonic features, the trends of which are generally N-S.

The Upper Mantle and Transition Zone

Continental Areas. Studies of spectral amplitudes and travel times of body waves make it possible to refine details of the structure of the upper mantle and transition zone (Archambeau et al., 1969; Hales et al., 1980; Patton 1980). The P-wave velocity structure beneath several crustal types is shown in fig 3.2. Most segments of the crust are characterized by a high-velocity lid ($V_P \geq 7.9$ km/sec) underlain by a low-velocity zone beginning at a depth as shallow as

60–70 km and extending to depths of 100–300 km. The top of this zone marks the base of the lithosphere and averages about 100 km in depth. Beneath the Basin and Range Province and probably beneath most continental rift zones, the low-velocity zone (anomalous upper mantle) extends to the Moho at depths as shallow as 25 km. S-wave low-velocity zones also characterize these regions, and may extend to depths of 200–300 km (Cara, 1979). Low-velocity zones are generally not detected beneath Precambrian shields. It is of interest to note that in those regions where a low-velocity zone is poorly developed or does not exist, free-air gravity anomalies are negative; where a well-developed zone exists, the gravity anomalies are weakly to strongly positive. Figure 3.3 shows an idealized model for the distribution of Q in the Earth for both body and surface waves. A prominent minimum of Q at 150–200 km characterizes the low-velocity zone.

From the base of the low-velocity zone to 400 km, P-wave velocities increase only slightly. At 400–420 km a major discontinuity occurs, which is commonly used to define the base of the upper mantle. A second velocity discontinuity is observed at about 650 km and a third one at about 1050 km, marking

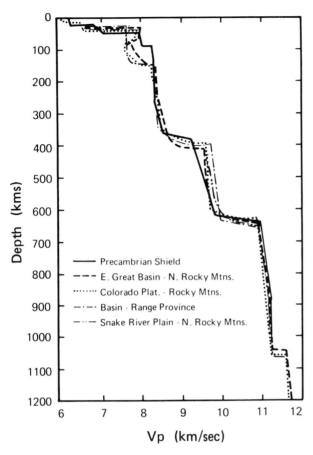

3.2. P-wave velocity distribution in the mantle (after Archambeau *et al.*, 1969; Hales *et al.*, 1980).

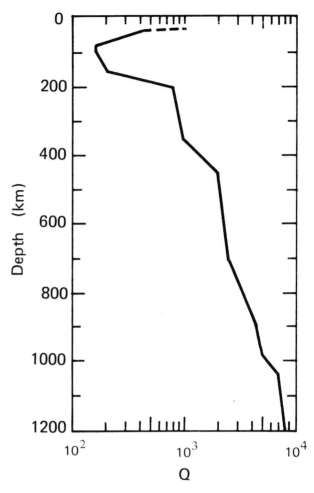

3.3. Typical variation of Q in the Earth as a function of depth as inferred for body waves (excluding Precambrian shield areas) (from Archambeau *et al.*, 1969).

the base of the transition zone. Beneath northern Australia, distinct discontinuities are recognized at 610, 630, 645, and 722 km with uncertainties of 5-10 km (Hales et al., 1980). Velocity profile models also suggest discontinuities between 500 and 550 km in some regions. Although P-wave velocity distributions beneath various crustal provinces at depths from 400 to 1000 km are very similar, S-wave distributions show lateral variability in this depth range (Patton, 1980). S-wave travel times in the upper mantle beneath continents that have undergone Phanerozoic tectonic activity are generally greater than times in the upper mantle beneath continents that were stabilized during the Precambrian. Also, travel times beneath oceanic regions decrease with crustal age away from oceanic spreading centers. Changes in the gradient of Q for body waves characterize the 400 km and 1050 km discontinuities (fig. 3.3).

Another way to study lateral variations in the upper mantle is by the use of travel-time anomaly distributions (Hales, 1972). Figure 3.4 is a P-wave

travel-time anomaly map of the United States. In general, P-wave arrivals are late in areas of young orogenic activity (0.2 to 0.8 secs) and early in shield and stable craton areas (0 to minus 0.5 secs). Both P_n velocity distributions (as discussed above) and travel-time anomaly distributions are consistent with fast upper-mantle P-wave velocities under shields and platforms and slow ones beneath young orogenic areas.

Oceanic Areas. Mantle structure in oceanic areas has been studied by P′dP′ seismic phases, which pass through the Earth and are reflected from the interior side of interfaces in the mantle and then return and are recorded on the opposite side of the Earth (Whitcomb and Anderson, 1970). Histogram plots of reflected waves from events summed in 20 km intervals for a portion of the Atlantic-Indian Ridge south

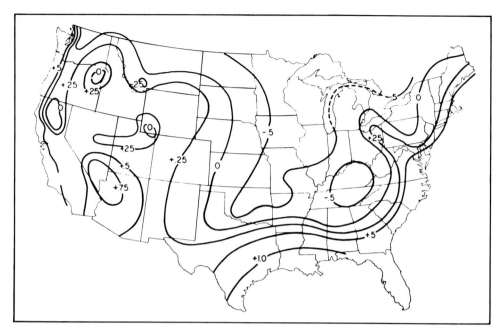

3.4. Travel-time anomaly map of North America (contours in seconds) (from Herrin and Taggart, 1968).

of Africa and for the Ninety-East Ridge southwest of Sumatra are given in fig. 3.5. Identification of waves reflected from depths greater than 100 km is not, on the average, as certain as that for shallower depths. Strong reflections (not resolved in fig 3.5) occur at 20–25 km, 50–60 km, and 130–140 km. The shallowest reflection may represent the Moho, although it seems deep in terms of available refraction data from the oceans. The two succeeding reflections are generally interpreted to represent the top and bottom of the low-velocity zone, respectively.

Deeper reflections listed in order of decreasing confidence are (in km): 630, 280, 520, 940, 410 (very weak), and 1250 (tentative). Reflections below 500 km show a consistent difference between the two oceanic areas of about 20 km. Possible explanations for this are: (a) temperatures beneath the Atlantic-Indian Ridge are higher, displacing phase changes to greater depths, or (b) the upwelling mantle beneath the Atlantic-Indian Ridge is richer in iron than the mantle beneath the Ninety-East Ridge.

The 630 km reflection corresponds well with the 650 km discontinuity recognized beneath the continents, and the 520 km reflection corresponds to the discontinuity recognized beneath part of the continents at 500–550 km. The fact that a 280 km discontinuity is not recognized beneath the continents but a strong reflection is produced beneath these two oceanic areas may be due to either of the following:

(a) there is a sharp change in density at 280 km, providing a strong reflecting interface but little change in refracted P-wave velocity; or (b) the upper mantle at 280 km beneath continental areas is different in composition, mineralogy, or thermal structure (or

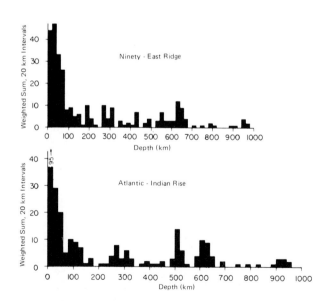

3.5. Histograms showing the distribution of P'dP' reflection depths for mantle regions beneath the Atlantic-Indian Rise south of Africa and beneath the Ninety-East Ridge southwest of Sumatra (from Whitcomb and Anderson, 1970).

some combination thereof) from that beneath the two oceanic areas. The occurrence of only a weak reflection at 410 km beneath these oceanic areas, compared to the major P-wave discontinuity recognized at this depth beneath continents, suggests that the 400 km discontinuity in these two areas is a poor reflector.

Recent seismic refraction studies in oceanic areas clearly recognize the 400 and 650 km discontinuities (England, Kennett, and Worthington; 1978; Cara, 1979). Existing data suggest that below about 250 km there is no difference in the seismic velocity distribution between oceanic and continental areas.

The Lower Mantle

Although seismic velocities, in general, have a rather uniform gradient in the lower mantle, several increased velocity gradients are recognized (fig. 3.6). Relatively high gradients occur at 1160–1220, 2180–2370, and perhaps at 2700–2750 km, while low gradients are recognized at 1260–1330, 1750–1850, and 2460–2600 km. A notable flattening of seismic-wave velocity and density gradients is also recognized just above the core-mantle interface (fig. 3.6), and extends for a distance to 30–150 km. Refinement of seismic data seems to rule out the presence of large bumps on the core-mantle interface, but will permit small ones with amplitudes not larger than 2 km (Cox and Cain, 1972). Although it appears to be small in magnitude

compared with that in the upper mantle and transition zone, some lateral variation in both P- and S-wave velocities in the lower mantle is indicated by existing data.

UPPER-MANTLE TEMPERATURE GRADIENTS

Although it is not possible to determine temperature gradients in the upper mantle directly, several sources of information allow estimates to be made. Estimates of temperature gradients between two oceanic and two continental areas are given in fig. 3.7. These are based on surface heat-flow measurements and models of heat productivity and thermal conductivity distributions with depth. Convection is assumed to be the dominant mode of heat transfer in the asthenosphere beneath oceanic areas, with conduction dominating in the lithosphere. Differences in gradients beneath various continental crustal provinces are further discussed in Chapter 4.

Although major differences in temperature distribution exist in the upper mantle, it is necessary that all temperature gradients converge at depths of a few hundred kilometers, or large unobserved gravity differences would exist between continental and oceanic areas as a result of thermally produced density differences. Most results indicate that temperatures

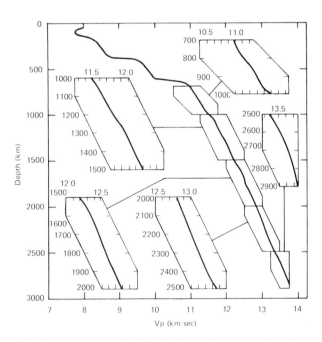

3.6. P-wave velocity distribution in the lower mantle (after Johnson, 1969).

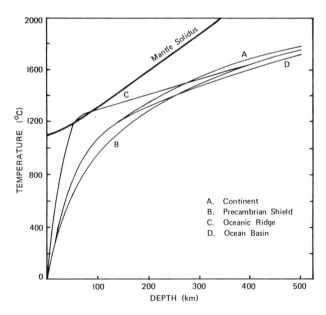

3.7. Temperature distributions beneath oceanic and continental areas (after Clark and Ringwood, 1964; Bottinga and Allegre, 1973).

beneath ocean basins and Precambrian shields converge at between 100 and 200 km depth. Heat-flow distribution, heat productivity models, and sea-floor spreading models suggest that temperature gradients to about 50 km depth range from 10° to 30°C/km. Data suggest that the mantle is partially molten beneath the axes of oceanic ridges (curve C, fig. 3.7) and that temperature gradient decreases with distance from ridges at a rate inversely proportional to lithosphere spreading rates (Bottinga and Allegre, 1973). With the exception of beneath oceanic ridge axes, temperature gradient must decrease significantly in the vicinity of 50–200 km in order to avoid large amounts of melting in the upper mantle, which is not allowed by seismic data. It is probable that the temperature gradients at depths of 100–200 km are close to the melting gradient of basalt, which is about 3°C/km at these depths.

ELECTRICAL CONDUCTIVITY DISTRIBUTION

Considerable uncertainty should be attached to estimates of the electrical conductivity distribution in the mantle due to poor resolution of data. Especially large variations occur in conductivity models for depths less than 500 km. An idealized electrical conductivity distribution in the mantle is shown by the solid line in fig. 3.8. The dashed lines indicate the combined range of uncertainty and real variations at depths less than 250 km. In most continental areas, the conductivity of the lower crust is in the range of 10^{-2} to 10^{-3} $\Omega^{-1}m^{-1}$. Conductivity appears to increase rapidly to the vicinity of 1 $\Omega^{-1}m^{-1}$ at the base of the transition zone and less rapidly to about 10 $\Omega^{-1}m^{-1}$ at the core-mantle interface. In regions where the seismic low-velocity zone occurs at shallow depths such as beneath the Basin and Range Province (beginning at $\leq$ 30 km), oceanic ridges, and beneath young continent rift systems, a region of shallow high electrical conductivity ($\leq 10^{-2}\Omega^{-1}m^{-1}$) also exists. Existing data indicate that the high conductivity zone is nearer the surface beneath oceans than beneath continents.

Significant lateral variations exist in terrestrial electrical conductivity. The best known variations occur at ocean-continent interfaces and are produced by the large conductivity of sea water. Water-saturated sediments can also produce shallow conductivity variations. Regional variations such as these are referred to as *conductivity anomalies* (CAs). Not all conductivity anomalies can be explained by sea

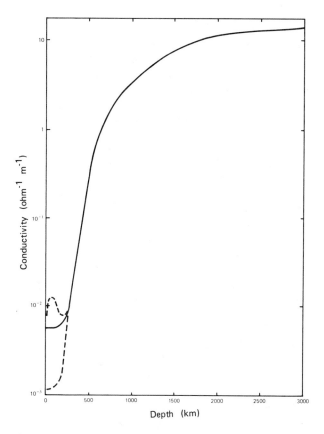

3.8. Idealized electrical conductivity distribution in the mantle (from Rikitake, 1973).

water or water-saturated sediments, and they appear to have their sources deep in the crust or upper mantle. Two types of such anomalies related to short-period magnetic variations have been identified. One is characterized by variations in the ratio of the vertical to the horizontal components of the field ($\Delta Z/\Delta H$), and the other by variations in the vertical component only (ΔZ). The first type of CA is characteristic of convergent plate boundaries and has been described in Japan and in the Andes in South America (Honkura, 1978). The second type results from shallowing of highly conductive layers and is found beneath both oceanic and continental rift systems, as well as within continental lithosphere. An example of the first type of CA in Japan is shown in fig. 3.9. The highly conducting layer increases in depth beneath eastern Japan to greater than 200 km, and then decreases to depths less than 30 km beneath the Japan Sea. Examples of the second type of CA are shown in fig. 3.10. In this figure three major anomalies are shown (intense stippling): one extends south from the north end of the southern Rockies beneath the Rio Grande rift; another trends in a

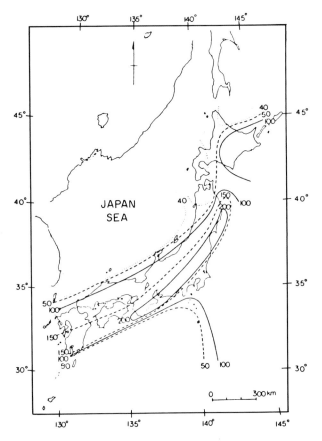

3.9. Contoured depth (in km) of a highly conducting zone beneath Japan as deduced from variations in the ratio of the vertical to the horizontal component ($\Delta Z / \Delta H$) of the Earth's magnetic field (from Rikitake, 1969).

possible to estimate the minimum amount of melt in the upper mantle from conductivity distribution and experimental results dealing with conductivity variation with degree of melting. The increase in depth of high conductivity zones beneath trenches appears to reflect depression of isotherms (fig. 6.22) produced by subduction (Honkura, 1978). The shallow depth of high-conductivity layers above descending slabs, such as beneath the Philippine and Japan seas, is related both to higher temperatures and to partial melting above such slabs. Seismic data also reflect these features (fig. 6.12). The shallowing of conductivity anomalies beneath rift systems and other parts of the continental crust probably reflects upwelling asthenosphere from the low-velocity zone or mantle plumes which have thinned the lithosphere (Chapter 6). Differences in depth to CAs along the East African rift system probably result from differences in degree of interconnection of magmas. The relatively high conductivity zone at shallow depths beneath the Basin and Range province corresponds to a shallow depth to anomalous upper mantle, which, in turn, reflects a widespread shallow depth of partial melting.

GLOBAL GRAVITY ANOMALIES

Global free-air gravity anomalies calculated from spherical harmonic coefficients of the gravitational field of degrees 2 through 16 are shown in fig. 3.11. In general, most of the large-scale gravity features of the Earth can be correlated with lithosphere-asthenosphere interactions or with mantle convection (Phillips and Lambeck, 1980). Positive anomalies in oceanic areas are located close to oceanic ridges but the converse is not true. In fact, part of the Pacific-Antarctic ridge system exhibits a negative anomaly. Large positive anomalies in oceanic areas probably reflect upwelling mantle convection currents or plumes (Chapter 6) that bring dense mantle material close to the surface. Some negative anomalies also appear to characterize plumes, however. This may be due to two competing processes of the same order of magnitude in the upper mantle: (a) dense material being brought close to the surface (favoring a positive anomaly), and (b) higher temperatures in the rising current causing a mass deficit (favoring a negative anomaly). Either process may dominate. Iceland is an example of the former case and the Galapagos Islands (anomaly not resolved in fig. 3.11) of the latter. Major negative anomalies characterize oceanic basins and appear to reflect maximum horizontal acceleration of lithosphere and asthenosphere away from

southwest direction along the Wasatch Front in Utah; and a third occurs beneath the southern Canadian Rockies. In general, upper mantle conductivities are higher in the western United States, and especially beneath the Basin and Range Province, than in the midwestern and eastern United States, as shown by the less dense stippling in the figure. The depths of conducting layers in the mantle and crust vary inversely with surface heat flow (Adam, 1976). The main conducting layers cluster around three curves of heat flow versus depth to conducting layer, with corresponding depth ranges of 300–700 km, 50–300 km, and 10–40 km.

The origin of high conductivity layers in the upper mantle and crust is related to increasing temperature, partial melting, increases in water content of rocks, changes in rock composition, or a combination of these. Temperature and partial melting appear to be most important in producing upper-mantle anomalies. For a given temperature distribution, it is

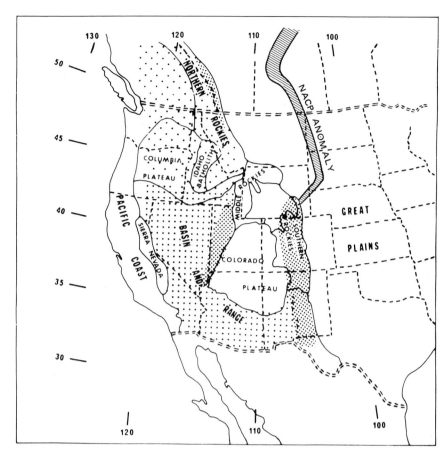

3.10. Conductivity distribution in the upper mantle in western North America (after Gough, 1973 and Camfield and Gough, 1976). Areas with no pattern are regions of very low conductivity; intensity of stippling in other areas is proportional to intensity of conductivity.

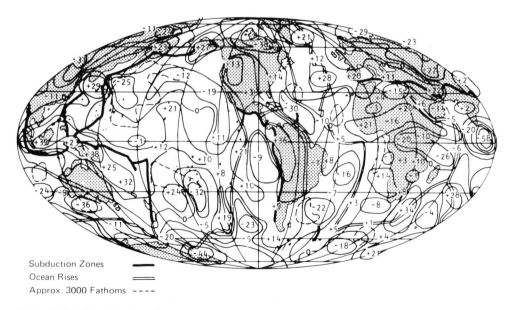

Subduction Zones ———
Ocean Rises ═══
Approx. 3000 Fathoms - - - -

3.11. Global free-air anomalies (in mgal) referred to an ellipsoid of flattening of 1/299.8 (after Kaula, 1972).

oceanic rises. There is not, however, a systematic relationship between the distribution of these anomalies and ocean depth after the cooling lithosphere effect (see Chapter 4) has been subtracted. Trench-arc systems also seem to be associated with positive anomalies, as is exemplified by the Indonesia area and Japan. Gravity model studies indicate that the downgoing slab contributes very little to these anomalies. Rather they appear to result from a combination of elevated back-arc lithosphere and a low topographic rise on the seaward side of most oceanic trenches. The low rise is produced by buckling of the oceanic lithosphere as it begins to descend into the mantle.

Major negative anomalies on the continents are commonly associated with regions that are isostatically rising after the retreat of Pleistocene glaciers, as for example northern Canada. In these areas relatively light lithosphere was depressed into the asthenosphere, producing the negative anomalies. Rapid recovery rates will greatly decrease or remove these anomalies in a few thousand years. The large negative anomalies in south-central Asia in the Hindu Kush–Tien Shan area and over Antarctica seem to have more complex explanations. Calculations indicate that thickening of the crust or lithosphere resulting from a continent-continent collision (see Chapter 9) cannot alone account for the negative anomaly in south-central Asia (Kaula, 1972). The negative anomaly in Antarctica, which is almost completely surrounded by active ocean ridges, cannot readily be explained in terms of either Pleistocene downwarping or lithosphere thickening produced by insufficient asthenospheric flow. It may reflect convection in deeper parts of the mantle.

COMPOSITION OF THE MANTLE

The mineralogical and chemical composition of the mantle can be approximated from the combined results of the study of seismic velocity distributions in the Earth, high-pressure-temperature experimental studies, shock-wave experiments, and geochemical studies of meteorites and ultramafic rocks. It is possible to attain static pressures in the laboratory of 300 kbar to more than 1.5 Mbar, which allows direct investigation of the physical and chemical properties of minerals that may occur throughout most of the Earth. Shock-wave studies also have been important in defining stability fields of possible mantle mineral assemblages at depths greater than 600 km.

Several lines of evidence indicate that *ultramafic rocks* compose large parts of the upper mantle. These rocks contain more than 70 percent of iron- and magnesium-rich minerals such as pyroxenes, olivine, and garnet. *Mafic rocks* are of basaltic composition and are composed of various mixtures of plagioclase, pyroxenes, olivine, amphibole, and garnet, and may also occur in parts of the upper mantle. Several specific ultramafic and mafic rock types that will be referred to hereafter are briefly described as follows. *Dunite* is an ultramafic rock composed principally of Mg-rich olivine, whereas *peridotite* is composed of Mg-rich olivine and pyroxenes. Some peridotites contain varying amounts of other minerals, such as spinel, garnet, amphibole, mica and plagioclase. These peridotites are referred to as spinel peridotite, garnet peridotite, and so forth. Ultramafic rocks are commonly altered to serpentine in varying degrees, and these are referred to as *serpentinized peridotites* (or dunites) or, if serpentinization is nearly complete, *serpentinites*. *Gabbro* is a mafic rock composed of plagioclase, pyroxenes, and sometimes olivine; it is the coarse-grained equivalent of basalt, a common volcanic rock on the Earth's surface. *Garnet granulite* is a mafic rock composed of plagioclase, pyroxenes, and garnet; *eclogite* is a mafic rock composed chiefly of Na-rich clinopyroxene (omphacite) and Mg-rich garnet (pyrope).

Measured P-wave Velocities at High Pressures

Laboratory measurements of V_P at pressures up to 20 kbar provide valuable data regarding possible mineral assemblages in the upper mantle (Birch, 1960; Christensen, 1966). Measured P-wave velocities in rocks increase with increasing pressure in response to (a) the intrinsic effect of pressure on individual minerals, (b) compression of minerals, and (c) progressive closing of pore spaces. An increase in temperature above room temperature will decrease the measured velocities. These effects are approximately self-canceling, such that V_P measured at room temperature and 10 kbar pressure will be only ≤ 0.1 km/sec larger than actual velocities in the Earth at 10 kbar (about 35 km) (Kumazawa et al., 1971). Ranges of P-wave velocity at 10 kbar and rock densities at 1 bar for various mafic and ultramafic rocks are summarized in table 3.1. The data indicate that dunite, peridotite, garnet peridotite, and eclogite or some mixture of these rock types is consistent with observed P_n velocities at the Moho in most areas.

Table 3.1. P-wave Velocities at 10 kbar

	ρ(at 1 bar) (gm/cm^3)	V_P(10 kbar) (km/sec)
Dunite	3.2–3.3	8.0–8.3
Peridotite	3.2–3.3	8.0–8.3
Garnet Peridotite	3.3–3.4	8.0–8.4
Eclogite	3.4–3.5	8.1–8.5
Mafic Garnet Granulite	3.2–3.3	7.8–7.9
Amphibole Peridotite	3.0–3.2	7.7–7.9
Plagioclase Peridotite	3.0–3.2	7.7–7.9
Serpentinized (13%) Peridotite	~ 3.2	~ 7.8
Serpentinite (100% serpentine)	~ 2.5	~ 5.2

Note: Velocities calculated if direct measurements not available.
After Birch (1960); Christensen (1966); Kumazawa, Helmstaedt, and Masaki (1971).

Higher than normal temperatures and/or some combination of garnet granulite, or amphibole, plagioclase, or serpentinized peridotite, may account for regions with anomalous upper mantle. Measured velocities of highly serpentinized ultramafic rocks, however, are too low for most mantle areas exhibiting low P_n velocities. The high temperatures characteristic of most of these regions also exceed the stability of serpentine, thus greatly limiting the extent of even slightly serpentinized peridotite in the mantle.

P-wave velocity measurements at various orientations to rock fabrics substantiate the idea that differences in mineral alignment can produce significant anisotropy (Ave'Lallemant and Carter, 1970; Kumazawa et al., 1971). Differences in V_P of more than 15 percent have been found in some ultramafic samples and are related primarily to the orientation of olivine grains. Anisotropy in the upper mantle beneath ocean basins may be produced by recrystallization of olivine and other minerals accompanying sea-floor spreading with the [100] axes of olivine statistically oriented normal to ridge axes (the higher V_P direction).

Experimental Data Related to the Moho

Two rival hypotheses exist for the origin of the density change that is responsible for the Moho. One is that it is produced by a phase change from a low- to a high-pressure mineral assemblage, and the other is that it represents a compositional (and phase) change. The complexities in defining the Moho in some areas as well as the differing temperatures and pressures of the oceanic and continental Moho suggest that it does not have a single origin.

One origin proposed for the Moho is that it represents a phase change from serpentinite in the lower crust to peridotite in the upper mantle (Hess, 1955). Existing experimental data, however, seem to eliminate this mechanism as a possible cause for either the oceanic or continental Moho. The experimentally determined univariant reaction curves for the formation of serpentine or serpentine plus brucite from Mg-rich olivine intersect oceanic geotherms at depths much greater than observed Moho depths. Also, average seismic-wave velocities in the lower crust are rather constant (see table 4.1) and would necessitate a fortuitously constant degree of serpentinization over much of the Earth.

The question as to whether or not the continental Moho represents a phase change from gabbro in the lower crust to eclogite in the upper mantle has been a subject of considerable research. Data indicate that a garnet granulite transition zone 5–10 kb wide exists between gabbro and eclogite at 1100°C (fig. 3.12). The upper and lower stability boundaries of the garnet granulite assemblage at continental Moho depths depend critically on the extrapolation of high-temperature data, since metastable plagioclase and garnet nucleation problems have, thus far, prevented accurate determination of these boundaries at temperatures lower than about 1000°C. Estimates of

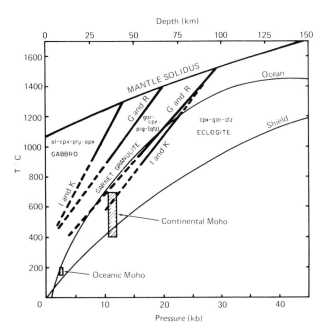

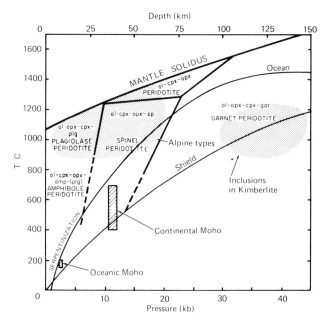

3.12. Pressure-temperature fields of mafic (basaltic) mineral assemblages. Also shown are oceanic and Precambrian shield geotherms and P-T ranges of the average oceanic and continental Moho. Oceanic geotherm is for a region 200 km from a rise and a spreading rate of 4 cm/yr. Boundaries of the garnet granulite field are after Ito and Kennedy, 1971 (I and K) and Green and Ringwood, 1972 (G and R). Abbreviations: ol = olivine; cpx = clinopyroxene; opx = orthopyroxene; gar = garnet; plg = plagioclase; qtz = quartz.

3.13. Pressure-temperature fields of ultramafic (periodotite) mineral assemblages (modified after Ringwood, 1969b). Also shown are oceanic and Precambrian shield geotherms and P-T ranges of the average oceanic and continental Moho. Oceanic geotherm is for a region 200 km from a ridge and a spreading rate of 4 cm/yr. Inferred minimum depths of formation of alpine-type ultramafics and of garnet periodotite inclusions in kimberlite (as deduced from clinopyroxene compositions) are also shown. Abbreviations: amp = amphibole; sp = spinel; others as given in Fig. 3.12.

the gradient for those boundaries range from 10 bar/°C (Ito and Kennedy, 1971) to 21 bar/°C (Green and Ringwood, 1972); extrapolations of each are shown in fig. 3.12, together with temperature and pressure estimates of the continental and oceanic Moho. The lower gradient extrapolation suggests that the transition from garnet granulite to eclogite may occur beneath some parts of the continental crust, while extrapolation of the steeper gradient suggests that eclogite is stable throughout the continental crust. The width of the extrapolated garnet granulite field at continental Moho temperatures using either gradient is too great to explain the sharpness of the Moho in most areas. Although it is not possible with existing data to determine which gradient is most accurate, the high seismic-wave velocities in eclogite severely limit its abundance in the crust. The absence of a correlation between surface heat flow and crustal thickness does not favor a gabbro-garnet granulite-

eclogite phase change at the Moho, since surface heat flow should reflect temperature distribution at depth and, hence, phase transition depths. Furthermore, it is difficult to see how eclogite could compose a major part of the near-Moho upper mantle, in that its average density (3.5) exceeds that of garnet peridotite (3.4) upon which it probably rests (fig. 3.13); hence, it would tend to sink. In subduction zones, however, phase transitions of mafic rocks to eclogite may enhance the sinking rate of the consumed plate (Chapter 6).

Experimental measurements of density changes that occur at the lower and upper boundaries of the garnet granulite field (reflecting the appearance of garnet and the disappearance of plagioclase, respectively) indicate that the density change at each of these boundaries is large enough to produce seismic discontinuities (Ito and Kennedy, 1971). It is possible that the two discontinuities found in some regions are produced by the gabbro-garnet granulite and garnet granulite-eclogite phase changes, respectively. Although this hypothesis is appealing, it encounters two major problems. First, experimental data of Green and Ringwood (1972) do not verify the existence of the density changes reported by Ito and Kennedy (1971). Second, it would seem to necessitate anomalously steep geothermal gradients to about 25 km in order for the geotherm to intersect both phase boundaries (fig. 3.12).

More experimental studies are needed to locate accurately the garnet granulite phase boundaries at lower temperatures. However, it seems clear that the Moho cannot represent a serpentinite-peridotite phase change, nor a single phase change from gabbro to eclogite. As will be discussed below and in the next chapter, a composition change from a lower crust of intermediate composition to an upper mantle composed dominantly of peridotite is most consistent with existing data.

Upper-Mantle Mineral Assemblages

Possible upper-mantle mineral assemblages of mafic and ultramafic compositions are shown in figs 3.12 and 3.13, respectively. These figures summarize phase equilibria studies at high pressures and temperatures. The mantle solidus—i.e., the temperature and pressure at which melting begins in the mantle—is assumed to be that of peridotite in both figures. Oceanic and shield geotherms and estimates of temperature and pressure ranges for the oceanic and continental Moho are also shown in the figures. Figure 3.12 for mafic compositions was discussed in the previous section.

The data in fig. 3.13 are consistent with an upper mantle beneath the continents composed chiefly of garnet peridotite. Stable ultramafic mineral assemblages beneath oceanic areas are dependent upon distance to active oceanic ridge systems. Most estimates of the range of oceanic geotherms are, however, consistent with an oceanic upper mantle at shallow depths (25–75 km) composed dominantly of spinel peridotite ($\pm$ amphibole or serpentinized peridotite) and, at greater depths, garnet peridotite. As previously mentioned, amphibole peridotite, plagioclase peridotite, and garnet granulite may occur in areas showing two discontinuities. Serpentinized peridotites appear to be produced at rather low temperatures and pressures (100–400°C; ≤ 25 km depth), probably during tectonic emplacement.

Ultramafic Rocks

There are six major occurrences of ultramafic rocks in the Earths's crust (after Wyllie, 1970):

1. Ultramafic zones of stratiform intrusions
2. Alkali intrusions and kimberlites
3. Dredge samples from ocean areas
4. Alpine-type ultramafics
5. Inclusions in basalt and kimberlite
6. Ultramafic flows and sills.

Stratiform intrusions are produced by fractional crystallization of basaltic magmas emplaced in the crust, and the ultramafic zones represent early differentiates. Alkali intrusions and kimberlite pipes appear to be highly fractionated (enriched in alkali, rare-earth, and volatile elements) and, in the case of kimberlites, contaminated in varying degrees by crustal rocks. For these reasons, it is unlikely that ultramafic rocks from either of these occurrences represent samples of the upper mantle. Some ultramafic rocks from the last four occurrences, however, probably do represent samples of the upper mantle.

Little is known about the occurrence of peridotites and serpentinites dredged from the ocean floors. Models for the oceanic crust, however (Chapter 4), suggest that oceanic ultramafic rocks occur as mantle diapirs that have penetrated the oceanic crust. Most of the samples are altered and serpentinized in varying degrees, and it seems doubtful that many or perhaps any represent fresh samples of the upper mantle beneath the oceans.

Alpine-type ultramafics are a rather heterogeneous group and include bodies of varying origins and modes of emplacement. Most appear to represent fragments of upper mantle (oceanic?) that have been emplaced on continental (or arc) crust by faulting or overthrusting (obduction—see Chapter 8) associated with subduction. Some may represent tectonically emplaced diapirs from the mantle. Concentric types appear to have been emplaced as magmas or crystal mushes of partially melted mantle material. A large proportion of alpine ultramafics have been altered, metamorphosed, deformed, and serpentinized in varying degrees. For this reason, extreme care must be used in interpreting the composition of these bodies in terms of the composition of the upper mantle.

Ultramafic inclusions are common in alkali basalts and in kimberlite pipes. Geochemical and mineralogical data indicate that inclusions of more than one origin may occur in the same lava flow or pipe. Some represent early fractionates of the host rock, while others appear to represent fragments of upper mantle that were picked up by the magmas at their source or during their ascent to the Earth's considerably less abundant than ultramafic inclusions, also occur in some alkali basalts and kimberlites.

Ultramafic flows and sills are rare. They occur in some Archean volcanic terranes and in some ophiolite complexes. Many of them appear to be Mg-rich basalts containing a large quantity of olivine crystals and hence should not be considered as complete melts of upper-mantle material.

Experimental studies of clinopyroxenes coexisting with olivine, orthopyroxene, and an Al-rich phase (such as spinel or plagioclase) indicate that Al, Mg, and Ca content of the clinopyroxene is sensitive to the temperature and pressure at which it crystallized (or recrystallized) (MacGregor, 1974). Hence, compositional data from clinopyroxenes in four-phase ultramafic rocks provide information on minimum depths of origin for these rocks. When this method is applied to a suite of ultramafic inclusions from a given locality, the results define a P-T curve which is thought to represent a paleogeotherm (Boyd, 1973). Such geotherms are similar to those estimated from geophysical and geochemical considerations, discussed previously. The depth ranges recorded by clinopyroxenes in alpine-type ultramafics and by garnet peridotite inclusions from kimberlites are outlined in fig. 3.13. Alpine-type ultramafic rocks record depths of at least 10–75 km (some of these are probably recrystallization depths), and those from garnet peridotites record depths greater than 100 km. Ultramafic inclusions from basalts fall between and significantly overlap these two groups.

Computerized Upper-Mantle Models

It is possible to construct computer models for the upper mantle using a large amount of available data (Cleary and Anderssen, 1979). One method involves the "mapping" of phase equilibria boundaries (from experimental studies) on a lithospheric plate for specified temperature distributions (Forsyth and Press, 1971). An example for the oceanic upper mantle, with a lithospheric plate 70 km thick and allowing both mafic and ultramafic mineral assemblages, is given in fig. 3.14. This model satisfies observed gravity, topography, Rayleigh- and Love-wave dispersion data, refraction data, seismic anisotropy, density distribution, temperature of surface lavas, and dredged

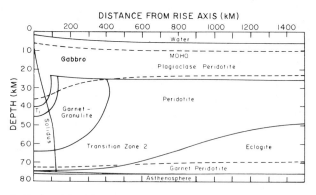

3.14. Model of the uppermost oceanic mantel for mixed mafic and ultramafic rocks (after Forsyth and Press, 1971). Model assumes a dry system, a spreading rate of 1 cm/year, and a solidus temperature of 1225°C. Boundaries between ultramafic assemblages are dashed lines; solid lines indicate boundaries between mafic assemblages T_1 = transition zone between garnet granulite and gabbro.

rock types from oceanic areas. It does not, however, explain a high-velocity zone detected in some parts of the oceanic upper mantle. Some pertinent features of the model are as follows: (1) the anomalous upper mantle beneath an oceanic ridge is explained in terms of garnet granulite and gabbro; (2) plagioclase peridotite occurs from the Moho down to about 20 km depth; (3) the lower part of the lithosphere is composed of about 50 percent eclogite (occurring probably as entrapped pockets in the peridotite); and (4) a large transition zone (Transition Zone 2) exists between garnet granulite and plagioclase eclogite.

Another approach involves generating a large number of random mantle models for both P- and S-wave data using Monte Carlo methods (Press, 1972) and then eliminating models by testing them against many different types of geophysical and geochemical data. This approach has been especially informative relative to the lower mantle. In general, solutions indicate that the upper mantle to about 150 km, although probably composed chiefly of garnet peridotite, must contain some eclogite, and that the lithosphere beneath the oceans is about 70 km thick and about 150 km thick beneath the continents.

Chemical Composition of the Mantle

Several approaches have been used to estimate the composition of the mantle: (1) using the compositions of various groups of ultramafic rocks; (2) using theoretical compositions calculated from geochemical considerations; (3) using compositions of various meteorite mixtures; and (4) using data from shock-wave studies and from ultra-high-pressure experimental studies. The mantle can be considered to be composed of depleted and undepleted regions. *Depleted mantle* is mantle in which most or all of the elements that do not readily substitute in mantle minerals (K, Rb, Ba, U, Th, Zr, rare earths, etc.) have been removed by partial melting episodes and concentrated into the crust (Chapter 7). *Undepleted mantle* still contains these elements, having escaped melting. Because of alteration, partial serpentinization, or complex metamorphic histories, it is hazardous to interpret ultramafic rocks found in the crust in terms of the composition of upper-mantle materials. Two estimates, however, of the major-element composition of the upper mantle based on various averages of ultramafic rocks are given in table 3.2 (columns 1 and 2) for comparison. Also shown is a theoretical undepleted upper-mantle composition (column 3), with "pyrolite" defined by the property that upon partial melting it will yield a basaltic magma and leave behind a refractory residue of alpine-type ultramafic material (Ringwood, 1975). It is constructed from a 1:3 mixture of Hawaiian tholeiitic basalt and synthetic alpine peridotite. Columns 4–6 show estimates of total mantle composition produced by mixing various meteorite classes in appropriate ratios to give a correct core/mantle mass ratio of 32/68.

Although variability exists between the estimates of mantle composition, it is clear that more than 90 percent of the mantle by weight can be expressed in terms of the system $FeO–MgO–SiO_2$ and that no other oxide exceeds 4 percent. The oxides CaO, Al_2O_3, and Na_2O collectively compose 5–10 percent of the mantle. More than 98 percent of the composition of the mantle can be expressed in terms of the six oxides listed in table 3.2. It is also noteworthy that the two estimates of upper-mantle composition based on ultramafic rocks exhibit lower Al_2O_3 and Na_2O and higher MgO than the meteorite or theoretical estimates. This may reflect a real depletion in aluminum in the upper mantle today, since the meteorite and theoretical estimates are designed to approximate more closely the composition of undepleted, primitive mantle. However, changes in composition during emplacement of the ultramafic rocks may also contribute to these apparent element differences.

The Low-Velocity Zone

As was previously discussed, the low-velocity zone is characterized by low seismic wave velocities, low Q, and high electrical conductivity. Any hypothesis about the origin of the low-velocity zone must explain these features. Among the causes considered for the low seismic wave velocities are: (a) an anomalously high temperature gradient, (b) a phase change, (c) a composition change, and (d) incipient melting. High temperature gradients seem unacceptable in terms of

Table 3.2. Estimates of the Composition of the Mantle

	1	2	3	4	5	6
SiO_2	44.5	44.2	45.2	46.0	48.1	43.2
MgO	41.7	41.3	37.5	38.1	37.1	38.1
FeO	8.7	8.3	8.4	8.6	12.7	9.3
Al_2O_3	2.6	2.7	3.5	3.6	3.0	3.9
CaO	2.2	2.4	3.1	3.1	2.3	3.7
Na_2O	0.3	0.3	0.6	0.6	1.1	1.8

1. Average of 168 ultramafic rocks (White, 1967).
2. Average of oceanic serpentinites (calculated water-free) and high-Ca–high-Al olivine inclusions (Harris, Reay, and White, 1967).
3. Pyrolite—a 1:3 mixture of basalt (Hawaiian tholeiite) with a synthetic alpine peridotite (Ringwood, 1975).
4. Mixture of 40% Type I carbonaceous chondrites, 50% ordinary chondrites, and 10% iron meteorites such that mantle/core ratio equals 68/32; mantle recalculated to 100% (after Murthy and Hall, 1970).
5. Mixture of 32.4% iron meteorite (with 5.3% FeS) and 67.6% oxide portion of bronzite chondrites; mantle recalculated to 100% (Mason, 1966).
6. Nonvolatile portion of Type I carbonaceous chondrites with FeO/FeO + MgO = 0.12 (Ringwood, 1975).

measured heat flow on the surface. Also, thermal gradients required to satisfy low P-wave velocities do not satisfy low S-wave velocities. No important phase changes are likely to occur in this depth range, and compositional changes (such as an increase in the amount of garnet granulite) either do not seem probable geochemically or do not readily explain the low Q and high electrical conductivity. Most data seem consistent with incipient melting as the cause for the low velocities, low Q, and high electrical conductivity in this zone. The rather sharp boundaries of the low-velocity zone and the high surface heat flow observed when it occurs at shallow depths are also compatible with incipient melting.

Experimental data show that to produce incipient melting in this region, a minor amount of water is required to depress silicate melting points (Wyllie, 1971b). Melting conditions for ultramafic and mafic mantle compositions under dry and slightly hydrous (0.1 percent water) conditions are shown in fig. 3.15. Ocean-basin and shield geotherms are also given for reference. It is clear that even with only 0.1 percent water, partial melting of both garnet peridotite and eclogite will occur in the appropriate depth range for the low-velocity zone. The source of water in the upper mantle may be from the breakdown of minor mantle phases that contain water such as hornblende (breakdown curve shown in fig. 3.15). mica, titan-clinohumite, or hydrated silicates. The amount of melt produced is very small and is almost a direct function of the water content. The theory of elastic wave velocities in two-phase materials indicates that only 1 percent melt is required to produce the lowest velocities measured in the low-velocity zone (Anderson et al., 1971). If, however, melt fractions are interconnected by a network of tubes along grain boundaries, the amount of melting may exceed five percent (Marko, 1980). The thickness of the zone of incipient melting varies with the geotherm and the ratio of peridotite to eclogite (for a given water content, the zone is thicker in eclogite than peridotite).

The downward termination of the low-velocity zone may be caused by one or a combination of several effects:

1. Rapid decrease in the amount of water available;

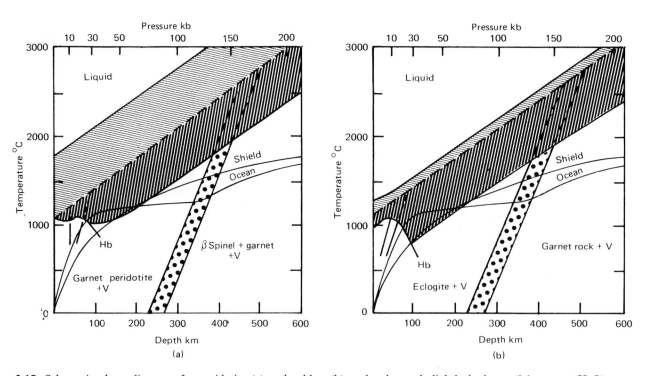

3.15. Schematic phase diagrams for peridotite (a) and gabbro (b) under dry and slightly hydrous (0.1 percent H_2O) conditions extrapolated to pressures of 200 kb (after Wyllie, 1971b). The melting interval consists of two parts: the bold-lined areas represent incipient hydrous melting intervals and the close-spaced line areas represent dry melting intervals. The bold dotted areas represent transition zones between mineral assemblages. Hb = upper stability limit of hornblende; V = water vapor.

2. Water enters high-pressure hydrous silicate minerals (such as hydrous pyroxenes and garnets);
3. The bottom of the zone coincides with the lower limit of eclogite (for the same amount of water there would be much less liquid in underlying peridotite than in eclogite-bearing peridotite);
4. The geotherm passes through the mantle solidus, as illustrated in fig. 3.15.

The low-velocity zone plays a major role in plate tectonics, providing a relatively low-viscosity region upon which lithospheric plates can slide with very little friction. Lateral motion of plates away from oceanic rises and consumption of those plates in subduction zones are offset by return flow deeper in the mantle. The fact that the low-velocity zone is absent or poorly developed beneath Precambrian shields suggests that the roots of shield areas may produce drag on lithospheric plates. Calculations, however, indicate that this drag is small compared to the drag produced in subduction zones.

The Transition Zone

The 400 km Discontinuity. High-pressure experimental studies document the breakdown of Mg-olivine to a high-pressure phase known as β-phase between 109 and 116 kbar (Ringwood and Major, 1970). Existing data indicate that mantle olivine is Fo_{89-90} in composition. Using the phase diagram for the system $Mg_2SiO_4-Fe_2SiO_4$ at 1000°C, Fo_{89} begins to transform to γ-spinel at 109 kbar. As pressure increases to 116 kbar, the amount of γ-spinel and its magnesium content also increase. At this point, it reacts to form β-phase by 118 kbar with the total increase in density of 10.6 percent. Assuming a temperature of 1600°C at 400 km depth in the Earth and a gradient of 30 bar/°C, the entire phase diagram is raised by 18 kbar, with the median pressure for the transformation occurring at a depth of 397 km and spread over a width of 27 km. This is in excellent agreement with the observed depth and width of the 400 km discontinuity. Since olivine is probably the most abundant mineral in the upper mantle, and the density increase of the phase change is in good agreement with the observed increase in seismic wave velocities, these data are consistent with the hypothesis that the olivine → β-phase transition is responsible for the 400 km discontinuity.

Other high-pressure experimental data indicate that at depths from about 350 km to 400 km,

aluminous orthopyroxenes are transformed into garnet, involving a density increase of about 10 percent (Ringwood, 1975). It is possible that the increase in velocity gradient commonly observed beginning at 350 km (fig. 3.2) and leading up to the 400 km discontinuity is caused by such pyroxene transformations.

The 650 km Discontinuity. The ultra-high-pressure experimental equipment of the last decade has made it possible to study phase transformations that may be responsible for the 650 km discontinuity. Data indicate that $(Mg,Fe)_2SiO_4$ spinel (Fo_{89}) undergoes a transition at about 260 kbar and 1000°C to a mixture of orthorhombic perovskite and periclase as follows (Liu, 1976):

$$(Mg,Fe)_2SiO_4 \rightarrow (Mg,Fe)SiO_3 + (Mg,Fe)O$$
spinel   perovskite   periclase

The reaction involves an increase in zero-pressure density of about 11 percent and corresponds closely with the observed depth of the 650 km discontinuity. At about the same pressure and temperature, pyrope garnet transforms to an ilmenite structure, which is followed by a transformation to an orthorhombic perovskite structure at about 300 kbar (Liu, 1979). These transformations involve increases in density of 7.9 and 7.7 percent, respectively. Existing data favor a combination of the spinel and garnet transformations to explain the 650 km discontinuity. The overall density increases for a peridotite mantle (~10%) and the depth interval over which these transformations occur (a few kilometers) correspond well with those observed at the 650 km discontinuity.

Summary of Mineralogical Changes to 1000 km. Possible mineral assemblages in the mantle to 1000 km depth and mineral transformations as determined largely from experimental data are summarized in table 3.3 The upper mantle to 400 km is composed chiefly of olivine, pyroxenes, and garnet. Incipient melting of these minerals, probably caused by the presence of a small amount of water, produces the low-velocity zone underlying all of the lithosphere except some Precambrian shield areas. The change in velocity gradient beginning at about 350 km may reflect transformation of Al-rich orthopyroxene to garnet. The 400 km discontinuity, recording a density change from 3.4 to 3.6, is probably caused by the olivine → β-phase transition. The uppermost part of the transition zone is a rather homogeneous region composed chiefly of β-phase, garnet and minor Na-

Table 3.3. Summary of Possible Mineral Assemblages of the Upper Mantle and Transition Zone

	Depth Range (km)	Mineralogy		Density (zero pressure)
Upper Mantle	up to 350	Olivine, Fo_{89} (ol)$(Mg,Fe)_2SiO_4$ Orthopyroxene (opx)$(Mg,Fe)SiO_3$ Clinopyroxene (cpx)$(Ca,Mg,Fe)_2(NaAl)(Si_2O_6)$ Pyrope garnet (gar)$(Mg,Fe,Ca)_3(Al,Cr)_2Si_3O_{12}$	*58% 17% 12% 14%	3.4
400 km Discontinuity	350–400 400–430	(Mg,Fe,Al) opx $\rightarrow$ (Mg,Fe) gar (Mg,Fe) ol $\rightarrow$ (Mg,Fe) β-phase		
Upper Transition Zone	400–550	(Mg,Fe) β-phase (Mg,Fe,Ca) gar $(NaAl)$ cpx (Ca) gar $\rightarrow$ (Ca) perovskite	57% 40% 3%	3.6
	500–550	(Mg,Fe) β-phase $\rightarrow$ (Mg,Fe) spinel		
	550–650	(Mg,Fe) spinel (Mg,Fe) gar (Ca) perovskite $(NaAl)$ cpx	57% 38% 2% 3%	3.7
650 km Discontinuity	650–700	(Mg,Fe) spinel $\rightarrow$ (Mg,Fe) perovskite + (Mg,Fe) periclase (Mg,Fe) gar $\rightarrow$ (Mg,Fe,Al) ilmenite $(NaAl)$ cpx $\rightarrow$ $(NaAl)$ Ca–ferrite + (SiO_2) stishovite		
Lower Transition Zone	700–1050	(Mg,Fe) perovskite (Mg,Fe) periclase (Mg,Fe) ilmenite (Ca) gar $(NaAl)$ Ca–ferrite (SiO_2) stishovite	70% 20% } 10% 	4.0

*Estimated weight percent of each mineral. After Ringwood (1975) and Liu (1979).

rich clinopyroxene. If excess Al_2O_3 is available, the Mg–Fe–Ca component of the clinopyroxene may be incorporated in the garnet solid solution; if not, it would transform to a high-pressure phase as yet unidentified. Ca garnet may also invert to cubic perovskite ($CaSiO_3$), which is stable in this depth range. At 500–550 km, β-phase will transform to spinel, involving only a 2.5 percent increase in density. This transition may account for mantle discontinuities observed in the 500 km depth range. The lower part of the transition zone is composed chiefly of (Mg,Fe) spinel and (Mg,Fe) garnet and has a density of about 3.7 gm/cm³.

The 650 km discontinuity would appear to be largely due to the breakdown of spinel to perovskite and periclase and the transition of garnet to an ilmenite structure. Na clinopyroxene may also break down to a phase with a Ca-ferrite structure plus stishovite at about this depth, although contributing less than one percent to the density increase. (Mg,Fe) ilmenite should invert to an orthorhombic perovskite structure at about 800 km. Hence, it would appear that the lower part of the transition zone in the mantle is comprised largely of an orthorhombic (Mg,Fe,Al) perovskite phase and $(Mg,Fe)O$ periclase.

The Lower Mantle

High-pressure data and studies of density distribution in the lower mantle indicate that this region is about 5 percent more dense than an oxide mixture with the approximate bulk composition of garnet peridote. Two explanations have been offered for this increase in density.

1. Further mineral transformations in the lower mantle to assemblages with a density greater than an isochemical oxide mixture, as mentioned above; or
2. An increase in the Fe/Mg ratio in the lower mantle.

In principle, a choice between these alternatives can be made by comparing seismic elastic ratios obtained from observed seismic velocities with those estimated from shock-wave studies. However, uncertainties in estimating elastic-ratio temperature corrections for the lower mantle allow either (or both) of the above interpretations at the present time. Recent studies suggest, however, that only phase assemblages dominated by $(Mg,Fe)SiO_3$ perovskite have sufficiently high densities and seismic parameters to be compatible with the observed density and seismic-wave velocity distribution in the lower mantle. One or more seismic discontinuities occur at about 1000 km and several velocity gradients are observed at greater depths (fig. 3.6). These may reflect further transformations to higher pressure phases or, alternately, changes in the spin state of Fe^{+2} in silicates. At very high pressures, Fe^{+2} may undergo a contraction in radius due to a coupling of d electrons. Evaluation of these alternatives awaits experimental studies with ultra-high-pressure equipment.

THE CORE

Physical properties

Seismological data indicate that the radius of the core is 3485 ± 3 km and that the outer core does not transmit seismic waves. This latter observation is generally interpreted to mean that the outer core is in a liquid state (Lilley, 1979). The inner core, with a radius of 1220–1230 km, transmits S waves at very low velocities, suggesting that it is a solid near the melting point or partly molten. Most recent estimates for the electrical conductivity of the core are in the range of $2–5 \times 10^5 \Omega^{-1} m^{-1}$. Recent estimates of the temperature at the inner core-outer core boundary are about 3700°C and, at the core-mantle boundary, about 3000°C.

A great deal of evidence, especially the time-dependent nature of the Earth's magnetic field, indicates that the magnetic field is generated within the core by dynamo action. Flow patterns in the outer core, which are responsible for the geodynamo, are poorly known, however. Proposed flows range from oscillations of a stable stratified fluid to turbulent convection. Failure to fit density and P-wave velocity data simultaneously by finite strain theory indicates that the outer core is not well mixed and may be crudely zoned (Anderson et al., 1971). The near coincidence of the magnetic and rotational poles of the Earth suggests that the daily rotation of the core strongly influences fluid motion. The geodynamo requires a great deal of energy, major sources of which are the cooling of the core, possible radioactivity in the core, and gravitational settling within the core.

Composition

Shock-wave experimental results limit the major constituents in both the outer and inner core to a mixture of such elements as Fe, Ni, V, and Co. Cosmochemical and nucleosynthesis considerations show that of these elements, only Fe is sufficiently abundant in the solar system to form a major part of the core. By analogy with meteorites, it would appear that an Fe–Ni mixture, dominated by Fe, comprises most of the Earth's core (table 3.4). Refinement of seismic velocity and shock-wave data, however, indicates that the outer core is 8 to 15 percent less dense than an Fe–Ni mixture and requires the presence of one or more elements with low atomic weight. The inner core, which is near its melting point or partially molten, need not contain light elements and may be enriched in Ni relative to the outer core. Shock-wave data indicate that the outer core must contain 5–15 percent of one or more of such elements as H, C, N, O, Si, and S (Brett, 1976). Experimental data are consistent with either sulfur or oxygen being the principal light element in the outer core (Murthy and

Table 3.4. Estimates of the Composition of the Earth's Core

	1	2	3
Fe	85.5	80	85
Ni	7.5	5	5
S	6.0	15	—
O	—	—	10

1. Iron meteorite composition (6% sulfur).
2. Iron meteorite composition (15% sulfur).
3. FeO model (Ringwood, 1979).

Hall, 1970; Ringwood, 1979) (table 3.4). One problem, if sulfur is the principal light element, is that of explaining why such a highly volatile element is not as depleted in the Earth as are other elements that are less volatile (such as Cr, Mn, Na, and K). Experimental high-temperature data indicate that FeO should be highly soluble in iron, and hence provide support for the idea that oxygen is the major light element in the core.

Origin

Heat generation and isotopic results indicate that the Earth's core formed rapidly, probably within 100–200 m.y. after accretion. This requires that the melting point of the iron alloy be exceeded throughout large volumes of the Earth's interior such that the metal can segregate and sink to the center. However, estimated temperature distributions in the Earth soon after accretion are considerably lower than the melting curve of iron at equivalent depths (Ringwood, 1979). The most likely solution to this problem is that either sulfur (in FeS) or oxygen (in FeO) entered the metal phase early and lowered its melting point. Experimental data are consistent with either possibility. The actual mechanism of segregation and core growth is poorly understood. Elsasser (1963) proposed one model whereby metal segregates into large "drops" in the upper mantle. These drops sink to the Earth's center as they heat their surroundings and lower the viscosity. If oxygen is the principal light element in the core, initial melting and segregation may have occurred in the lower mantle, thus depleting the lower mantle in iron. This would lead to a gravitationally unstable situation where less dense mantle underlies more dense mantle. Hence, the metal-free light material would rise and displace the overlying iron-silicate mixture, which would sink and undergo metal-phase separation.

SUMMARY STATEMENTS

1. The Moho occurs from 10 to 12 km beneath oceans and generally from 30 to 50 km beneath continents. Mantle P-wave (P_n) velocities at the Moho are generally in the range of 7.9–8.1 km/sec. Low P_n velocities (≤ 7.8 km/sec) define anomalous upper mantle and in some areas two discontinuities can be detected. Significant P_n velocity anisotropy has been found in some oceanic areas.

2. The mantle beneath the continents is commonly characterized by a high-velocity lid ($V_P \geq 7.9$ km/sec) underlain by a low-velocity zone ($V_P < 7.8$ km/sec). This zone is also characterized by low Q and high electrical conductivity. The high-velocity lid is absent under some segments of the crust (such as continental rifts and oceanic ridges) and a P-wave low-velocity zone is absent beneath Precambrian shields. Major discontinuities beneath the continents occur at 400 km and 650 km, and a minor discontinuity occurs at 1050 km.

3. Geophysical data indicate that the mantle is laterally inhomogeneous to the base of the low-velocity zone (and to greater depths for S waves), and that it is coupled to the crust down to the top of the low-velocity zone.

4. Most large-scale gravity anomalies in the Earth can be correlated with lithosphere-asthenosphere interactions or with mantle convection patterns.

5. Measured seismic velocities at high pressures and the results of high-pressure phase equilibria studies indicate that the upper mantle is composed chiefly of olivine, pyroxenes, and garnet and that garnet peridotite is probably the dominant rock type. Neither a peridotite-serpentinite phase change nor, at least in most areas, a gabbro-garnet granulite-eclogite phase change can readily explain the Moho. Most data are consistent with a composition change from a lower crust of diorite composition to a dominantly ultramafic upper mantle.

6. Estimates of the chemical composition of the mantle indicate that it is more than 90 percent Mg, O, Fe, and Si and that more than 98 percent of its composition can be described in terms of the six oxides MgO, FeO, SiO_2, CaO, Al_2O_3, and Na_2O.

7. The upper-mantle low-velocity zone is probably produced by incipient melting of ultramafic rocks ($\pm$ eclogite) in the presence of small amounts of water.

8. The 400 km discontinuity appears to be caused primarily by the $(Mg,Fe)_2SiO_4$ olivine $\rightarrow \beta$-phase transition, with a minor contribution from the Al-rich orthopryoxene $\rightarrow$ garnet transition.

9. Ultra-high-pressure experimental data suggest that the 650 km discontinuity is produced by a combination of the breakdown of $(Mg,Fe)_2SiO_4$ spinel to $(Mg,Fe)SiO_3$ perovskite plus (Mg,Fe) periclase and the transformation of pyrope garnet to an ilmenite structure.

10. Shock-wave studies indicate that the lower mantle is denser than an oxide mixture with the bulk composition of garnet peridotite. This may be accounted for by either phase transformations to more closely packed phases than oxides or an increase in the Fe/Mg ratio in the lower mantle, or both.

11. Fluid motions within the outer core appear to act as a geodynamo and to generate the Earth's magnetic field. Seismic-wave velocity data and shock-wave experimental studies indicate that the outer core is liquid Fe-Ni, it is not homogeneous, and it contains 5–15 percent of one or more low-atomic-weight elements. The inner core is comprised of an Fe–Ni alloy that is near its melting point or partially molten.

SUGGESTIONS FOR FURTHER READING

Jacobs, J.A., Russell, R.D., and Wilson, J.T. (1974) *Physics and Geology*. New York: McGraw-Hill. 622 pp.

Jacobs, J.A. (1975) *The Earth's Core*. London: Academic Press. 253 pp.

McElhinny, M.W., editor (1979) *The Earth: Its Origin, Structure and Evolution*. London: Academic Press. 597 pp.

Ringwood, A.E. (1975) *Composition and Petrology of the Earth's Mantle*. New York: McGraw-Hill. 618 pp.

Robertson, E.C., editor (1972) *The Nature of the Solid Earth*. New York: McGraw-Hill. 677 pp.

Wyllie, P.J. (1971) *The Dynamic Earth*. New York: Wiley. 416 pp.

Chapter 4

The Crust

CRUSTAL TYPES

As was discussed in the last chapter, the crust is that region of the Earth above the Moho. There are three major crustal divisions—oceanic, transitional, and continental—of which only oceanic and continental divisions are of major importance. Typically, *oceanic crust* ranges from 5 to 15 km thick and comprises 59 percent of the total crust by area. Islands, island-arcs, and continental margins are examples of *transitional crust* that exhibit thicknesses of 15–30 km. *Continental crust* generally ranges from 30 to 50 km thick, with thicknesses up to 80 km reported in some areas. The continental crust (including most transitional crust) comprises 79 percent of the total crust by volume.

The Earth's crust can be further subdivided into crustal types (Brune, 1969). A *crustal type* is herein defined as a segment of the crust exhibiting similar geological and geophysical characteristics. The crust can be adequately described in terms of 12 major crustal types, listed in table 4.1 with some of their physical properties. The data in this table will be referred to throughout the chapter. The first two columns of the table, together with fig. 4.1, summarize the areal and volume abundances of various crustal types. Column 3 describes tectonic stability, and reflects chiefly the prevalence of earthquake activity and recent deformation. Crustal types are categorized into stable (S) (little or no earthquake activity), intermediate or variably stable (I), or unstable (U) areas. When crustal types have more than one stability entry in the table, the first entry is most characteristic. Brief geological descriptions of each major crustal type are given below. Reference should

also be made to the tectonic map of the world in the pocket (Plate I).

Shields

Shields are stable parts of the crust composed of Precambrian rocks with little or no sediment cover. Constituent rocks within a specific shield may range in age from 0.5 to >3.5 b.y. Metamorphic and plutonic rock types dominate, and temperature-pressure regimes recorded in the exposed rocks record burial depths ranging from as shallow as 5 km to as deep as 35 km or more. Shield areas, in general, exhibit very little relief and have remained tectonically stable for long periods of time. They comprise about 12 percent of the total crust by volume, with the largest shields occurring in Africa, Canada, and Antarctica.

Platforms

Platforms are also stable parts of the crust with little relief. They are composed of Precambrian basement rocks similar to those exposed in shields, which are typically overlain by 1 to 3 km of relatively undeformed sedimentary rocks. Shields and the Precambrian basement of platforms are collectively referred to as a *craton*. A craton is generally considered as an isostatically positive portion of the continent that is tectonically stable relative to adjacent mobile belts (see Chapter 5). The age of the sedimentary cover in platforms may range from Precambrian to Cenozoic and may locally reach thicknesses up to 5 km, for instance as found in the Williston basin of the north-central United States. Platforms comprise most of the crust in terms of volume (35 percent) and most of the continental crust in terms of both area and volume (fig. 4.1 and Plate I).

Paleozoic Orogenic Belts

Paleozoic orogenic belts are long, curvilinear belts of Paleozoic folding, faulting, and igneous activity. They range from several hundred to several thousand kilometers in width, and extend for thousands of kilometers in length (Plate I). They are composed of a variety of rock types and are characterized by low, often highly eroded mountain ranges or regions with little relief. They are generally characterized by intermediate or stable tectonic conditions.

Mesozoic-Cenozoic Orogenic Belts

Mesozoic and Cenozoic orogenic belts are similar in size and in rock types present to Paleozoic orogenic belts. They differ from the latter in that they are tectonically unstable. The most extensive Mesozoic belts are the Cordilleran in western North and South America and similar belts in eastern Asia (Plate I). The two most extensive orogenic belts of Cenozoic age are the Cordilleran and the Alpine-Tethyan belt crossing southern Eurasia. *Plateaus* occur in both Paleozoic and Mesozoic-Cenozoic belts and are characterized by uplifted crustal areas that have escaped the intensive deformation characterizing most of an orogenic belt. Examples are the Colorado and Tibet plateaus.

Continental Rift Systems

Continental rift systems are fault-bounded valleys ranging in width from 30 to 75 km and in length from a few tens to thousands of kilometers. They are characterized by a tensional tectonic setting. The longest rift system on the Earth today is the East African system, which extends over 6500 km from the western part of Asia Minor to southeastern Africa (Plate I). Rift systems range from a simple *graben* (a down-dropped block between two normal faults) or *half-graben* (a tilted block with one or more normal faults on one side) to a *complex graben system* (a down-dropped or tilted block cut by many normal faults, many or most of which are buried by sediments in the graben). The last is most common. The *Basin and Range Province* in western North America is a multiple rift system composed of a complex series of alternating graben valleys and ridges (*horsts*). Rift systems may occur in rocks of any age, and young

systems (≤ 30 m. years) are characterized by unstable tectonic conditions. Some large volcanic plateaus such as the Columbia River Plateau in the northwestern United States and the Deccan Traps in India also appear to be related to extensive tensional faulting (see Chapter 7).

Volcanic Islands

Oceanic islands are of diverse types and origins. Some occur as island arcs, others as volcanic peaks in oceanic areas, and still others result from shallow flooding of the continental shelves. Only islands of primarily volcanic origin not associated with subduction zones are included here. Island-arcs are considered a separate crustal type. Islands formed by flooding of continental shelves (such as the Arctic archipelago) or by continental drift (such as Greenland and Madagascar) are considered continental crust belonging to one of the crustal types described above. Volcanic islands occur on or near oceanic ridges (e.g., St. Paul Rocks and Ascension in the Atlantic) and in ocean basins (such as the Hawaiian Islands). They range in tectonic stability from intermediate or unstable inactive volcanic islands to stable (such as Easter Island in the South Pacific). Their size may range from very small (< 1 km^2) to the size of Iceland.

Island-Arc Systems

Island-arcs occur above active or recently extinct subduction zones. They are generally arcuate chains of islands ranging in size from < 1 km^2 to islands the size of Honshu in Japan and New Guinea in Indonesia. Arc systems may be continuous with peninsular areas (such as the Kurile Islands and Kamchatka). They are composed dominantly of young calcalkaline volcanic and plutonic rocks and derivative sediments. Modern arcs are characterized by intense earthquake activity and volcanism and by variable physical properties. Continental analogues of island-arcs occur in regions (e.g., Central America and the Andes in South America) where subduction zones plunge beneath continental borderlands.

Trenches

Oceanic trenches mark the beginning of subduction zones which are associated with intense earthquake activity. Trenches parallel island-arcs or volcanic chains above subduction zones along continental margins. They range in depth from 5 to 8 km, repre-

Table 4.1. Physical Properties of the Earth's Crust

		1	2	3	4	5	6
	Major Crustal Types	Percentage	Abundance	Tectonic Stability[1]	Heat Flow[7] (μ cal/cm^2 sec)	Bouguer Anomaly (mgal)	Total Thickness[3] (km)
		area	volume				
1.	Shield	6	12	S	1.0	−10 to −30	35
2.	Platform	18	35	S,I	1.3	−10 to −50	41
3.	Paleozoic orogenic belt	8	14	I,S	1.5		43
4.	Mesozoic-Cenozoic orogenic belt[2]	6	13	U,I	1.8	} −200 to −300	40
5.	Continental rift system[2]	<1	<1	U	⪆2.5		28
6.	Volcanic island[2]	<1	<1	I,U,S	⪆2.5		
	a. Hawaii					+250	14
	b. Iceland					−30 to +45	12
7.	Island-arc[2]	3	3	U	⪆2.0	−50 to +100	22
8.	Trench[2]	3	2	U	1.2	−100 to −150	8 (14)
9.	Ocean basin	41	11	S	1.3	+250 to +350	7 (11)
10.	Oceanic ridge[2,5]	10	5	U	>5	+200 to +250	5 (6)
11.	Marginal-sea basin[2]	4	3	U,I	⪆1.5	+50 to +100	9 (13)
12.	Inland-sea basin[2]	1	2	I,S	1.3	0 to +200	22 (25)
	Average Continent	41	79		1.3	−100	40
	Average Ocean	59	21		1.6 (2.3)	+250	7 (10)

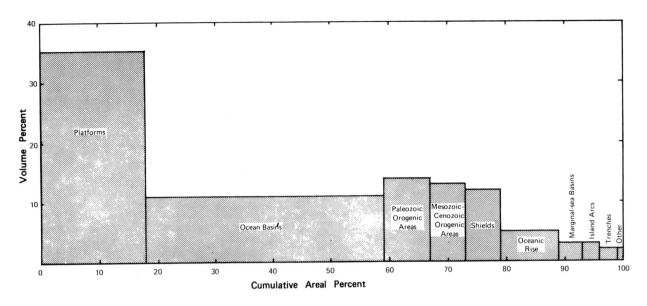

4.1. Areal and volume distribution of major crustal types.

Table 4.1 (cont'd.)

	7	8	9	10	11	12
Major Crustal Types	V_{P_n} (km/sec)	V_P Upper Layer (km/sec)	Thickness Upper Layer (km)	V_P Lower Layer (km/sec)	Thickness Lower Layer (km)	Average Crustal V_P[4] (km/sec)
1. Shield	8.1	6.1	16	6.6	19	6.3
2. Platform	8.1	6.1	17	6.8	21	6.5
3. Paleozoic orogenic belt[2]	8.1	6.1	16	6.7	23	6.4
4. Mesozoic-Cenozoic orogenic belt[2]	≤ 8.2	6.0	17	6.8	19	6.4
5. Continental rift system[2]	≤ 7.8	6.1	17	6.9	11	6.4
6. Volcanic island[2]						
a. Hawaii	8.2	5.1	6	7.1	6	6.1
b. Iceland	7.2	5.0	5	6.5	7	6.1
7. Island-arc[2]	≤ 8.0	5.5	12	6.6	12	6.1
8. Trench[2]	8.0	5.0	3	6.6	5	6.4
9. Ocean basin	8.2	5.1	1.5	6.8	5	6.4
10. Oceanic ridge[2,5]	≤ 7.5	5.0	1	6.5	4	6.4
11. Marginal-sea basin[2]	≤ 7.9	5.0	2	6.6	5	6.2
12. Inland-sea basin[2]	8.0	4.0[6]	12[6]	6.5	10	6.0
Average Continent	8.0	6.1	17	6.8	21	6.4
Average Ocean	8.1	5.1	1.5	6.6	5	6.4

[1] S = stable
   I = intermediate or variable stability
   U = unstable
[2] Properties vary considerably.
[3] Depth to Moho from sea level given in parentheses.
[4] Excluding sedimentary layer except for inland sea.
[5] Excluding Iceland.
[6] Sedimentary layer.
[7] To express as mW/m^2 multiply by 41.8

senting the deepest parts of the oceans, and contain a small amount of sediment derived chiefly from nearby arc or continental areas.

Ocean Basins

In terms of area, ocean basins comprise most of the Earth's surface (see fig. 4.1). They are tectonically stable and are characterized by a thin, deep-sea sediment cover (approximately 0.3 km thick) and linear magnetic anomalies. The sediment layer thickens along continental slopes where detrital sediments are supplied from the continents. Although ocean basins are rather flat, they contain abyssal hills, seamounts, guyots, and oceanic islands. *Abyssal hills*, which are the most abundant topographic feature on the ocean floor, are sediment-covered hills ranging in width from 1 to 10 km and in relief from 50 to 1000 m.

Seamounts are submarine volcanoes (both active and extinct), and *guyots* are flat-topped submarine volcanoes produced by erosion at sea level followed by submersion, probably due to sinking of the ocean floor. Coral reefs grow on some guyots as they sink, producing *atolls*.

Oceanic Ridges

Oceanic ridges are linear belts occurring near the centers of the oceans (Plate I). They are topographic highs and are usually tectonically unstable. A rift valley system commonly occurs near their crests in which new oceanic lithosphere is produced by intrusion and extrusion of basaltic magmas as sea-floor spreading continually moves the lithosphere away from the ridge axis. The relief in the rift zones varies from rough, mountainous terranes, such as those

characterizing the Mid-Atlantic Ridge, to more subdued relief, such as that characterizing the East Pacific
Rise. In general, rough topography occurs on
slow-spreading ridges, while smooth topography is
more common on fast-spreading ridges. Fast-spreading ridges also commonly lack axial rifts. Iceland is
the only known example of a surface exposure of a
modern oceanic ridge system. The worldwide oceanic
ridge system is interconnected from ocean to ocean
and is more that 80,000 km long. Ridge crests are cut

by numerous transform faults (see Chapter 6), which
may offset ridge segments by thousands of kilometers.

Marginal-Sea Basins

Marginal-sea basins are segments of the oceanic crust
located between island-arcs (as the Philippine Sea) or
between an island-arc and a continent (as the Japan

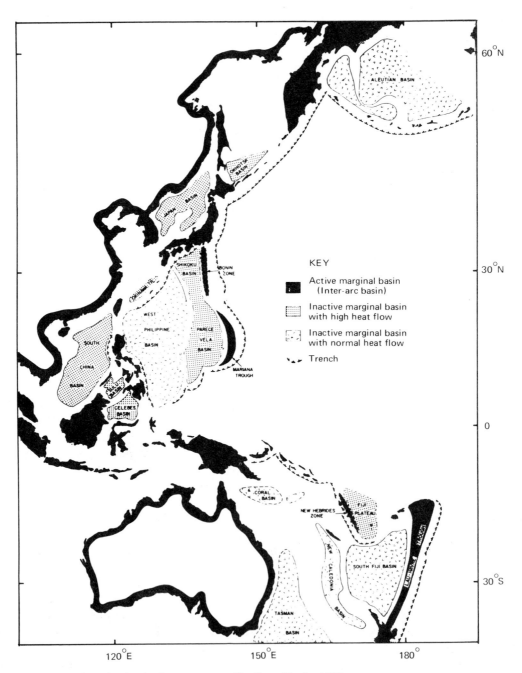

4.2. Marginal-sea basins in the western pacific (from Karig, 1971).

Sea and Sea of Okhotsk). They are most abundant today in the western Pacific (fig. 4.2), and are characterized by a horst-graben topography (similar to the Basin and Range Province) with major faults subparalleling the adjoining arc system(s) (Karig, 1971). The thickness of sediment cover is variable, and sediments appear to be derived chiefly from continental or arc areas. Marginal-sea basins are classified depending on whether they are tectonically *active* or *inactive* (fig. 4.2). Tectonic activity is inferred from thin sedimentary cover, rugged horst-graben topography, and high heat flow. Examples of active (or interarc) basins are the Lau-Havre and Mariana troughs and typical inactive basins are the Tasman and West Philippine basins.

Inland-Sea Basins

Inland-sea basins range from basins that are completely surrounded by continent (e.g., the Caspian Sea) to marginal seas not associated with active arc systems, which are nearly completely surrounded by continents (such as the Black Sea and the Gulf of Mexico). Tectonically, they vary from regions of intermediate stability to stable regions. Sediment thicknesses in inland seas vary from as little as a few kilometers in parts of the Mediterranean Sea to over 20 km in the Caspian Sea. Parts of the Mediterranean Basin are most appropriately classified as marginal-sea basins since they occur adjacent to active arc-trench systems (viz., the Balearic and Aegean basins). Mud and salt diapirs occur in sediment layers of some inland-sea basins.

SEISMIC FEATURES

The structure of the crust is determined chiefly from body-wave refraction studies and surface-wave dispersion studies. Crustal models based on seismic data indicate that the oceanic crust can be broadly divided into three layers, which are, in order of increasing depth: the *sediment layer* or layer 1 (0–1 km thick), the *basement layer* or layer 2, (0.7–2.0 km thick), and the *oceanic layer* or layer 3 (3–7 km thick). Models for the continental crust show a greater range in both number and thicknesses of layers. Although two- or three-layer models for the continental crust are most common, one-layer models and models with more than three layers are proposed in some regions. Generally, the continental crust can be considered to be

composed of a *sediment layer* (0–5 km thick) followed by an *upper layer* (10–20 km thick) and a *lower layer* (15–25 km thick). The boundary between the upper and lower layers in the continental crust has been referred to as the *Conrad discontinuity*. This discontinuity is poorly defined in some regions and is probably gradational over several to many kilometers.

P-wave velocities in crustal sedimentary layers range from 2 to 4 km/sec, depending upon such factors as degree of compaction, water content, and lithologic types. P-wave velocities in the upper oceanic crust are about 5 km/sec, while those in the upper continental crust are about 6 km/sec (see table 4.1). Lower crustal layers in both areas are characterized by $V_P = 6.5$ to 6.9 km/sec. The average crustal P-wave velocity (excluding the contribution of the sediment layer) in both oceanic and continental areas is 6.4 km/sec (ranging from 6.1 to 6.4 km/sec). With the exception of continental borderlands and island-arcs, the average thickness of continents is between 35 and 40 km, while the average thickness of the oceanic crust is only 5 to 7 km. Corresponding average upper-layer thicknesses are about 17 km and 1.5 km, respectively, and average lower layer thicknesses are 20 km and 5 km, respectively. Existing data indicate that mean P_n velocities roughly increase with both mean crustal V_P and depth to Moho in continental areas (Woollard, 1968).

A generalized contour map of the thickness of the North American crust is given in fig. 4.3. The thickness distribution appears to be related to the distribution of crustal types as discussed in more detail in the next section. The central and eastern United States are characterized by rather thick crust (40–50 km), the Canadian Shield by crust of average thickness (30–40 km), and the western part of the continent by crust of variable thickness (25–50 km). The crust shows a notable thickening beneath Lake Superior, along the Grenville Front in Quebec (see Chapter 5), beneath the Sierra Nevada Range in California, beneath eastern Montana, and beneath the southern Appalachian Mountains in the southeastern United States. Thin crust (< 30 km) occurs in continental borderlands and beneath the Basin and Range Province in Nevada and southern Arizona. Crust of variable thickness, variable P_n velocities (see fig. 3.1), and positive upper-mantle travel-time anomalies (see fig. 3.4) all seem to characterize Mesozoic-Cenozoic orogenic areas, while moderate to thick crust, average P_n velocities (8.1–8.2 km/sec), and negative upper-mantle travel-time anomalies characterize pre-Mesozoic crustal types.

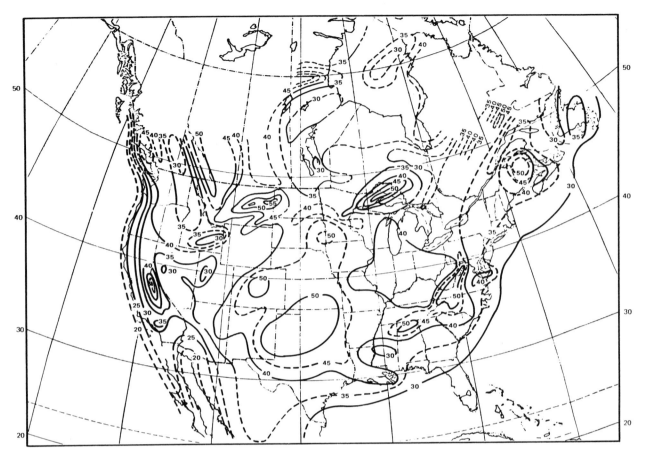

4.3. Contour map of the thickness (in km) of the North America crust (data cheifly from Pakiser and Zietz, 1965; McConnell *et al.*, 1966; Prodehl, 1970; Goodwin *et al.*, 1972; Hall and Hajnal, 1973; Warren and Healy, 1973; and Warren *et al.*, 1973).

Crustal Types

Compilation of Data. The average seismic structure and velocity distribution of various crustal types are compiled in table 4.1 and illustrated diagrammatically in fig. 4.4. The results represent median values of total crustal thicknesses, layer thicknesses, and P-wave velocities as compiled from many published seismic cross sections of the continental and oceanic crust. Most of the continental data come from North America, Europe, and the Soviet Union. The majority of the models used are for a two-layer crust (excluding the sediment layer). In those few cases where more-than-two-layer models are presented, layers are combined (if velocities are close) or the models are rejected (only three out of more than 150 sections were rejected on this basis).

Shields and Platforms. Shields and platforms have similar upper- and lower-layer thicknesses and velocities. The difference in their median thickness (column

6, table 4.1) appears to reflect primarily the presence of the sediment layer in the platforms. Upper-layer thicknesses range from about 10 to 25 km, lower-layer thicknesses from 16 to 30 km. Velocities in both layers are rather uniform, generally ranging from 6.0 to 6.2 km/sec in the upper layer and 6.5 to 6.9 km/sec in the lower layer. P_n velocities are typically in the range of 8.1 to 8.2 km/sec, rarely reaching 8.6 km/sec. Shield areas with high-grade metamorphic rocks dominating at the surface may be characterized by a homogeneous, one-layer crust (with $V_p = 6.5$–6.6 km/sec) as exemplified, for instance, by the Grenville Province in eastern North America (Taylor et al., 1980). This crustal model is consistent with the idea that the Conrad discontinuity, when present, reflects a change from medium- to high-grade metamorphic rocks. Relief up to 10 km per 100 km distance has been reported on the Moho in some platform and shield areas, and refinements in the interpretation of seismic and gravity data indicate the presence of significant lateral variations in crustal

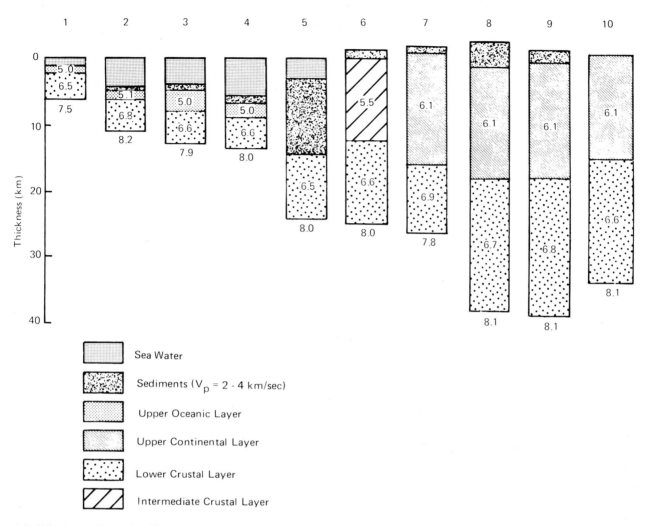

4.4. Seismic sections of various crustal types. P-wave velocities in km/sec. Key: (1) oceanic ridge (2) ocean basin; (3) marginal-sea basin; (4) trench; (5) inland-sea basin; (6) island-arc; (7) Basin and Range Province; (8) Phanerozoic orogenic area; (9) platform; and (10) shield.

layer and velocity distributions. An example of such lateral variation is given in fig. 4.5, which is a seismic cross section extending from west of the Rocky Mountains in British Columbia into the platform of Alberta and Saskatchewan. The data suggest that this region is composed of juxtaposed blocks with varying velocities and densities. Some of the near-vertical contacts may represent large fault zones that penetrate the entire crust and part of the upper mantle. Also, the data suggest the existence of a high-velocity zone (~ 7.2 km/sec) in the lower crust (or, alternately, a thin zone of anomalous upper mantle). Similar complex lateral variations have been proposed for platform areas in Eurasia.

Phanerozoic Orogenic Areas. Crustal thickness in young, post-Paleozoic orogenic areas is extremely

variable, ranging from about 20 km in island-arc and continental borderland areas to 80 km beneath the Himalayas. Paleozoic orogenic areas exhibit about the same range of thicknesses and velocities (including V_{P_n}) as platform areas. With the exception of continental rift systems, it is noteworthy that the average layer thicknesses and velocities (including V_{P_n}) of Phanerozoic orogenic areas are about the same as platforms (table 4.1, fig. 4.4). In areas of very thick crust, such as beneath the Andes (70 km) and Himalayas (80 km), the thickening occurs primarily in the lower crustal layer. P_n velocities in post-Paleozoic orogenic areas range from normal (8.1 to 8.2 km/sec) to low (< 7.8 km/sec). Lower crust high-velocity zones (7.0 to 7.4 km/sec) or, alternately, rather thin anomalous upper mantle zones, are also more frequent in post-Paleozoic orogenic areas than

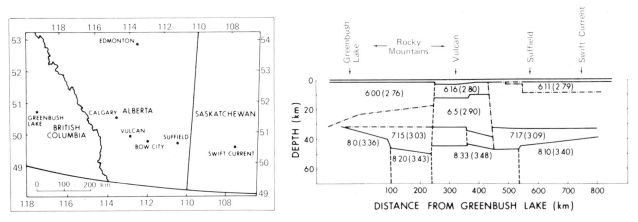

4.5. Crustal model of southwestern Canada based on seismic and gravity data (after Chandra and Cumming, 1972). Number outside the parentheses gives P-wave velocity and that inside the calculated density.

in older crust. The velocity contrast (ΔV_P) between the lower crust (V_P) and upper mantle (V_{P_n}), defined as $\Delta V_P = V_{P_n} - V_P$ is also commonly smaller beneath young orogenic areas (0.5 to 1.5 km/sec) than beneath cratons and shields (1 to 2 km/sec) (Oxburgh, 1969).

Continental rift systems are characterized by thin crust (< 30 km) and low P_n velocities (< 7.8 km/sec). It is of interest in regard to the origin of continental rift systems that thinning of the crust in these regions is accomplished by thinning of the

lower crustal layer, which ranges only from 4 to 14 km thick in the Basin and Range Province.

Island-Arcs and Volcanic Islands. Resolution of seismic data is poor in island-arc systems, and hence a fair degree of uncertainty exists regarding arc crustal structure. Seismic sections of several arc systems are given in fig. 4.6 to illustrate the large range in proposed velocity and layer thickness distributions. Crustal thickness ranges from about 5 km in the Lesser Antilles to 35 km in Japan, averaging about 22

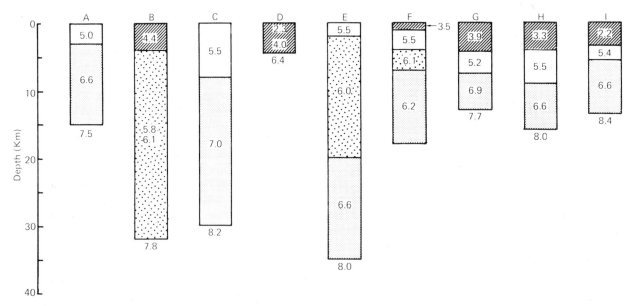

4.6. Seismic sections of island arcs. Key: (A) Southern Kurile Islands (Gainanov *et al.*, 1968); (B) Sakhalin (Gainanov *et al.*, 1968); (C) Puerto Rico (Talwani *et al.*, 1959); (D) Lesser Antilles (Fisher and Hess, 1963); (E) Northern Honshu, Japan (Rikitake *et al.*, 1968); (F) New Zealand, near Wellington (McConnell *et al.*, 1966); (G) Tonga Islands (Shore *et al.*, 1971); (H) Mariana Islands (Murauchi *et al.*, 1968; (I) Fiji Islands (Shor *et al.*, 1971).

km. Upper crustal velocities range from about 4.4 to 6.0 km/sec (median = 5.5 km/sec) and lower crustal velocities from 6.2 to 7.0 km/sec (median = 6.6 km/sec). P_n velocities range from normal (8.0 to 8.2 km/sec) to low (< 7.8 km/sec). All values given in table 4.1 and fig. 4.4 have large standard deviations and, as is indicated by the sections in fig. 4.6, arc crustal structure cannot be fitted to a simple two-layer crustal model. There is also evidence in some arc systems of an intermediate-velocity layer (5.0 to 6.0 km/sec) of varying thickness.

Most seismic data for volcanic islands come from the Hawaiian Islands or Iceland (Jacoby et al., 1980). Crustal thickness ranges from 10 to 20 km beneath the Hawaiian Islands, with upper crustal velocities ranging from 4.7 to 5.3 km/sec and lower crustal velocities from 6.4 to 7.2 km/sec (Furumoto et al., 1973). V_{P_n} ranges from 7.8 to 8.5 km/sec. A NW-SE cross section across part of Iceland is given in figure 4.7, and indicates a crustal thickness ranging from about 8 to 15 km. Low P_n velocities (7.2 km/sec) characterize the upper mantle beneath Iceland as they do oceanic ridges in general. The thinning of the crust in central Iceland chiefly involves thinning of the upper layer, and it corresponds, in general, with the active rift zones and with zones of recent volcanic activity that appear to represent the landward expression of the Mid-Atlantic Ridge.

Trenches. As with island-arcs, the crustal structure of oceanic trenches is poorly known. Data suggest a wide range in crustal thickness beneath trenches (3 to 12 km), with most thickening and thinning occurring in the lower layer, which ranges from 4 to 8 km thick. The upper layer in trenches ranges from 2 to 4 km thick and exhibits a variable V_p in the range of 3.8 to 5.5 km/sec (averaging about 5 km/sec). Most trenches have a sediment layer less than 3 km thick.

Ocean Basins. Crustal structure in ocean basins is rather uniform, not deviating greatly in either velocity or layer thickness distributions from that shown in fig. 4.4.

Total crustal thickness ranges from 6 to 8 km, and P_n velocities are generally uniform in the range of 8.1 to 8.2 km/sec (except as previously discussed). The sediment layer averages about 0.3 km in thickness and exhibits several strong reflecting zones, some of which are probably produced by cherty layers, as is suggested by cores retrieved from the Deep Sea Drilling Project. The most prominent reflecting horizon is the interface between the sediment layer and layer 2, which is characterized by a rough topography. The thickness of layer 2 (basement layer), which averages about 1.5 km, decreases toward oceanic ridges and, in general, is inversely proportional to the calculated spreading rates for a given ocean basin or segment thereof. Layer 3 is generally rather uniform in both thickness (4 to 6 km) and velocity (6.7 to 6.9 km/sec), as was previously mentioned. However, a high-velocity layer (7.1 to 7.7 km/sec), averaging about 3 km thick, has been detected beneath the 6.8 km/sec layer in parts of the Pacific crust (Sutton, Maynard, and Hussong, 1971). The lateral extent of this layer is at present unknown.

Recent developments in seismic experimental design and analysis techniques require some modifications of the simplified model described above (Spudich and Orcutt, 1980). Layer 2, for instance, is a layer in which velocity increases rapidly with depth and in which fine structure can be identified. Layer 3 is more homogeneous than layer 2 but exhibits occasional low-velocity zones. Also, the width of the oceanic Moho is variable (0–2 km). Seismic velocity distributions in layers 2 and 3 are consistent with the presence of porous, water-saturated mafic rocks at elevated pressures.

Oceanic Ridges. Crustal thickness ranges from 3 to 6 km beneath oceanic ridges, most of which is accounted for by the lower crust that ranges from 2 to 4 km in thickness (Le Pichon et al., 1965). Layers 1 and 2 thin or disappear entirely on most rises. Unlike other oceanic areas, the velocities in the lower crustal layer are quite variable, ranging from 4.4 to 6.9 km/sec. Such a range of velocities is probably produced by seismic anisotropy and by varying but, on the whole, high temperatures beneath ridges. Anomalous upper mantle ($V_{P_n} < 7.8$ km/sec) and a low-Q zone also occur beneath ridge axes. Surface-wave data indicate that the lithosphere increases in thickness from ≲ 35 km beneath oceanic ridges to 65 km at a crustal age of 40 m.y. The thin crust and lithosphere, common absence of layers 1 and 2, presence

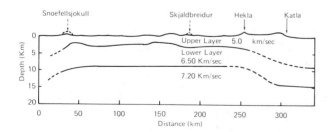

4.7. NW-SE seismic profile through southwestern Iceland (after Palmason, 1971).

of anomalous upper mantle, and high heat flow (which will be discussed later) that characterize oceanic ridges appear to reflect the youthfulness of the crust and intrusion of magmas at shallow depths beneath ridge crests.

Marginal-Sea Basins. The major difference between the crustal structure of marginal-sea basins and typical ocean basins is the slightly greater total thickness of the former (10 to 15 km). This thickening results from a thicker sediment layer in most marginal seas. Most seismic models show two to four layers above the lower crustal layer, which exhibits velocities of 6.6 to 6.9 km/sec. The upper crustal layers commonly range in velocity from about 2 km/sec at the top (unconsolidated sediments) to about 5 km/sec just above the lower crustal layer. It is noteworthy that an intermediate velocity layer (6.0 to 6.2 km/sec) occurs just above the lower crustal layer in the Caribbean and New Caledonia basins (Fox, Schreiber, and Heezen, 1971).

Inland-Sea Basins. Inland-sea basins show a considerable range in crustal thickness and layer distribu-

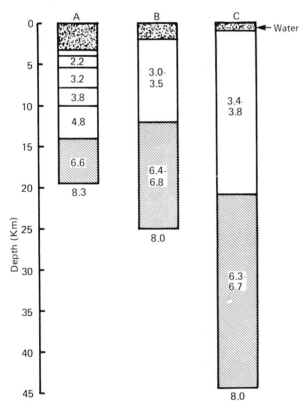

4.8. Seismic sections of inland-sea basins (after Menard, 1967). P-wave velocities in km/sec. Key: (A) Gulf of Mexico; (B) Black Sea; (C) Caspian Sea.

tions (table 4.1, fig. 4.8). Crustal thickness can range from about 15 km, as exhibited by the Gulf of Mexico, to 45 km in the Caspian Sea basin. In general, the sedimentary layer or layers ($V_P = 2-5$ km/sec) rest directly on the lower crust ($V_P = 6.3-6.7$ km/sec), with little or no upper crustal layer. Differences in crustal thickness among inland-sea basins are accounted for by differences in thickness of both sedimentary and lower crustal layers. Increasing velocities in sediment layers with depth, as for instance shown by the Gulf of Mexico (fig. 4.8A), appear to reflect an increasing degree of compaction and diagenesis.

Crustal Low-Velocity Zones

Crustal low-velocity zones have been reported in some young orogenic provinces and platform areas. Two examples are shown in fig. 4.9, one for the Basin and Range Province in northwestern Utah and one for the platform in eastern New Mexico. The latter also includes an electrical conductivity profile. A low-velocity zone is detected at about 10 km in northwestern Utah and another at 20 to 25 km in both areas. In other parts of the Basin and Range Province, seismic data suggest the existence of a widespread low-velocity layer (>3.5 km thick), beginning at a depth of 5 to 6 km. One or two crustal low-velocity layers have also been reported beneath the Rhine graben in Germany and beneath the Alps (Landisman et al., 1971). It is pertinent that hypocenters of most crustal earthquakes occur in the same depth range as that in which crustal low-velocity layers are reported, suggesting the presence of a region that readily deforms at these depths. The origin of crustal low-velocity zones will be further considered in the discussion of electrical conductivity of the crust.

Crustal Structure and Mean Crustal Age

Seismic velocity and thickness data from the continents are now considered as a function of mean crustal age. The results are shown in fig. 4.10, and include upper (T_{uc}), lower (T_{lc}), and total (T_c) crustal thickness, upper (P_{uc}) and lower (P_{lc}) crustal velocities, and upper mantle velocity (P_n). Data are grouped into five age categories: <225, 225–600, 600–1200, 1200–2500, and >2500 m.y. Each point on the figure represents a mean value and the vertical bar shows one standard deviation. Because crustal thickness is quite variable in Mesozoic-Cenozoic regions, seismic thicknesses are weighted according to areal distribu-

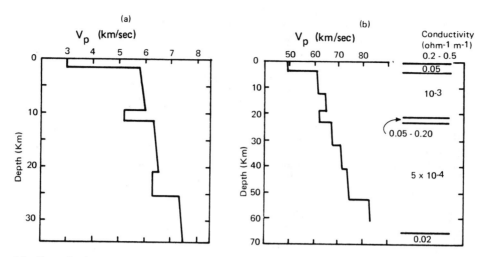

4.9. Crustal seismic and electrical conductivity profiles. (a) northwestern Utah (Landisman *et al.*, 1971); (b) eastern New Mexico (after Mitchell and Landisman, 1971).

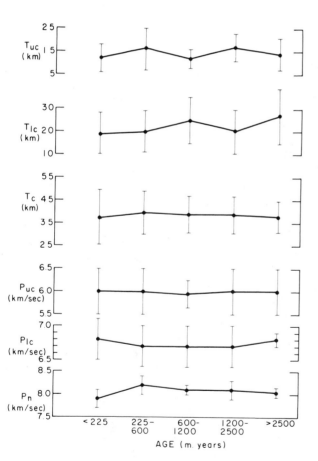

4.10. Variation of crustal thickness (T_c), layer thicknesses (T_{uc} = upper layer; T_{lc} = lower layer), crustal P-wave velocities (P_{uc} = upper layer velocity; P_{lc} = lower layer velocity), and P_n velocity) with mean crustal age (after Condie, 1973).

tion in these areas. Continental shelves and island-arcs (other than New Zealand and Japan) are excluded from the compilation, and mean crustal ages ignore the contributions of late Precambrian and Phanerozoic sediments on shields and platforms. Precambrian terranes that have been extensively reworked during younger orogenies are included with the younger crust.

The most striking feature of both the thickness and velocity patterns are their approximate constancy with respect to mean crustal age. The results indicate that crustal thickness (T_c) averages about 38 km regardless of age. The standard deviation of T_c, however, appears to decrease with increasing age, probably reflecting progressive cratonization, which tends to stabilize the crust and produce more uniform thicknesses. Young crust (<225 million years) ranges from very thick beneath mountain ranges (60–80 km) to thin along continental margins and in arcs (20–30 km). Craton-forming processes would appear to eliminate preferentially thick and thin crust and to stabilize crust with an equilibrium thickness of 35–40 km.

A systematic trend in the spread about the mean is not observed for the other thicknesses or for the velocity parameters. The apparent decrease in T_{lc} between the >2500 and 1200–2500 m.y. crust may not be real, since an unknown portion of the Precambrian crust in the 1200–2500 category (now mostly covered with platform sediments) is reworked >2500 m.y. crust. The low P_n velocity in the <225 m.y. category probably reflects the common occurrence of warmer upper mantle in young orogenic areas. If only those areas <225 m.y. in age with normal heat flow ($\sim 1.3 \mu$cal/cm^2 sec) are used, P$_n$ velocity is about the same as is observed beneath older crust.

Crustal Reflection Profiles

Continental Areas. Seismic reflection profiles from the continental crust have been extremely helpful in enhancing our knowledge of crustal structure. The most detailed studies are those by the COCORP group in the United States (Brewer and Oliver, 1980). An example of one such profile across the Wind River Range in Wyoming is given in fig. 4.11. The most striking feature of the profile is a series of dipping reflectors which originate where the Wind River thrust crops out at the surface. These reflectors trace the thrust to a depth of about 24 km, with an average dip angle of 30–45 degrees. This structure suggests that compression was largely responsible for the formation of the Wind River Mountains during

the Laramide orogeny. Also shown on the section are arcuate reflections (A) and complex composite reflections (B). The arcuate reflections are similar to patterns shown on synthetic seismograms for multiply deformed folds, and are interpreted as complexly deformed high-grade metamorphic rocks similar to those exposed in portions of the Wind River Range (Smithson et al., 1980). These results suggest the lower crust in this area may be similar to the deeply eroded Precambrian terranes now exposed at the surface in portions of the continents.

Profiles in the southern Appalachian Mountains indicate that a low-angle decollement underlies the Blue Ridge, Piedmont, and Atlantic Coastal Plain. The metamorphic rocks of the Blue Ridge and Piedmont appear to have been thrust onto a late Paleozoic continental margin (Chapter 9). Profiles across the Rio Grande Rift exhibit strong reflections at 19–20 km which are in agreement with other seismic data that are interpreted to define the top of

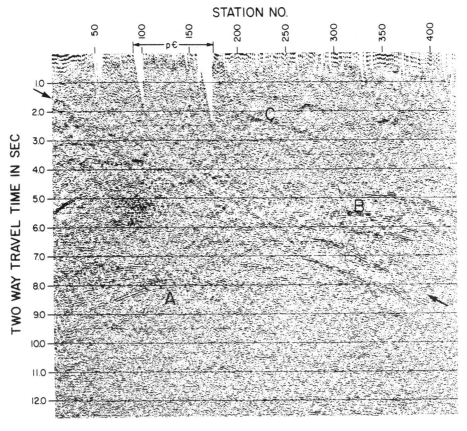

4.11. COCORP reflection profile across the southern Wind River Range, Wyoming (after Smithson, *et al.*, 1980). Arrows mark reflection from Wind River thrust. Deep complex crustal structure marked by arcuate reflections (A) and numerous composite reflections. (B). Strong shallow reflection at C. Note: Depth (km) $\cong$ 3 times the 2-way travel time in sec.

a magma body at this depth (Brown et al., 1979). A recent profile across the San Andreas Fault in California shows a zone of shallow brittle fracturing underlain by a featureless zone perhaps created by ductile flow (Long et al., 1978).

Although there are still many problems with the interpretation of continental seismic profiles, existing data allow tentative conclusions regarding crustal structure (Brewer and Oliver, 1980):

1. Short, discontinuous, and arcuate reflections in the middle and lower crust appear to record complexly folded metamorphic rocks.

2. Strong continuous events can be interpreted in terms of major low-angle faults or, at shallow levels, perhaps as layering in sediments and volcanics.

3. Blank or transparent regions in profiles may represent highly folded rocks with dips too steep to return reflected energy or massive granitic plutons.

4. Reflections from the lower crust that exhibit low dips ($< 5°$) may represent either horizontal metamorphic layering or stratiform igneous intrusions.

5. There is no evidence for a single mid-crustal reflection representing the Conrad discontinuity.

6. Reflections from the Moho are of variable intensity and character. Results indicate the boundary ranges from a simple reflector to a completely layered structure or, in some instances, to a nonreflecting or highly deformed boundary.

7. Large faults, such as the Wind River thrust and the boundary faults of the Rio Grande Rift, may penetrate most or all of the crust.

Oceanic Areas. Reflection profiling in oceanic areas has been especially important in revealing detailed structure of continental margins. Interpretation of reflecting horizons in some instances has been verified from cores of the Deep Sea Drilling Project. Typical reflection profiles of two types of continental margins are given in fig. 4.12. The profile from the western margin of the Gulf of Mexico exhibits detailed stratigraphy and structure in the upper few kilometers. The strong discontinuous reflections of the Sigsbee–Cinco de Mayo and Mexican ridges and the Campeche units are interpreted as alternating fine-grained pelagic sediments and turbidites. Reflections in the Challenger unit may represent, in part, deep-water carbonates. The most striking features of the profile are the nearly symmetrical folds with wave lengths of the order of 10 km and topographic relief of 500–700 m. These folds act as barriers to young sediments being transported down the continental slope. Thrust faults also occur in the section and dip chiefly toward the continent at steep angles. The folds are probably not caused by salt doming, in that the salt-bearing unit (the Challenger unit) continues relatively undisturbed beneath the folds. The folds

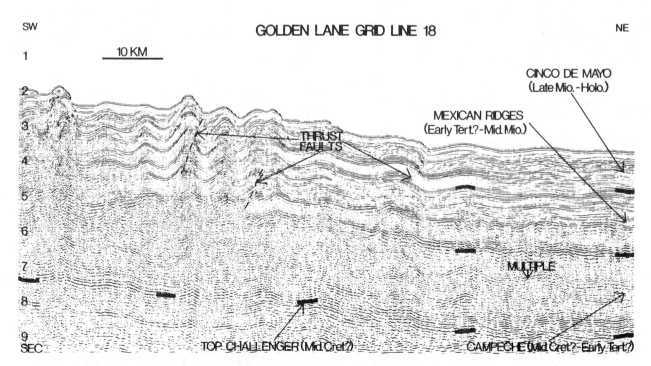

4.12. a) Seismic reflection profile across the western continental margin of the Gulf of Mexico (from Buffler *et al.*, 1979). Depth (km) $\cong$ 3 times the 2-way travel time (secs).

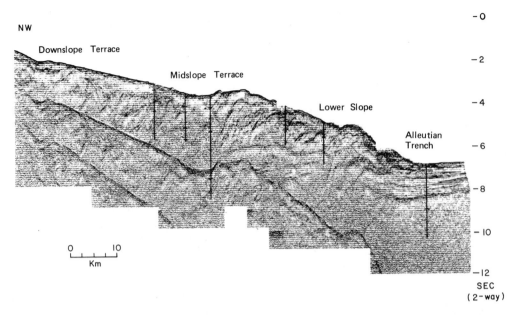

4.12. b) Seismic reflection profile across the Aleutian trench off Kodiak Island, Alaska (from Von Huene, 1979). Depth (km) ≅ 3 times the 2-way travel time (secs).

are interpreted to have developed in a seaward-sliding decollement triggered either by regional uplift to the west followed by gravity sliding into the Gulf of Mexico, or by compressive forces from the west.

A profile across the Aleutian Trench off southern Alaska exhibits extremely complex structure (fig. 4.12b). The strong continuous reflection near the middle of the section appears to represent the contact of deformed ocean sediments with metamorphic basement. The upper strong reflector in the trench (which looks like an unconformity) represents the contact between trench turbidites and oceanic sediments, whereas the strongest reflection in the trench occurs between the oceanic sediments and layer 2 of the oceanic crust. These reflections can be followed for about 10 km toward the trench slope before they are cut by faults. Although they are deformed by steep faults which dip toward the continent, the reflections can be traced as far as the midslope terrace, where they appear to plunge into the continent. The curved features in the lower slope are diffractions caused by complex structure related to downslope mass movement and faulting. Topographic ridges in this region may represent the surface expression of thrust faults dipping landward. The downslope terrace (forearc basin) appears to be comprised of undeformed turbidites at shallow depth.

HEAT FLOW

Heat-Flow Distribution

Surface heat flow in continental and oceanic crust is controlled by many factors, and for any given segment of crust decreases with time. One of the most important sources of heat loss from the Earth appears to be the hydrothermal circulation associated with oceanic ridges. Model calculations suggest that approximately 25 percent of total global heat flow can be accounted for by such hydrothermal transport (Davies, 1980). Heat flow on continents and islands varies as a function of the age of the last magmatic event, the distribution of heat-producing elements, and erosion level. Considering all sources of heat loss, the total heat loss from the Earth today appears to be about 10^{13} cal/sec, which is comprised of 3×10^{12} cal/sec from the continents and 7×10^{12} cal/sec from the oceans (Sclater et al., 1980). The equivalent heat flows are 1.3 and 2.3 HFU, respectively, for a worldwide average heat flow of 1.95 HFU. The difference between the average measured oceanic heat flow (1.6 HFU) and the calculated value (2.3 HFU) (table 4.1) is due to heat losses at ocean ridges by hydrothermal circulation. Models indicate that 70 percent of the average surface heat flow is lost from convection in the mantle by plate creation at oceanic ridges, 20 percent by conduction, and the remainder by radioactive decay of heat-producing elements in the crust.

Estimated average heat flows of major crustal types are tabulated in column 4 of table 4.1. Shield areas exhibit the lowest and least variable continental heat-flow values, generally in the range of 0.9 to 1.1 HFU, averaging about 1 HFU. Platforms and Paleozoic orogenic provinces are more variable—usually falling between 1 and 2 HFU and averaging about 1.5 HFU. The difference of about 0.3 HFU between the shield and platform averages appears to be real, although it reflects North American data for the most part. A significant amount of this difference can be accounted for by the fact that most platform heat-flow measurements are from areas with thicker crust (40–50 km) than those in which most shield measurements have been made (30–40 km).

Post-Paleozoic orogenic areas and island-arcs exhibit high and quite variable heat flow, in the range of 1.5 to 2.5 HFU. Measurements available from continental rift systems are high and variable. The high heat-flow in some island-arcs and oceanic volcanic islands reflects recent volcanic activity in these areas. Most evidence supports a correlation of high heat flow with thin crust, low P_n velocities, positive travel-time anomalies, and shallow depth to the zone of seismic low-velocity, low Q, and high electrical conductivity in the upper mantle.

Heat flow in ocean basins generally falls between 1.0 and 1.5 HFU, averaging about 1.3 HFU. Oceanic ridges, on the other hand, are characterized by extremely variable heat flow, ranging from less than 1 HFU to much greater than 5 HFU. Heat flow decreases as a function of increasing distance and age from oceanic ridges (see fig. 4.16). Active and some inactive marginal-sea basins are also characterized by high heat flow (1.5 to 2.5 HFU), while other older, inactive basins exhibit typical ocean-basin heat flows (fig. 4.2) (Watanabe et al., 1977). Inland-sea basins as a group exhibit variable heat flow (0.5 to 2.0 HFU), which reflects, in part, Cenozoic sedimentation rates in the basins.

Crustal Heat Production and Heat Flow

Heat-flow values are significantly affected by the age and intensity of the last magmatic event(s), the distribution of radioactive elements in the crust, and the amount of heat flowing up from the mantle. Information from the continents has made it possible to evaluate quantitatively the relative contributions of each of these effects. Data indicate that continental surface heat flow (q_o in $\mu cal/cm^2$ sec) is linearly related to average radiogenic heat production (A_o in 10^{-13} cal/cm^3 sec) of near-surface granitic and gneissic rocks by the following equation (Roy, Blackwell, and Birch, 1968):

$$q_o = q_r + A_o D. \qquad (4.1)$$

q_r and D are constants characteristic of crustal types; q_r, commonly referred to as the *reduced heat flow*, is the intercept value for rocks with zero heat production; and D, the slope of the line relating q_o to A_o, has units of depth and is commonly referred to as the *characteristic depth*.

Examples of the linear relationship between q_o and A_o for several crustal types are given in fig. 4.13. The line representing the Appalachian Mountains appears to characterize most continental platform areas and pre-Mesozoic orogenic areas. The slopes of the lines range from 4.5 to 14 km. The chief factors distinguishing the individual crustal types are the reduced heat-flow values (q_r) and, to a lesser extent, the range of surface heat flow and heat productivity

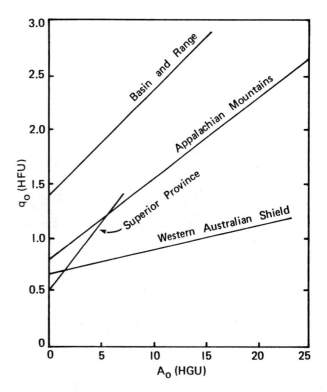

4.13. Surface heat flow (q_0) versus heat production of near-surface rocks (A_0) for several crustal provinces.

Table 4.2. Summary of Parameters for Reduced Heat-Flow Provinces

Province	Age (m.y.)	q_o (HFU)	q_r (HFU)	D (km)
Basin and Range (U.S.A.)	0–65	2.20	1.41	9.4
Eastern Australia	0–65	1.72	1.37	11
Appalachians (U.S.A.)	100–400	1.36	0.79	7.5
United Kingdom	300–1000	1.40	0.56	16
Western Australian Shield	≥ 2500	0.93	0.63	4.5
Indian Shield (India)	≥ 1800	1.52	0.93	15
Superior Province (Canada)	≥ 2500	0.96	0.50	14
Baltic Shield (Scandinavia)	≥ 1800	0.86	0.53	8.5

q_o and q_r are average surface and reduced heat flow, respectively; D is the characteristic depth.

values. Reduced heat flow distribution can be used to define *reduced heat flow provinces* (table 4.2).

Two vertical distributions of radioactivity in the crust have been proposed that are consistent with the linear q-A relationship. The first model (illustrated in fig. 4.14a) involves a heat-producing surface layer of approximately constant thickness D—the slope from equation (4.1)—and variable heat production, with q_r representing the combined heat coming from the lower crust and upper mantle (Blackwell, 1971). The second model involves an exponential decrease in radiogenic heat production according to the expression,

$$A = A_o e^{-x/D}, \qquad (4.2)$$

where A_o is the near-surface heat production and A is the heat production at depth x (Lachenbruch, 1968).

The following lines of evidence tend to favor the exponential model as most nearly representing the actual distribution of radioactivity in the crust:

1. Radiogenic heat-production data from a few deep drill holes (> 3 km) suggest decreasing radioactivity with depth.

2. Radioactivity decreases in granitic rocks as a function of increasing inferred depths of emplacement in the crust (Swanberg and Blackwell, 1973).

3. Studies of the distribution of U, Th, and K in rocks of varying regional metamorphic grade indicate that granulite-facies terranes, which probably formed at lower crustal depths (20–40 km), have concentrations of radioactive elements much lower than upper-crustal metamorphic and plutonic terranes (Hyndman et al., 1968).

4. Thermodynamic models based on an equilibrium distribution of heat-producing elements in a gravitational field are consistent with an exponential

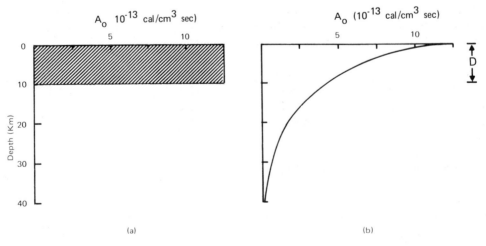

4.14. Models of crustal heat production that satisfy the linear relationship between q and A (after Blackwell, 1971). (a) Constant heat generation to depth D; (b) heat generation decreasing exponentially with depth.

decrease of these elements with depth (Turcotte and Oxburgh, 1972a).

5. The exponential model readily preserves the linear relationship of q and A under differential erosion, whereas the constant thickness model does not (Lachenbruch, 1970). The low heat flow in shield areas, for instance, is readily explained in the exponential model by erosion of much of the surface layer which is rich in radioactive elements.

Temperature Distribution in the Crust

Temperature distribution in the crust is dependent upon surface and mantle heat flow and the distribution of thermal conductivity and radioactivity with depth. Estimates of temperature-depth distributions in four crustal types are given in fig. 4.15. Models are calculated assuming steady state conditions. The shaded areas for each crustal type represent the combined uncertainties of the factors listed above. Both the exponential and constant-thickness radioactivity distribution models give temperature-depth curves within the shaded areas. It is noteworthy that the

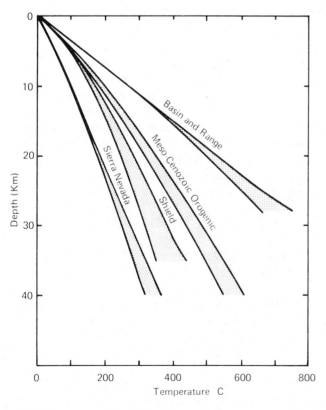

4.15. Temperature-depth distributions in four continental crustal types (after Blackwell, 1971; Hyndman *et al.*, 1968).

temperature differences between the crustal types are, in general, larger than the combined uncertainties of data used in calculating the temperature-depth curves. The results suggest the following Moho temperatures: Sierra Nevada, 325–375°C; Precambrian shield, 350–450°C; Mesozoic–Cenozoic orogenic area, 500–650°C; and Basin and Range Province, 750–800°C.

Age Dependence of Heat Flow

Existing data indicate that average surface heat flow decreases with the average age of crustal rocks in both oceanic and continental areas (Sclater et al., 1980). The range of variation of average heat flow in continental and oceanic crust is plotted as a function of increasing mean crustal age in fig. 4.16. The data suggest that continental heat flow falls off with age to an approximately constant value of 1 HFU in about 1 b.y. In oceanic areas, on the other hand, constant heat-flow values of about 1.3 HFU are reached in only 50–100 m.y. The continental heat flow can be considered in terms of three components, all of which decay with time (Vitorello and Pollack, 1980). Radiogenic heat in the upper crust contributes about 40 percent of the total heat flow in terranes of all ages. The absolute amount of heat from this source decreases with time in response to erosion of the upper crust. The second component of heat, contributing about 30 percent to the heat flow in Cenozoic terranes, comes from the residual energy of igneous activity associated with orogeny. These two components decay rapidly in a few hundred million years. The third and generally minor component comes from convective heat from within the mantle.

In oceanic areas, heat flow (q) falls off with the square root of the age of the crust (t), according to

$$q = 11.3t^{-1/2} \qquad (4.3)$$

to approximately 120 m. years (Sclater, Jaupart, and Galson, 1980). As the lithosphere cools it contracts, and thus the depth to the ocean floor increases going away from oceanic ridges. The depth of the ocean floor (d) back to about 70 m.y. can be approximated by

$$d(\text{meters}) = 2500 + 350t^{1/2} \qquad (4.4)$$

if the lithosphere is considered the upper thermal boundary of the mantle.

The age of the oceanic crust increases as a function of distance from oceanic ridges and as a function of decreasing mean elevation (as a consequence of sea-floor spreading—see Chapter 6). The

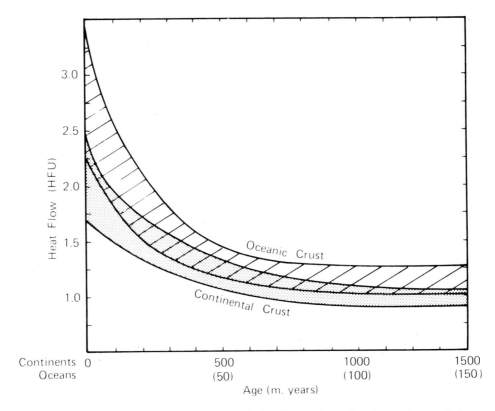

4.16. Age dependence of average heat flow variation for continental and oceanic crust (after Sclater, 1972). Numbers in parentheses on horizontal axis refer to ages in oceanic areas.

decrease in heat flow away from oceanic ridges is readily explained by plate tectonics, in which new crust is created at the axial rift zones by intrusion of magma that cools and solidifies as it moves laterally by sea-floor spreading. Models assuming a lithospheric plate 50–100 km thick with basal temperatures in the range of 550–1500°C produce a good match for observed heat-flow distributions away from oceanic ridges decaying to equilibrium levels of about 1.0 to 1.3 HFU in 50–100 m.y.

Reduced Heat Flow Provinces

It is clear from the results in table 4.2 that reduced heat flow values also decrease with the average age of the continental crust. No relationship, however, appears to exist between the characteristic depth D and mean crustal age. The reduced heat flow decays within 200–300 m.y. to a value between 0.5 and 0.6 HFU, which is about a factor of three faster than surface heat flow decays. Corresponding equilibrium q_r values for oceans are 0.6–0.9 HFU. D is generally equated with the depth of a surface layer in which most of the heat-producing elements are concentrated. Two interpretations of the change in D

between reduced heat flow provinces have been proposed (Jessop and Lewis, 1978): D values reflect differences in crustal erosion level or differences in degree of fractionation of the crust. Because the crust level exposed at the surface as deduced from metamorphic mineral assemblages does not correlate with D, nor does D correlate with mean crustal age, the first interpretation seems unlikely. The alternate interpretation necessitates variable intensities of fractionation of U, Th, and K into the upper crust which is not related to the age of the crust.

Continental and Oceanic Equilibrium Heat Flow

As described above, both continental and oceanic heat flow decay to about the same equilibrium value of 1.0–1.3 HFU. One model which accomodates this similarity is illustrated in fig. 4.17. The model assumes a 10 km thick oceanic crust and a 40 km thick continental crust, with an exponential distribution of radiogenic heat sources in the continental crust. Equilibrium reduced heat flow values (q_r) are selected from the range of observed values in both crustal types. Both models have similar temperatures by

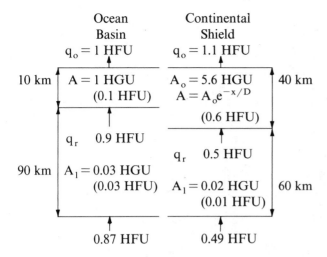

4.17. Heat flow model of Precambrian shield and ocean basin consistent with approximate equivalency of heat flow from both types of crust (after Sclater, et al. 1980). A_1 = heat productivity of the lower lithosphere.

depths of 100 km, the average thickness of the lithosphere. The model clearly implies that more mantle heat is entering the base of the oceanic lithosphere (0.87 HFU) than of the continental lithosphere (0.49 HFU). Hence, when a plate carrying a Precambrian shield moves over oceanic mantle, the surface heat flow should rise until mantle convective systems readjust themselves so they are again liberating most heat beneath oceanic areas. This may account for the relatively high surface heat flow of the Indian Shield (table 4.2), which has moved over oceanic crust in the last 50 m.y. (Chapter 9).

GRAVITY ANOMALIES

Continental Areas

Gravity features in continental regions are generally discussed in terms of Bouguer anomaly distributions. Average Bouguer anomalies expressed in milligals are given in table 4.1 for each major continental and oceanic crustal type. Shield and platform areas are characterized by broad regional anomalies of −10 to −50 mgal and occasional sharp anomalies of local importance. Anomalies with widths of hundreds of kilometers reflect inhomogeneities in the lower crust or upper mantle, and those of smaller size reflect near-surface rock types or fault zones. Gravity anomalies can be used to trace structural trends or rock units in Precambrian shields beneath the sedi-

mentary cover in platform areas. One example is the Keweenawan basalts (1 b.y. in age), which can be traced southwest from Lake Superior beneath the sedimentary cover as the Mid-Continent gravity high (fig. 4.18). Structural contacts between Precambrian crustal provinces in shield and platform areas (such as between the Grenville and Superior provinces in Canada) are also often expressed by gravity anomalies.

Phanerozoic orogenic areas are commonly characterized by large negative Bouguer anomalies of −200 to −300 mgal. Anomalies parallel near-surface structural and lithologic elements in these areas and may reflect thicker crust and warmer (low density) upper mantle. Young orogenic belts near continental borderlands and island-arc areas exhibit smaller anomalies (−50 to +100 mgal). Large negative anomalies in the Basin and Range Province and over most continental rifts reflect the thinning of the lower crust and the presence of shallow, low-density upper mantle in these areas. Small positive gravity anomalies superimposed on the regional negative anomalies occur in the centers of some rifts and appear to reflect near-surface intrusions of mafic magma (Darracott et al., 1972).

Oceanic Areas

Bouguer anomalies in oceanic volcanic islands range from large and positive (+250 mgal over the Hawaiian Islands) to small and variable (−30 to +45 mgal over Iceland). The negative Bouguer anomalies observed over trenches reflect chiefly the descent of the less dense lithosphere into the upper mantle. Some gravity profiles suggest thinning of the crust on the oceanward side of trenches. Bouguer anomalies range from +250 to +350 in ocean basins, from +200 to +250 on ridges, and from 0 to +200 over marginal and inland-sea basins. Bouguer and free-air anomaly profiles across the northern Mid-Atlantic Ridge and East Pacific Rise are given in Figs. 4.19 and 4.20. Crustal structures are also shown from available seismic data. In the North Atlantic, the Bouguer anomaly increases away from the Mid-Atlantic Ridge (+200 mgal) toward the continents (to values of +350 mgal). A similar although much less pronounced and more variable trend exists across the East Pacific Rise. The minimum in the Bouguer anomaly over oceanic ridges suggests that they are approximately isostatically compensated by the low density, anomalous upper mantle beneath them. Free-air anomalies in oceanic areas reflect, for the most part, topography on the ocean floor, as can be

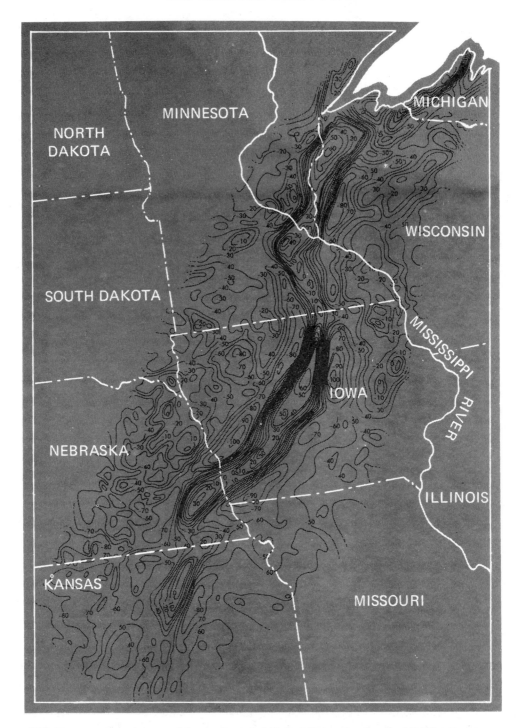

4.18. Bouguer anomaly map of the north-central United States showing the Mid-Continent
Gravity High (from Bouguer Gravity Anomaly Map of the United States, U.S. Geol.
Survey, 1964). Contours in milligals.

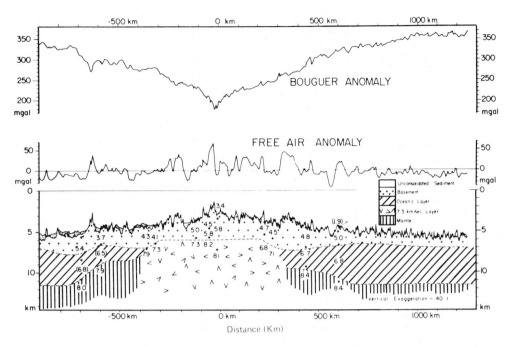

4.19. Gravity anomaly profiles and seismic structure across the northern Mid-Atlantic Ridge (from Talwani *et al.*, 1965). P-wave velocities are indicated.

seen from inspection of the profiles in Figs. 4.19 and 4.20. Smaller Bouguer anomalies in marginal and inland seas result from thicker sediment layers in most of these basins.

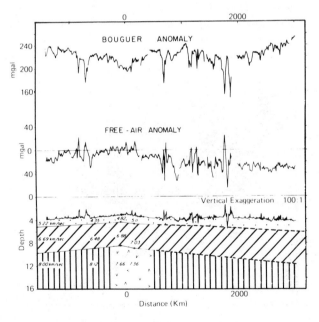

4.20. Gravity anomaly profiles and seismic structures across the East Pacific Rise (from Talwani *et al.*, 1965). P-wave velocities are indicated. Symbols as in Fig. 4.19.

Discussion

It has long been known that Bouguer anomalies roughly increase with the thickness of the crust. The relationship can be expressed by the following equation (after Grushinsky, 1967);

$$C = C_o - x\Delta g, \qquad (4.5)$$

where C is crustal thickness (in km), Δg is the Bouguer anomaly (in mgal), and C_o and x are variable constants depending on the crustal type. Generally, C_o ranges from 31 to 38 km and x from 0.06 to 0.03. Oceanic crust has low values of C_o and continental crust has high values. Because of the variation in C_o and x, however, it is not possible to calculate crustal thickness from a knowledge of only the Bouguer anomaly.

Isostatic anomalies are small over most crust, indicating a close approach to isostatic equilibrium. The major exceptions are arc-trench systems where lithosphere is being consumed by the asthenosphere. Thick crust or "roots" exist under some mountains (e.g., the Sierra Nevada) but not under others (e.g., the Rocky Mountains), suggesting that the Airy hypothesis of isostatic compensation (see Chapter 1) is not always the dominant compensation mechanism. Lateral variations in the density of the crust and

Table 4.3. Mean velocities, isostatic anomaly, and depth to Moho

V_{P_n} (km/sec)	Mean Crustal Velocity (km/sec)	Isostatic Anomaly (mgal)	Depth to Moho (km)
7.7	6.1	-30	25
8.0	6.4	-5	36
8.3	6.6	$+20$	46

After Woollard (1972).

upper mantle as implied by seismic data tend to support the Pratt compensation mechanism. It is likely that both mechanisms operate in varying degrees in different areas. Isostatic compensation must occur on a time scale of less than 50 m.y., since both direct and indirect estimates of continental erosion rates indicate that otherwise, continents would be eroded to sea level in 25–50 m.y. A rough correlation exists between P_n velocity, mean crustal velocity, isostatic anomaly, and depth to Moho as shown by the data in table 4.3. In terms of existing data, it is probable that isostatic transfer of mass occurs in the asthenosphere.

MAGNETIC ANOMALIES

Continental Areas

Magnetic anomalies over the continents usually have amplitudes of a few hundred gammas, widths of tens to hundreds of kilometers, and lengths of hundreds to thousands of kilometers (MacLaren and Charbonneau, 1968; Zietz et al., 1969; Zietz et al., 1971). They reflect upper crustal rock types or isothermal surfaces of Curie point distributions. In both shield and platform areas, magnetic anomalies characteristically exhibit broad "swirling" patterns. In Archean shield areas granitic-gneiss terranes usually correlate with magnetic highs and mafic volcanic belts with lows (fig. 4.21). Since granitic rocks do not exhibit large magnetizations, the magnetic highs of these terranes must reflect terranes beneath the granites and gneisses that are more magnetic than those beneath the mafic volcanic belts (Hall, 1968). Major near-surface intrusions, both mafic and siliceous in composition, usually exhibit magnetic signatures. Structural discontinuities such as major faults and crustal-province boundaries also usually produce magnetic anomalies. Magnetic anomalies, like gravity anomalies, provide a means of tracing major rock units and structural discontinuities beneath sediment cover in platform and continental shelf areas.

Magnetic anomaly patterns over Phanerozoic orogenic areas and island-arcs are more complex and broken up and show a wide range of amplitudes. Broad anomalies appear to be related to deeper crustal inhomogeneities and shallow, narrow anomalies to near-surface faults or mafic and ultramafic plutons. Overall magnetic grain generally parallels major orogenic structures. However, in some young orogenic and island-arc areas and, in particular, in those areas with high heat flow, thin crust, and anomalous upper mantle, the magnetic grain does not parallel surficial geologic features, and anomaly amplitudes are generally less than 200γ. For instance, the Basin and Range Province, which exhibits overall N–S structural trends, is characterized by dominantly E–W-trending, small-amplitude magnetic anomalies. The small amplitude of magnetic anomalies in these regions is generally interpreted to indicate that the lower crust and/or upper mantle are above the Curie point of magnetite (578°C)—the most common magnetic mineral. The peculiar anomaly trends may be

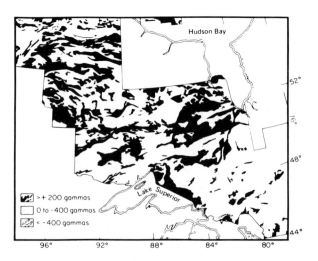

4.21. Magnetic anomaly patterns in the southern Canadian Shield (after Morley *et al.*, 1967). Magnetic lows reflect mafic volcanic belts and highs (in black) areas of granite and gneiss.

related to older lithologic or structural elements in the crust at intermediate depths.

Oceanic Areas

Oceanic crust is characterized by linear to curvilinear magnetic stripes that roughly parallel ridge crests (Raff and Mason, 1961) (fig. 4.22). The anomalies are typically 5 to 50 km wide, hundreds of kilometers long, and range from about 400γ to 700γ in amplitude (relative to sea level). The fact that they are roughly symmetrical about oceanic ridges is an important feature that led to the sea-floor spreading hypothesis (discussed in Chapter 6). Magnetic anomalies are offset along transform faults such as the Blanco and Mendocino fracture zones shown in fig. 4.22. Existing data indicate that the magnetic

anomalies are produced in basalts emplaced in the central rift zones of oceanic ridges by reversals in the Earth's magnetic field. Magnetic anomalies are less distinct in marginal-sea basins and are very weak to absent in inland-sea basins. In marginal-sea basins this is caused in part by the comparatively high temperatures beneath the basins (see fig. 4.2), above or close to Curie temperatures. Linear anomalies have been mapped in the East Scotia and Tasman Sea basins, indicating the presence of now-extinct spreading centers in these basins (Barker, 1972; Hayes and Ringis, 1973). In some inland-sea basins, the thick sediment cover appears to mask or partially mask magnetic anomalies. Weak anomalies varying from linear and subparallel to irregular-shaped have been detected in the Gulf of Mexico (Yungul, 1971).

ELECTRICAL CONDUCTIVITY DISTRIBUTION

Models for the electrical conductivity distribution in the continental crust are nonunique and complex (Keller, 1971). One or more high-conductivity layers are recognized between 20 and 40 km depth in some post-Paleozoic orogenic provinces and at shallower depths in some younger orogenic areas. Highly conducting layers commonly correlate with low-velocity zones in the crust, as found in eastern New Mexico for instance (see fig. 4.9). Crustal conductivity anomalies have also been described from some continental regions. An example is the North American Central Plains (NACP) anomaly, which extends over a distance of 1500 km from southeastern Wyoming to the edge of the Canadian Shield in Saskatchewan (fig. 3.10) (Camfield and Gough, 1977). This anomaly lines up with major shear zones exposed in the Canadian Shield in northern Saskatchewan and with the Mullen Fork–Nash Creek shear zone in southeastern Wyoming (fig. 5.7). It is possible that this highly conducting zone is produced by dehydration reactions in a shear zone now chiefly covered by cratonic sediments extending from Wyoming to central Saskatchewan. Such a shear zone, if actually present, may represent a collisional plate boundary extending across north-central North America.

Two origins are generally considered for high-conductivity (low-velocity) zones in the crust: the presence of water, derived perhaps from dehydration reactions, and partial melting. A third possibility, the presence of conductive graphite zones, may also be of local importance. Experimental data, combined with geothermal results, indicate that partial melting is

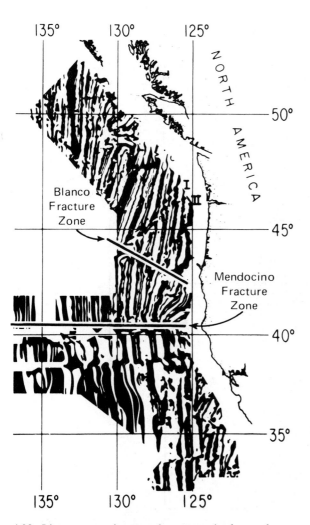

4.22. Linear magnetic anomaly patterns in the northeastern Pacific (after Raff and Mason, 1961). Positive anomalies are black.

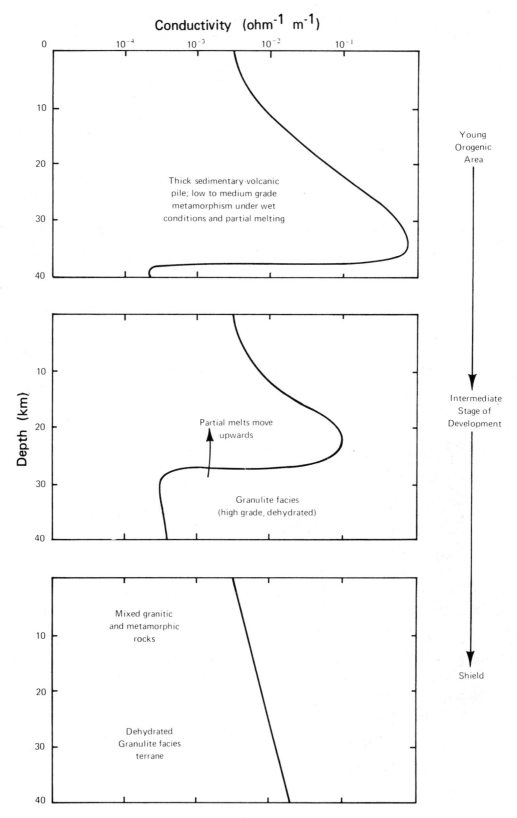

4.23. Schematic electrical conductivity profiles of three stages in crustal evolution (modified after Hyndman and Hyndman, 1968).

unlikely in the crust at depths less than 20 km (Wyllie, 1971a). Some combination of water and partial melting is responsible for conducting zones found in the crust at depths greater than 20 km, while water is probably responsible for high-conducting layers at shallower depths. The importance of water and partial melting in controlling the electrical conductivity in evolving continental crust is illustrated in fig. 4.23. During the first stage, rapid burial of clastic sediments and volcanics, perhaps in a back-arc basin, leads to dehydration reactions and partial melting near the base of the succession, thus increasing the conductivity with depth. As magmas and water move to shallower depths (stage 2), the lower crust is dehydrated to granulite-facies mineral assemblages and the zone of maximum conductivity moves upward. The final stage illustrates a stable continental craton relatively dry and free from partial melting.

METAMORPHISM

Burial of rocks in the Earth's crust results in progressive metamorphism as the rocks are subjected to increasing pressure and temperature. Given that metamorphic reactions are more rapid with rising than with falling temperature, metamorphic mineral assemblages generally record the highest P-T regime to which rocks have been subjected. Later uplift and erosion make it possible to directly study rocks that were once buried to various depths in the crust but are now exposed at the Earth's surface. Such rocks contain metamorphic mineral assemblages "arrested" at the maximum burial depth or maximum metamorphic grade. Minor changes in mineralogy known as retrograde metamorphism may occur during uplift, but such changes can commonly be identified without great difficulty by studying textural features of the rocks (Miyashiro, 1972). Progressive metamorphism is accompanied by losses of H_2O, CO_2, and other volatile constituents, with some high grade terranes being almost anhydrous. Other geochemical changes may also accompany metamorphism, although changes in major-element concentrations are generally considered to be of minor importance.

Zones of increasing metamorphic grade can be classified into *metamorphic facies*, which represent limited ranges of load pressure, temperature, and water content in the crust. Five major facies of region metamorphism are recognized (fig. 4.24). The zeolite facies is characterized by the development of zeolites in sediments and volcanics and reflects temperatures generally less than 200°C and burial depths less than

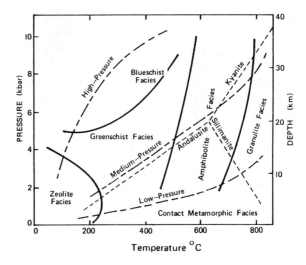

4.24. Approximate P-T regimes of regional metamorphic facies and facies series (after Miyashiro, 1973; Ernst, 1973).

5 km. The greenschist facies is characterized by the development of chlorite, actinolite, epidote, and albite in mafic volcanics and of muscovite and biotite in pelitic rocks. Blueschist-facies assemblages form at high pressures (>5 kbar) yet low temperatures (<400°C), and are characterized by such minerals as glaucophane, lawsonite, and jadeite. The amphibolite facies is characterized by kyanite, staurolite, and sillimanite in quartzo-feldspathic rocks and by intermediate plagioclase and hornblende in mafic rocks. The granulite facies is characterized by a sparsity or absence of hydrous minerals and the appearance of pyroxenes.

By employing experimental petrologic results and oxygen isotope data it is possible to estimate the temperature and pressure at which metamorphic mineral assemblages crystallized. Hence, it is possible to estimate the burial depths of various portions of metamorphic belts now exposed at the Earth's surface. Because the geothermal gradient varies from place to place in the Earth, the succession of metamorphic zones should also vary. A succession of metamorphic zones from a given geographic area is called a *facies series*. Three major facies series are recognized (fig. 4.24). The low-pressure series reflects a relatively steep geothermal gradient (≳25°C/km) and is characterized by such minerals as andalusite, cordierite, and sillimanite. The medium-pressure series reflects intermediate geotherms (15–25°C/km) and is characterized by kyanite and garnet, and the high-pressure series (≲10°C/km) is characterized by glaucophane, jadeite, and lawsonite. Low-pressure metamorphic belts of all ages are known, but they are particularly common in Archean terranes. Medium-pressure belts have also been described from crustal provinces of all ages; however, they are uncommon in the Archean. Blueschist terranes make their first appearance in the late Proterozoic and do not become common until after mid-Paleozoic. These terranes appear to have formed on the trench side of arc systems in response to depressed geotherms in subduction zones. Their absence in pre–late-Proterozoic terranes may reflect higher geothermal gradients beneath arc systems during the Precambrian.

COMPOSITION

Several approaches have been used to estimate the chemical and mineralogical composition of the crust. One of the earliest methods was based on chemical analysis of glacial clays, which were assumed to be representative of large portions of the continents (Goldschmidt, 1954). Later estimates were based on mixing average basalt and granite compositions in ratios generally ranging from 1:1 to 1:3 (Taylor, 1964a) or by weighting the compositions of various igneous, metamorphic, and sedimentary rocks according to their inferred abundances in the crust (Poldervaart, 1955; Ronov and Yaroshevsky, 1969). More recently, it has been possible to estimate the composition of the continents based on the results of extensive sampling of rocks of various metamorphic grade (Eade and Fahrig, 1971; Holland and Lambert, 1972). Such rocks were formed at various depths in the crust and have been subsequently uplifted and exposed at the surface. From the results of experimental studies it is possible to outline roughly the P-T conditions under which metamorphic mineral assemblages form (fig. 4.24).

Estimates of rock and mineral abundances in the crust are given in table 4.4. These estimates are based on weighting rock types in the sediment layer and upper layer of continents according to their exposed abundances (assuming shield areas to represent the upper layer) and assuming a gradual change from granitic rocks in the upper crust to gabbro near the Moho (Ronov and Yaroshevsky, 1969). The sediment layer composition in oceanic areas is estimated from observed sediment abundances and layer 2 is assumed to be composed of half sediments and half low-K tholeiite. The data indicate that granitic-gneissic rocks and mafic rocks or granulite are the most abundant rock types in the crust, with sedimentary rocks composing only about 8 percent. Mineralogically, plagioclase is the most abundant mineral in the crust, followed by quartz and K-feldspar. Pyroxenes and olivine compose about 14 percent and hydrous silicates about 15 percent. All other minerals compose less than 9 percent of the crust.

Continental Crust

An estimate of the composition of the sedimentary layer in the continents is given in column 1 of table 4.5. It is based on a weighted average of common sedimentary rock types as observed in drillcores and in measured stratigraphic sections. The abundance of SiO_2, Al_2O_3, and CaO in the estimate reflects the

Table 4.4 Rock Type and Mineral Abundances in the Crust

Rocks	Vol. Percent	Minerals	Vol. Percent
Sandstone	1.7	Quartz	12.0
Clays and shales	4.2	K-feldspar	12.0
Carbonates	2.0	Plagioclase	39.0
Granites	10.4	Micas	5.0
Grandiorite and quartz diorite	11.2	Amphiboles	5.0
Syenites	0.4	Pyroxenes	11.0
Basalts, gabbros, amphibolites, granulites	42.5	Olivine	3.0
Ultramafic rocks	0.2	Sheet Silicates	4.6
Gneisses	21.4	Calcite	1.5
Schists	5.1	Dolomite	0.5
Marbles	0.9	Magnetite	1.5
		Other	4.9
	100.0		100.0

After Ronov and Yaroshevsky (1969).

Table 4.5. Estimates of the Chemical Composition of the Crust*

	CONTINENTAL CRUST							OCEANIC CRUST				TOTAL CRUST		INTERMEDIATE IGNEOUS ROCKS		
	Seds.	Upper		Lower		Total		Seds.	Upper	Upper and Lower	Total					
	1	2	3	4	5	6	7	8	9	10	11	12	13	14	15	16
SiO$_2$	50.0	63.9	65.2	58.2	64.0	60.2	63.3	40.6	45.5	49.3	48.8	61.3	57.9	59.5	58.7	51.9
TiO$_2$	0.7	0.6	0.6	0.9	0.5	0.7	0.6	0.6	1.1	1.5	1.4	0.8	0.9	0.7	0.8	1.5
Al$_2$O$_3$	13.0	15.2	15.8	15.5	16.8	15.2	16.0	11.3	14.5	17.0	16.3	16.3	15.4	17.2	17.3	16.4
Fe$_2$O$_3$	3.0	2.0	1.2	2.9	1.5	2.5	1.5	4.6	3.2	2.0	2.0	1.5	2.4	2.9	3.0	2.7
FeO	2.8	2.9	3.4	4.8	3.5	3.8	3.5	1.0	4.2	6.8	6.6	4.2	4.4	3.9	4.0	7.0
MgO	3.1	2.2	2.2	3.9	2.2	3.1	2.2	3.0	5.3	7.2	7.0	3.2	4.0	3.4	3.1	6.1
CaO	11.7	4.0	3.3	6.1	3.7	5.5	4.1	16.7	14.0	11.7	11.9	5.3	6.8	7.0	7.1	8.4
Na$_2$O	1.6	3.1	3.7	3.1	4.1	3.0	3.7	1.1	2.0	2.7	2.7	3.7	3.0	3.7	3.2	3.4
K$_2$O	2.0	3.3	3.2	2.6	2.6	2.9	2.9	2.0	1.0	0.2	0.2	2.3	2.3	1.6	1.3	1.3
H$_2$O	2.9	1.5	0.8	1.0	0.7	1.4	0.9	5.0	2.7	0.8	1.0	0.8	1.3	1.2	1.2	0.8
Rb	90	100		90	70	95	85	60	30	1	4	70	75	30		
Sr	300	300		375	400	340	350	1000	570	130	170	310	300	390		
Ba	300	1000		400	1000	670	1150	1000	510	15	65	920	540	270		
U	3	3		3	0.5	3	1.8	1	0.6	0.1	0.2	1.4	2.5	0.7		
Th	8	10		10	2	100	6	5	2.5	0.2	0.5	4.8	8.0	2.2		
Ni	40	20		75	20	50	20	100	100	100	100	35	60	20		

*Major element oxides in weight percent; trace elements in ppm.

NOTE: All references are for major element data only. Trace element data from various sources.

1. Average continental sedimentary rocks (Ronov and Yaroshevsky, 1969, Table 2).
2. Average upper crust ("granitic" layer) (Ronov and Yaroshevsky, 1969, p. 47).
3. Average Canadian Shield (Eade and Fahrig, 1971, p. 42).
4. Average lower crust ("basaltic" layer)—mixture of 1 part basaltic rocks and 1 part average granitic rocks (Ronov and Yaroshevsky, 1969, table 2).
5. Average granulite-facies terrane, New Quebec (Canadian Shield) (Eade and Fahrig, 1971, p. 44).
6. Average continental crust (Ronov and Yaroshevsky, 1969, p. 47)—8% sedimentary (column 1), 23% basaltic and 69% granitic rocks; 1:1 basalt-granite mixture assumed in lower crust (column 4).
7. Average continental crust assuming lower crust is granulite (8% sediments—column 1; 46%—column 3; 46% —column 5).
8. Average deep-sea sediments (Ronov and Yaroshevsky, 1969, p. 47).
9. Average oceanic layer 2 assuming low-K ridge tholeiite and deep-sea sediments mixed in 1:1 ratio (Ronov and Yaroshevsky, 1969, p. 47).
10. Average oceanic layers 2 and 3 assuming ridge tholeiite composition (Engel et al., 1965).
11. Total oceanic crust assuming 16% layer 2 (column 10), 79% layer 3 (column 10), and 5% sediments (column 8).
12. Total earth's crust assuming granulite in lower continental crust (79%—column 7; 21%—column 11).
13. Total earth's crust assuming 1:1 basalt-granite mixture in lower continental crust (79%—column 6; 21%—column 11).
14. Average andesite (Taylor, 1969, p. 60).
15. Calcic andesites from island-arcs (McBirney, 1969a, p. 503).
16. Average diorite (Nockolds, 1954, p. 1019).

abundances of shales and carbonates, respectively. Two estimates of the composition of the upper crustal layer in continental areas are also given in the table. In column 2 is an estimate based on mixing average igneous, metamorphic, and sedimentary rock compositions in the ratios they are thought to occur in continental shield areas. The second estimate (column 3) is based on systematic widespread sampling of the Canadian Shield. Both estimates are remarkably similar, and suggest a composition of the upper continental layer approximately the same as granodiorite or tonalite. Measured P-wave velocities at

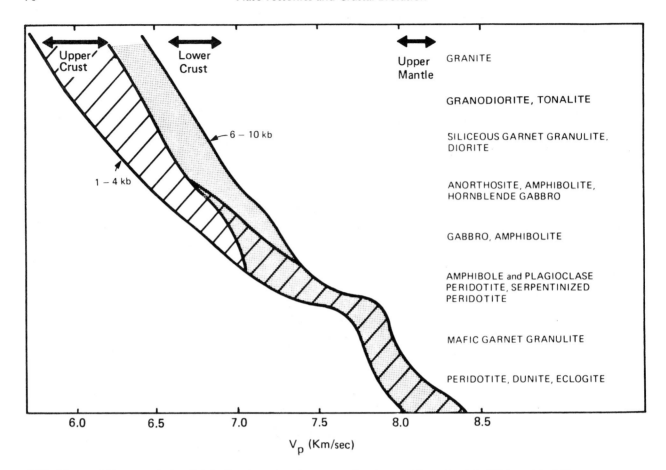

4.25. Measured P-wave velocity distribution in various igneous and metamorphic rocks at 6–10 kb and 1–4 kb pressure (data from Birch, 1960; Christensen, 1965). The general range of continental crust and upper mantle P-wave velocities are indicated by the arrows.

upper crustal depths (1–4 kbar) are also consistent with an upper crust composed of rocks having an overall composition of granodiorite or tonalite (fig. 4.25).

The composition of the lower continental crust is much less certain. Measured P-wave velocities at lower crustal pressures (6–10 kbar) are consistent with a lower crust composed of siliceous garnet granulite, diorite, anorthosite, amphibolite, or some combination of these rock types (fig. 4.25). Although seismic data and experimental data under hydrous conditions indicate that amphibolite could be an important constituent in the lower continental crust, studies of progressive regional metamorphic terranes indicate that water is driven out as metamorphic grade (burial depth) increases. Hence, it is unlikely that sufficient water is available in the lower continental crust for amphibolite to be an important component. A lower crust composed of gabbro, as proposed by many early crustal models, should have

P-wave velocities (7.0–7.3 km/sec) greater than those commonly observed in the lower crust. It is clear, however, that a mixture of granitic and gabbroic rock could account for the lower crustal velocities. Two estimates for the composition of the lower crust are given in table 4.5. The first (column 4) is constructed from a 1:1 mixture of average basaltic and granitic rocks from the continents. The second (column 5) is based on the average composition of high-grade metamorphic rocks from analyses of large numbers of samples from granulite-facies terranes of the eastern Canadian Shield. These siliceous granulites appear to have formed at lower crustal depths, as suggested by their mineral assemblages, and may be representative of rocks composing large parts of the lower continental crust today (Smithson and Brown, 1977). A granulite lower crust would be richer in Si and Ba and poorer in Fe, Mg, Ca, Ti, U, Th, and Ni than a lower crust composed of the basalt-granite mixture.

Two estimates of total continental crustal composition are given in columns 6 and 7. They assume that 8 percent of the crust is composed of sediments of the composition given in column 1 and lower crustal compositions as given in columns 4 and 5, respectively. Both estimates are similar, on the whole, and suggest an overall composition for the continents similar to intermediate igneous rock such as andesite or diorite (columns 14–16). However, the continental crust appears to be enriched in K_2O, Rb, Ba, U, and Th and depleted in CaO relative to these average intermediate rocks.

Oceanic Crust

An average composition of the sedimentary layer in oceanic areas is given in column 8 (table 4.4), based on dredging and coring of this layer. With the exception of being higher in CaO and lower in SiO_2, the average major-element composition of the oceanic sediment layer is similar to that of the continental sediment layer. The composition of layer 2 is less certain. The study of ophiolites (Chapter 8) and the data obtained by dredging and drilling in areas where the sediment layer is thin or absent suggest that layer 2 is composed principally of basalt, with perhaps minor amounts of altered and metamorphosed sediments. Seismic velocity distributions also support this interpretation (Lewis, 1978). Data also indicate that the linear magnetic anomalies of the ocean floors originate in the upper part of this layer, and hence support a basaltic composition. Two estimates of the composition of layer 2 are given in columns 9 and 10. The first assumes a 1:1 mixture of deep-sea sediments (column 8) and low-K ridge tholeiite (i.e., the type of basalt produced at oceanic ridges—see Chapter 7); the second (column 10) assumes an entirely ridge tholeiite composition.

Seismic velocity distributions in the lower oceanic crust are consistent with hornblende gabbro and diabase and possibly serpentinized ultramafic rocks (Christensen, 1970; Lewis, 1978). Gabbro may be an important constituent in the high-velocity (> 7.0 km/sec) layer observed in the lower part of the oceanic crust in some areas (Peterson et al., 1974). Column 10 is an estimate of the composition of the lower oceanic layer, assuming it to be composed of hornblende gabbro of ridge tholeiite composition. A final estimate of the composition of the entire oceanic crust is given in column 11, assuming both layers 2 and 3 are composed of rocks of ridge tholeiite composition. Since the sediment layer contributes only 5 percent, the estimate suggests a composition for the oceanic crust similar to that of the low-K tholeiite generated at oceanic ridges.

Total Crust

Two estimates of total crustal composition are given in columns 12 and 13 of table 4.5. The first assumes a lower continental crustal layer of siliceous granulite composition (columns 7) and the second, a lower continental crustal composition produced by mixing basalt and granite in a 1:1 ratio (column 6). Three averages of intermediate igneous rocks are given in columns 14–16 for comparison. The total crustal averages are similar to the continent averages (columns 6 and 7), and reflect the volumetric abundance of continental crust. Furthermore, the two total crust estimates are similar and suggest an overall crustal composition similar to andesites and diorites. Like the continents, however, the total crust appears to be enriched somewhat in K_2O, Rb, Ba, U, and Th compared to intermediate igneous rocks.

SUMMARY STATEMENTS

1. On the basis of thickness, the Earth's crust is broadly divisible into continental (30–80 km), transitional (15–30 km), and oceanic (5–15 km) types. Continental plus transitional crust comprises 41 percent by area and 79 percent by volume of the total crust. The crust can be further divided into the 12 crustal types in table 4.1.

2. In general, both the oceanic and continental crust are composed of three seismic layers. The contact between the upper and lower layers in continental crust (i.e., the Conrad discontinuity) may be poorly defined and probably gradational over several kilometers. The lower crustal layer ($V_P = 6.5$ to 6.9 km/sec) ranges in thickness from 3 to 7 km in oceanic areas and 15 to 25 km in continental areas. Both thicknesses and velocities of the upper crustal layer differ in the two crustal types: (a) oceanic, $V_P = 4.5$ to 5.5 km/sec; t = 0.7 to 2.0 km and (b) continental, $V_P = 5.8$ to 6.2 km/sec; t = 10 to 20 km. Sediment-layer velocities in both crustal types range from 2 to 4 km/sec and thicknesses from 0 to 1 km in oceanic areas and 0 to 5 km in continental areas. In continental areas, total crustal thickness, layer thicknesses, and layer

velocities do not vary significantly with mean crustal age.

3. Seismic and electrical conductivity data suggest the presence of one or more low-velocity, high-conductivity layers (usually within the depth range of 10 to 30 km) in some continental areas. The presence of water and/or partial melting are probably responsible for producing such layers.

4. Seismic profiling of the crust provides a means of characterizing major stratigraphy and structure of the crust.

5. Heat is lost from the oceanic crust primarily at oceanic ridges where losses by hydrothermal transport are important. Continental heat loss depends upon age of the last magmatic event, distribution of radiogenic heat sources with depth, and erosion level. Approximately 70 percent of the Earth's heat loss is from oceanic ridges.

6. The average heat flow of the crust decreases with mean crustal age. In the continents a decrease from 1.7 HFU to 1 HFU occurs in about 1 b.y. and in the oceans a decrease from 2.5 HFU to 1.3 HFU occurs in 50–100 m.y. Reduced heat flow in the continents also decreases with mean crustal age.

7. Surface heat flow from the continents (q_o) is linearly related to crustal near-surface heat productivity (A_o) by $q_o = q_r + A_o D$ where q_r is the reduced heat flow and D the characteristic depth. A model in which radiogenic heat sources decrease in abundance exponentially with depth best explains this relationship.

8. Bouguer anomalies in most continental areas reflect near-surface rock types and structural discontinuities, whereas in oceanic areas they reflect, for the most part, the presence or absence of anomalous upper mantle. Large negative anomalies over young orogenic areas reflect thicker crust and/or warmer, less dense upper mantle. Small isostatic anomalies in the crust indicate a close approach to isostatic equilibrium.

9. Magnetic anomalies in continental areas reflect near-surface rock types and structural discontinuities. Except in regions of thin crust and high heat flow, magnetic grain generally parallels orogenic structures. Oceanic crust is characterized by roughly linear magnetic anomalies paralleling oceanic ridge systems. Such anomalies are less distinct and more variable in marginal-sea and inland-sea basins.

10. Crustal types have similar geologic and geophysical characteristics, some of which are summarized in table 4.1. A few examples are as follows: (a) Shields and cratons are characterized by average crustal thicknesses (35–40 km), negative upper mantle travel-time anomalies, low to moderate heat flow (1.0 to 1.5 HFU), V_{P_n} 8.1 to 8.2 km/sec, small Bouguer anomalies) (-10 to -50 mgal), pronounced magnetic anomalies, absent or moderately well-defined low-velocity (low Q) zone in the upper mantle, and low reduced heat flow (< 0.8 HFU). (b) Post-Paleozoic orogenic areas are characterized by variable crustal thickness (20–80 km), positive upper mantle travel-time anomalies, variable V_{P_n} (7.7 to 8.2 km/sec), high to variable heat flow (1.3 to 2.5 HFU), large negative Bouguer anomalies (-200 to -300 mgal), variable magnetic anomalies, pronounced upper mantle low-velocity (low Q) zone, occasional low-velocity, high electrical conductivity layers in the lower crust, and high reduced heat flow (> 1.3 HFU). (c) Continental rifts are characterized by thin crust (< 30 km), positive upper mantle travel-time anomalies, anomalous upper mantle ($V_{P_n} < 7.8$ km/sec) and shallow low-velocity (low Q) zone, high heat flow (≥ 2.5 HFU), small, poorly developed magnetic anomalies, large negative Bouguer anomalies (-200 to -250 mgal), and high reduced heat flow (~ 1.4 HFU). (d) Ocean basins are characterized by uniform crustal thickness (6–8 km), typical V_{P_n} (8.1 to 8.2 km/sec), low to moderate heat flow (1.0 to 1.5 HFU), linear magnetic anomalies, and large positive Bouguer anomalies ($+250$ to $+350$ mgal). (e) Oceanic ridges are characterized by thin crust (3–6 km), anomalous upper mantle ($V_{P_n} < 7.8$ km/sec), variable to commonly high heat flow (> 5.0 HFU), linear magnetic anomalies paralleling ridge crests, and positive Bouguer anomalies (about $+200$ mgal).

11. Detailed sampling of Precambrian shields and measured seismic velocities in a variety of igneous and metamorphic rocks indicate that the upper layer in the continents is of granodiorite

or tonalite composition. The lower crust is probably composed of rocks of diorite or siliceous garnet granulite composition.

12. Geological and geophysical data from the ocean floors suggest that oceanic crustal layer 2 is composed of mixed deep-sea sediments and basalts (ridge tholeiite) or of rocks entirely of ridge tholeiite composition. The lower oceanic layer is probably composed of hornblende gabbro and diabase of ridge tholeiite composition.

13. The average composition of the total crust is similar to intermediate igneous rocks (andesites and diorites). It is, however, enriched in K_2O, Rb, Ba, U, and Th compared to most intermediate igneous rocks.

SUGGESTIONS FOR FURTHER READING

Hart, P.J., editor (1969) The Earth's crust and upper mantle. *Amer. Geophys. Union Mon.* No. 13, 735 pp.

Heacock, J.G., editor (1971) The structure and physical properties of the Earth's crust. *Amer. Geophys. Union Mon.* No. 14, 348 pp.

Keen, C.E., editor (1979) *Crustal Properties Across Passive Continental Margins.* Amsterdam: Elsevier. 390 pp.

Prodehl, C. (1979) Crustal structure of the western United States. *U.S. Geol. Survey Prof. Paper* 1034. 74 pp.

Riecker, R.E., editor (1979) *Rio Grande Rift: Tectonics and Magmatism.* Washington, D.C.: Amer. Geophys. Union. 438 pp.

Rona, P.A. (1980) *The Central and North Atlantic Ocean Basin and Continental Margins.* New York: Pergamon Press. 99 pp.

Sclater, J.G., Jaupart, C., and Galson, D. (1980) The heat flow through oceanic and continental crust and the heat loss of the Earth. *Revs. Geophys. Space Physics* **18**, 269–311.

Chapter 5

Crustal Provinces

RADIOMETRIC DATING

Radiometric dating is based on the decay of radiogenic nuclides to stable nuclides. The following conditions must be satisfied in order to interpret radiometric dates in terms of specific geologic events:

1. The time interval over which a rock or mineral formed must be short compared to its age. This problem can be important in dating young rocks (≤ 10 m.y.), especially metamorphic and plutonic rocks, and for authigenic minerals in young sediments.

2. Radioactive decay constants must be accurately known. Decay constants for ^{40}K, ^{235}U, ^{238}U, and ^{232}Th are known to better than 5 percent, and to better than 10 percent for ^{87}Rb and ^{147}Sm. Because older values of the decay constants were used to calculate dates prior to 1977, care should be taken when comparing dates to recalculate results using the new decay constants given in Steiger and Jäger (1977).

3. The initial isotopic concentration of the daughter isotope (^{87}Sr, ^{206}Pb, etc.) must be determined before an accurate date can be calculated. There are several methods for obtaining this information, some of which are discussed below.

4. No losses or gains of parent or daughter isotopes can have occurred since the time of the event being dated or, if they have, adequate corrections must be made for such losses or gains.

K-Ar Dates

The K-Ar dating method is one of the most widely used (Dalrymple and Lanphere, 1969). It has served to delineate major crustal stabilization in orogenic belts as well as to date young magmatic events. However, because of the mobility of argon and imprecise knowledge about just when argon becomes immobilized in a rock system, individual K-Ar dates are of limited value in bracketing the age of orogenies or regional metamorphism. Also, some minerals (such as pyroxenes) commonly accept excess argon into their structures and, if not corrected for, give ages that are too old. Today K-Ar dates are used in studying the geochronology of young, unaltered volcanic rocks and sediments, in detecting incipient metamorphism or alteration, and in defining regional epeirogenic uplift.

Rb-Sr Isochron Dates

The following expression is commonly used to calculate the age of a geologic event using the Rb-Sr dating method (Faure, 1977):

$$\left(^{87}Sr/^{86}Sr\right)_t - \left(^{87}Sr/^{86}Sr\right)_i = \left(^{87}Rb/^{86}Sr\right)_t(e^{\lambda t} - 1)$$

(5.1)

where $\left(^{87}Sr/^{86}Sr\right)_t$ and $\left(^{87}Sr/^{86}Sr\right)_i$ are the atomic ratios of the two Sr isotopes at the present time and at the time of the event being dated, respectively; $\left(^{87}Rb/^{86}Sr\right)_t$ is the atomic ratio at the present time; λ is the decay constant; and t is the age. This is the equation of a straight line with intercept $\left(^{87}Sr/^{86}Sr\right)_i$ and a slope of $e^{\lambda t} - 1$, which is defined as an *isochron*. All portions of a rock system that formed at the same time, had the same initial $^{87}Sr/^{86}Sr$ ratio $[\left(^{87}Sr/^{86}Sr\right)_i]$, and have not lost or gained either parent or daughter isotope since the rock formed will lie on this line. Extrapolation of the line defined by samples of differing Rb/Sr ratios to ^{87}Rb/^{86}Sr $= 0$ gives the $(^{87}$Sr/^{86}Sr$)_i$ ratio. In order to obtain accurate estimates of both the age and $(^{87}$Sr/^{86}Sr$)_i$ ratio, it is necessary to have a group of samples with a reasonably wide distribution of ^{87}Rb/^{86}Sr ratios. If an accurate isochron can be determined, the method provides a means of identifying single samples that have lost or gained parent or daughter since the system became closed, since such samples will probably plot off the isochron.

It is sometimes possible to distinguish two events using the isochron method. An example for Precambrian granitic rocks from central Colorado is illustrated in fig. 5.1. Whole-rock samples of the granites define an isochron of 1650 m.y. with an initial ^{87}Sr/^{86}Sr ratio of 0.703. When individual minerals in one sample are analyzed, an isochron of 1384 m.y. [$(^{87}$Sr/^{86}Sr$)_i = 0.726$] is obtained. The latter isochron is interpreted as dating an isotopic exchange on a small scale between minerals, and probably records a time of regional metamorphism. The 1650 m.y. whole-rock isochron is interpreted as the original formation date of the granites.

Sm-Nd Isochron Dates

The Sm-Nd isochron dating method is similar to the Rb-Sr method. ^{147}Sm decays by alpha particle emission to ^{143}Nd with a half-life of about 10^{10}y ($\lambda = 6.54 \times 10^{-12}$y^{-1}), and the standard isochron expression can be written by normalizing to ^{144}Nd, a stable isotope of Nd,

$$\left(^{143}\text{Nd}/^{144}\text{Nd}\right)_t - \left(^{143}\text{Nd}/^{144}\text{Nd}\right)_i$$

$$= \left(^{147}\text{Sm}/^{144}\text{Nd}\right)_t(e^{\lambda t} - 1) \qquad (5.2)$$

where each term has an analogous definition to terms in the Rb-Sr isochron equation 5.1. Because both Sm and Nd are rare-earth elements, they are not fractionated from each other, as are Rb and Sr, and hence the range in Sm/Nd ratios is generally much smaller than the Rb/Sr range on isochron plots. This lack of Sm-Nd fractionation also applies to secondary geological processes such as alteration and metamorphism, and hence the Sm-Nd dating method has the advantage over the Rb-Sr method that it can see through secondary events to primary igneous

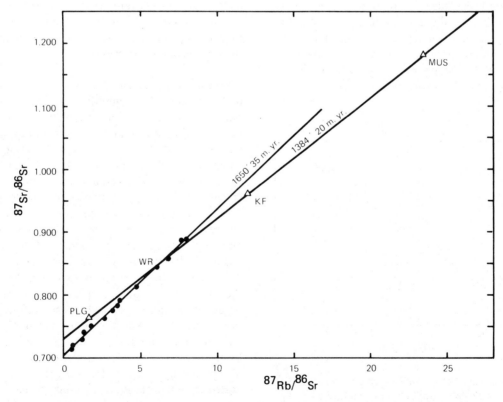

5.1. Rb-Sr isochron diagram for Precambrian granitic rocks from central Colorado (after Wetherill and Bickford, 1965). Key: ● = whole-rock samples; Δ = mineral separates from whole rock WR. PLG = plagioclase; KF = K-feldspar; MUS = muscovite.

events (O'Nions et al., 1979). Also, because the Sm/Nd ratio is generally higher in mafic and ultramafic rocks than in felsic rocks, the Sm/Nd method provides the only accurate method of dating mafic and ultramafic rocks.

U-Th-Pb Dating Methods

Concordia Dates. When $^{238}U/^{206}Pb$ and $^{235}U/^{207}Pb$ dates for a given rock are the same, they are said to be *concordant*. If the $^{206}Pb/^{238}U$ and $^{207}Pb/^{235}U$ ratios are plotted against each other, concordant samples of various ages define a single curve known as *concordia*. In fig. 5.2, $^{206}Pb/^{238}U$ and $^{207}Pb/^{235}U$ ratios in zircons from the Boulder Creek batholith in Colorado define a chord that intersects concordia at 1725 m.y. and 65 m.y. The zircon ages are referred to as *discordant*, and the chord upon which they fall is a line of lead loss (or uranium gain). The upper intersection with concordia is generally interpreted as the initial formation age of the zircons (in this case assumed to be equal to the crystallization age of the Boulder Creek batholith). If an episodic model for lead loss is assumed, the lower intercept gives the age of the thermal event that resulted in lead loss (65 m.y.). The data can also be fitted to a continuous diffusion model for the loss of lead in which the lower end of the chord curves down toward the origin (zero age). Unless independent geologic evidence favors one of these models, it is not possible to decide which model is most suitable for a given rock terrane (Faure, 1977).

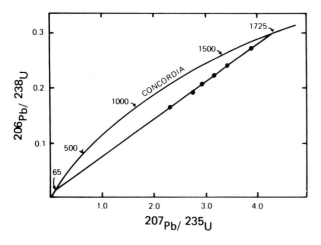

5.2. U-Pb concordia relationships of zircons from the Boulder Creek Batholith, Colorado, interpreted in terms of an episodic lead-loss model (after Stern *et al.*, 1971). Individual zircon analyses = ●; ages on concordia in m. years.

As with Rb-Sr isochron dates, discordant U-Pb dates necessitate a reasonably wide range of $^{206}Pb/^{238}U$ and $^{207}Pb/^{235}U$ ratios (or nearly concordant ages) to define accurately the intersection(s) with concordia. Zircon is the most common mineral used in U-Th-Pb dating, in that the initial amount of lead incorporated in zircon is generally negligible compared to uranium and thorium, and zircon is extremely refractory. Zircon dates by this method have potentially the greatest resolving power for closely spaced Precambrian events (approximately 5 m.y. at 3.0 b.y.).

Isochron Dates. Equations similar to 5.1 and 5.2 can be written for $^{238}U-^{206}Pb$, $^{235}U-^{207}Pb$, and $^{232}Th-^{208}Pb$ normalizing to the stable isotope ^{204}Pb. In addition, a third isochron expression relating $^{207}Pb/^{204}Pb$ to $^{206}Pb/^{204}Pb$ can be formulated by combining the $^{238}U-^{206}Pb$ and $^{235}U-^{207}Pb$ expressions, thus providing a total of four isochron dating relations in the U-Th-Pb system. Because of U loss during secondary geologic processes, the U-Pb isochron methods have not been very successful. The Th-Pb isochron method can be definitive in some geologic systems (Faure, 1977). The Pb-Pb isochron method has the advantage that it is not sensitive to recent losses of uranium or lead but depends only on the requirement that all samples have remained closed to early secondary processes.

Common Lead Dates. Common lead is lead in a rock or mineral at the time of formation. Those minerals or rocks with extremely low U/Pb and Th/Pb ratios, such that no significant radiogenic lead has accumulated since they crystallized, are used in common lead dating methods (Russell and Farquhar, 1960). Galenas, K-feldspars, and some whole-rock volcanic samples are examples of samples relatively free from such radiogenic contamination. Lead model ages are calculated from a knowledge of present and initial lead isotope ratios ($^{206}Pb/^{204}Pb$ and $^{207}Pb/^{204}Pb$), the age of the Earth, and various assumptions difficult to prove regarding the history of the Earth. The initial isotopic ratios are assumed to be equal to those observed in the sulfide phases in iron meteorites, which contain negligible uranium and thorium, and the age of the Earth is inferred from meteorite ages. The model dates calculated are interpreted by some as the ages that lead was separated from mantle source regions, probably by melting, followed by extrusion or intrusion of magma (or ore solutions) into the crust. Another interpretation is that the dates represent the last crust-mantle mixing event in a given area. Because of assumptions that are difficult to evaluate, lead model ages are generally considered the least accurate of the U-Th-Pb radiometric dates.

Disequilibrium Dating. The decay chains of ^{235}U and ^{238}U are often broken by the loss of one or more daughter isotopes due to geological processes. Alteration, weathering, chemical or biochemical precipitation, and magma crystallization are examples of processes that disturb the uranium decay chains. It is possible to make use of these separations to date relatively young geologic events by (1) monitoring the activity of the radiogenic nuclide separated from a decay chain, or (2) monitoring the rate of growth of a daughter nuclide produced from a parent previously removed from the decay chain. These methods are particularly important in dating events during the last million years and fill a gap between ^{14}C and K-Ar dating methods. Applications have been made to measure sedimentation rates in the oceans using the ^{230}Th and ^{231}Pa methods and to the age of carbonate-precipitating organisms using the ^{234}U and ^{230}Th methods. Growth of glaciers has also been studied employing ^{210}Pb.

Fission Track Dates. Natural fission of ^{238}U leaves damage tracks in glasses and minerals which can be enlarged by etching. The number of tracks, which is a function of the age of the sample and its uranium content, can be counted with a petrographic microscope and used to date the sample, provided certain conditions are met. The two most important conditions are that the samples must have been cooled rapidly and that the track density must not have been affected by secondary processes. Because tracks anneal in minerals at different rates as a function of falling temperature, the fission-track method can also be used to study the thermal history of rocks and rates of uplift (Wagner and Reimer, 1972).

Radiogenic Isotope Tracer Studies

Radiogenic daughter isotopes continually increase in rocks at a rate proportional to the parent/daughter element ratios (Rb/Sr, Sm/Nd, U/Pb, etc.). Employing this relationship, it is possible to place constraints on the composition of the source rocks from which magmas are derived. For instance, ultramafic source rocks have lower Rb/Sr and U/Pb ratios but higher Sm/Nd ratios than granitic rocks. Hence, magmas produced by partial melting of ultramafic rocks in the mantle will have lower $^{87}Sr/^{86}Sr$ and

$^{206}Pb/^{204}Pb$ ratios and higher $^{143}Nd/^{144}Nd$ ratios than magmas produced by partial melting of granitic crustal rocks. This assumes that the isotopic ratios are not changed during melting or fractional crystallization; experimental and theoretical considerations support this assumption. Small differences in parent/daughter element ratios within the same type of source rock can also be detected, and these results can be interpreted in terms of source heterogeneity. It is possible to extend the application of isotopic tracer studies to older geologic terranes by subtracting the amount of radiogenic daughter isotope that has grown in igneous rocks since their crystallization. The isochron equations (5.1 and 5.2) can be used to make this correction by solving for the initial daughter isotope ratios ($^{87}Sr/^{86}Sr_i$, $^{143}Nd/^{144}Nd_i$, etc.).

Examples of applications of radiogenic isotope tracer studies are given in Chapters 7 and 11.

EVENTS THAT CAN BE DATED

A radiometric date gives the age that a system was last open to migration of daughter isotopes in and out of a rock terrane or between adjacent minerals. This date, in effect, records an *isotope immobilization temperature* for a given mineral or rock. An immobilization temperature may be defined as that temperature at which diffusion rates of a given daughter isotope in a mineral (or rock) change from geologically rapid to imperceptible. The cooling rate for a rock is also an important factor bearing on element immobilization temperatures. The temperature at which Ar is immobilized is lower than that for Sr, and the immobilization temperature of Sr is lower than that for Pb in zircons. Hence, various dates in different minerals and rocks record different parts of the thermal history of a rock or mineral. It is not always obvious which dates (if any) correlate with specific tectonism, magmatism, or other geologic phenomena. The problem is further complicated by the fact that daughter isotopes may be only partially retained if cooling rates are slow, thus giving radiometric dates that may be meaningless in terms of specific geologic events. For these reasons, a rather complete knowledge of geologic field relationships is necessary if radiometric dates are to be accurately interpreted.

Magmatic Events

Magmatic crystallization is one of the most common types of events that can be dated. Felsic igneous rocks are generally more satisfactory for dating than mafic or intermediate igneous rocks since they contain greater concentrations of most radiogenic parent isotopes. An exception is the previously mentioned Sm-Nd system. Magmatic events may record tensional or compressional stresses in the crust or they may be completely unrelated to any known crustal deformation. Their interpretation in terms of tectonics is critically dependent upon field relationships.

Orogenies

The term *orogeny* is used in different ways. In its broadest sense it means mountain building. However, in a more restricted way, as it will be used here, it refers to crustal deformation involving major folding, which occurs during a limited interval of geologic time. It is also commonly accompanied by volcanism, plutonism, and regional metamorphism. Regions that have undergone only faulting and broad warping are excluded from this definition. Orogenies can be dated by dating syn-tectonic granitic plutons or regional metamorphism. As pointed out by Gilluly (1973), however, one should not equate orogeny with plutonism, since some orogenies occur with little if any magmatic activity (such as the Antler orogeny in the western United States). Also, plutonism may be post-tectonic or, as with the 1.4–1.5 b.y. plutons in the central and southwestern United States, *anoro-genic* (i.e., emplacement without regional deformation).

The exact times that various daughter isotopes become immobilized during an orogeny are not well known, but they occur after the thermal intensity peak. U-Pb zircon dates and Sm-Nd dates probably coincide most closely with maxima of plutonism, while K-Ar dates appear to record regional uplift during the final stages of orogeny. It is possible with Rb-Sr, U-Th-Pb, and Sm-Nd dating methods to resolve 10–50 m.y. at 3 b.y., and in some instances, with U-Pb zircon dates, to resolve as little as 5 m.y. at 3 b.y. (Krogh and Davis, 1971). Dating of Phanerozoic orogenies suggests that orogenies last from 10 to 50 m.y. Dates from Precambrian mobile belts, on the other hand, commonly span 100–200 m.y., and it is possible or even probable that several individual orogenies are "hidden" in such a spread of dates. Such periods of time over which several closely spaced orogenies may have occurred will be referred to here as *orogenic periods*, to distinguish them from Phanerozoic-type orogenies.

Phanerozoic and late Precambrian orogenies are of two types: those related to convergent-plate boundaries and those related to continental collisions (Chapter 8). Whether or not Precambrian orogenies can be interpreted in a similar manner is a subject discussed in Chapter 10.

Regional Uplift

Because of the low immobilization temperature of argon, K-Ar dates from orogenic terranes are generally 10–20 percent lower than Rb-Sr whole-rock isochron and U-Pb zircon dates. Argon-retention ages, however, appear to record the termination of denudation and uplift during the final stages of orogeny or at any other time. K-Ar mineral ages record uplift in the Appalachian Mountains during the Paleozoic and in the Alps during the late Tertiary.

Diagenetic Dates

Diagenetic dates sometimes may be estimated using K-Ar and Rb-Sr isochron methods. Glauconite, diagenetic micas or feldspars, or whole-rock samples are generally used. Several problems are encountered in attempting to obtain diagenetic ages: (1) Ar and, to a lesser extent, Sr leak from diagenetic minerals; (2) it is difficult to separate detrital and diagenetic components in sedimentary rocks; and (3) Sr may not equilibrate within diagenetic phases, resulting in poorly defined isochrons. Extreme care should be used in interpreting radiometric ages obtained from sediments or sedimentary rocks in terms of diagenetic ages unless the above problems have been evaluated.

Relict Dates

Minerals that are resistant to weathering, erosion, and metamorphism, like zircon, may be recycled through more than one orogeny. U-Pb dates obtained from recycled zircons give ages somewhere between the initial age of the zircon and the age of the last metamorphic or partial melting episode. Such dates are referred to as relict dates, and they may or may not record any particular geologic event. Their chief importance is in identifying older crustal terranes or segments thereof that are isolated or partially isolated in younger terranes. Such terranes are known as *relict-age provinces.*

Defining Crustal Provinces

Stockwell (1965) suggested that the Canadian Shield can be subdivided into structural provinces based on differences in structural trends and style of folding

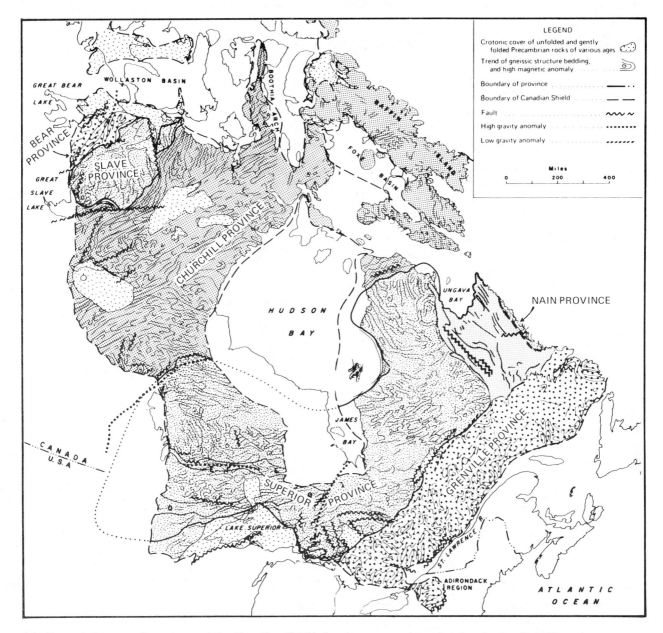

5.3. Precambrian crustal provinces of the Canadian Shield showing structural trends (after Stockwell, 1965).

(fig. 5.3). Structural trends are defined by foliation, fold axes, bedding, and sometimes, magnetic anomalies. Boundaries between the provinces are drawn where one trend cross-cuts another, along either unconformities or structural-metamorphic breaks. Large numbers of radiometric dates from the Canadian Shield indicate that the structural provinces are broadly coincident with age provinces. Similar relationships have been described on other continents and lead directly to the concept of a crustal province. A *crustal province* is herein defined as a segment of the continental crust which records a similar range of

radiometric dates and exhibits similar deformation styles. Crustal provinces are of two general types: orogenic and anorogenic. *Orogenic provinces* record complex polyphase deformational, magmatic, and metamorphic histories; *anorogenic provinces* exhibit only minor deformation. Orogenic provinces can be divided into *subprovinces*, some of which have been designated mobile belts. Subprovinces are defined chiefly on the basis of characteristic structural and metamorphic styles. *Mobile belts* are linear to curvilinear subprovinces that range from tens to hundreds of kilometers in width and up to several thousand

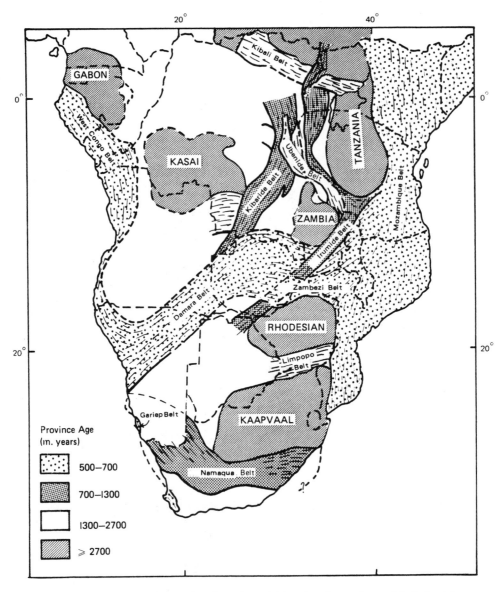

5.4. Crustal province map of central and southern Africa showing major mobile belts (after Cahen and Snelling, 1966).

kilometers in length (fig. 5.4). Mobile belts of the same crustal province may have sharp contacts with each other, or they may merge with each other, as is so well exemplified by the Pan-African belts in southern Africa. The Churchill Province in Canada is comprised of many subparallel mobile belts containing remnants of Archean crust (Goodwin, 1974). Anorogenic provinces are comprised of igneous or sedimentary rocks and have undergone only slight warping and extensional faulting, and these provinces may later be deformed and become part of orogenic provinces.

OROGENIC PROVINCES

Orogenic provinces appear to underlie most of the continental crust today (fig. 5.5). Most of these provinces have now evolved into stable cratons and are covered partly or entirely by anorogenic sedimentary or volcanic provinces (Plate I). Some mobile belts cross-cut older mobile belts at steep angles without offsetting the older belts (fig. 5.4). This feature is a major constraint on the origin of this type of mobile belt (Chapter 10). Structural trends within orogenic provinces range from linear to exceedingly complex

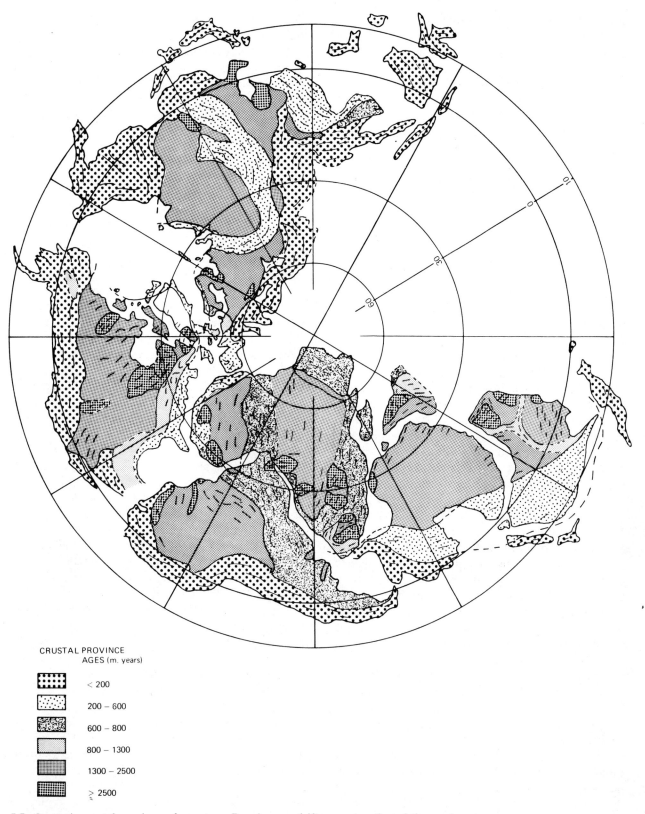

CRUSTAL PROVINCE
AGES (m. years)

< 200

200 − 600

600 − 800

800 − 1300

1300 − 2500

≥ 2500

5.5. Orogenic crustal provinces shown on a Permian pre-drift reconstruction of the continents.

and swirling (fig. 5.3). Because many structural trends are the result of multiple stages of deformation, careful and detailed structural studies are necessary to reconstruct the deformational significance of structural trends. Structural trends may change rapidly in small areas or they may be persistent over great distances. Relict-age terranes are abundant in many orogenic provinces, for instance in the late Proterozoic mobile belts of southern Africa (fig. 5.4) and in the Churchill Province of Canada. Such terranes, together with widespread relict dates, indicate that much of these provinces is comprised of reworked Archean crust. Most of Africa classified as orogenic provinces less than 2.5 b.y. in age appears to represent reworked older crust (Kröner, 1977).

The dominant rock type in most orogenic provinces is tonalite to granodiorite gneiss. Such gneisses contain remnants of older supracrustal rocks and may also contain infolded successions of unconformably overlying supracrustal rocks. *Supracrustal rocks* are rocks that formed at or near the Earth's surface (sediments, volcanics, shallow intrusions) and are subsequently buried and metamorphosed, then once again exposed at the surface by uplift and erosion. Precambrian supracrustal rocks can be classified into one of several assemblages, as will be discussed in Chapter 10. The gneissic complexes in orogenic provinces are also intruded by gabbros, stratiform complexes, anorthosites, and a variety of granitic rocks. Granitic plutons range from pre- to post-tectonic. Mafic dike swarms occur in most orogenic provinces, and again are represented by both pre- and post-deformational examples.

Metamorphic grade is highly variable in orogenic provinces ranging from prehnite-pumpellyite to granulite facies. The highest-grade zones may accompany extensive migmatite development, and occur in the central parts of mobile belts. Metamorphic facies series are also variable within mobile belts. On the whole, however, there seems to be a time-dependent trend from low-pressure to high-pressure series in going from Archean provinces to younger provinces. Mesozoic-Cenozoic orogenic provinces commonly exhibit paired metamorphic belts, as is discussed in Chapter 8. Mobile belts that evolve at relatively low temperatures and high pressures are characterized by extensive thrusting and nappe formation and appear to record substantial shortening of the crust. Examples are the Nagssugtoqidian belt in Greenland and the Alpine-Tethyan belt in Eurasia (Bridgewater et al., 1973). On the other hand, mobile belts that evolve at relatively high temperatures and low pressures like the Ketilidian in southern Greenland and the Hercynian in Europe are characterized by extensive

granitic plutonism and dominantly vertical deformation patterns.

The maximum age of sediments unconformably overlying orogenic provinces can be used to estimate the time involved in uplift and erosion leading to cratonization of an orogenic province (Watson, 1976). Existing data indicate that uplift and stabilization lasted no more than 200 m.y. in provinces older than 700 million years. Most Phanerozoic orogenic provinces, however, appear to be characterized by periods of sporadic uplift and erosion lasting greater than 200 million years.

Major orogenic provinces are continuous from one continent to another, as can be illustrated on a Permian pre-drift reconstruction of the continents (fig. 5.5). Archean provinces occur as small equidimensional remnants within and surrounded by younger provinces. As was mentioned above, however, the widespread distribution of relict dates and relict-age terranes within younger orogenic belts indicates that the Archean continental crust was much more extensive prior to 2 b.y. ago (see fig. 10.8). The widespread distribution of early Proterozoic provinces (fig. 5.5) may be misleading, in that most of these areas are covered with Phanerozoic cratonic sediments and their ages are not well known. The best-documented orogenic provinces of this age (chiefly 1.7–1.9 b.y.) are the Churchill Province in North America and the Svecofennian Province in northern Europe. Provinces between 800 and 1300 m.y. in age are of only minor extent. They are represented by the Grenville Province in eastern North America and western Europe, the Namaqua Province in southern Africa, and minor, less well-defined provinces in India and Australia. The Pan-African provinces (500–700 m.y.), as is discussed below, are widespread in Africa and South America but of very minor extent or unknown elsewhere. Paleozoic provinces are preserved around the North Atlantic (Appalachian-Caledonian belt), in western Europe (Hercynian belt), in eastern Australia and Antarctica, and in central and southeastern Asia. Mesozoic-Cenozoic provinces are represented chiefly by the Cordilleran belt extending from Siberia and Alaska, through western North and South America perhaps to Antarctica and New Zealand, and the Alpine-Tethys belt, extending from western Europe to southeastern Asia (fig. 5.5; Plate I).

Province Boundaries

Boundaries between crustal provinces or mobile belts may be parallel to the structural trends within adjacent provinces, such as the boundary between the

Irumide and Mozambique mobile belts (fig. 5.4). Alternately, one province may cross another at steep angles, as is represented by the contact of the Zambezi belt where it crosses the Irumide belt. The actual contacts between crustal provinces or between mobile belts fall into one or a combination of four categories: (1) faults or shear zones; (2) unconformities; (3) rapid changes in metamorphic grade; (4) intrusive contacts. Magnetic and gravity anomalies also characterize provincial boundaries (fig. 5.3), reflecting juxtaposition of rocks of differing densities and magnetic susceptibilities. Changes in crustal thickness may also occur at province boundaries, represented, for instance, by the increase from 35–40 km to 45 km at the Grenville Front between the Grenville and Superior provinces (fig. 4.3).

The Grenville Front has been the subject of numerous geological and geochronological studies. This boundary varies from a fault contact to a metamorphic transition zone along strike (Wynne-Edwards, 1972). It ranges in width from a few meters to several kilometers and involves the overprinting of one metamorphic episode (the Grenville at ~ 1.0 b.y.) on another (the Kenoran at 2.7 b.y.). The Grenville isograds cut across the Kenoran isograds at steep angles near the Grenville Front (fig. 5.6). K-Ar dates

are reset at rather low temperatures (~ 200°C) and gradually decrease from 2.7 b.y. to about 1.0 b.y. over a zone of variable width across the front. At some locations the Grenville Province is thrust over the Superior Province, transporting older gneisses (as recorded by Rb-Sr dates) with completely reset K-Ar dates into contact with Archean rocks of the Superior Province.

Shear zones between provinces or mobile belts may be accompanied by extensive mylonite zones ranging up to tens of kilometers in width. A cross section of such a boundary between the Wyoming Archean Province and the Proterozoic province to the south is shown in fig. 5.7. This contact, which is exposed in the Medicine Bow Mountains in southeastern Wyoming, is a near vertical shear zone separating Archean gneisses from a gneissic complex (probably ~ 1.8 b.y.) and granitic rocks (1.4 b.y.). The contact is complicated by the fact that deformed metasedimentary rocks (1.7–1.8 b.y. in age) rest unconformably on the Archean gneisses and are also cut by the shear zone. The shear zone, which is up to several kilometers wide, is composed chiefly of mylonitized quartzo-feldspathic gneisses. Some boundary fault zones exhibit transcurrent motions, as is exemplified by the northern shear zone of the

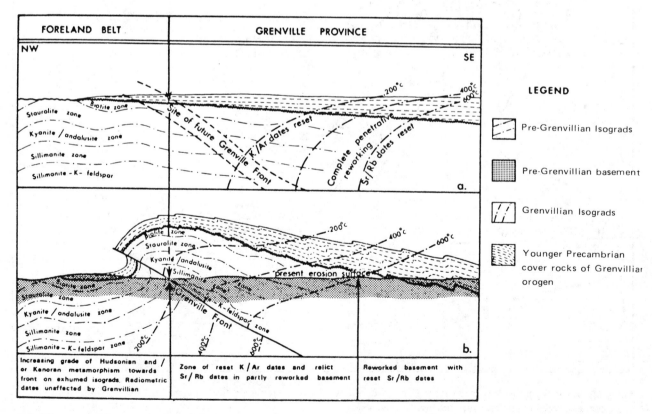

5.6. Diagrammatic cross-section of the Grenville Front before (a) and after (b) uplift (after Wynne-Edwards, 1972).

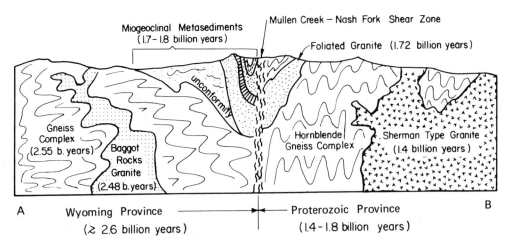

5.7. Schematic cross-section of the boundary between the Wyoming Province and Proterozoic
      provinces in southeast Wyoming (from Hills, *et al.*, 1968). Location of section shown in
      Fig. 5-8.

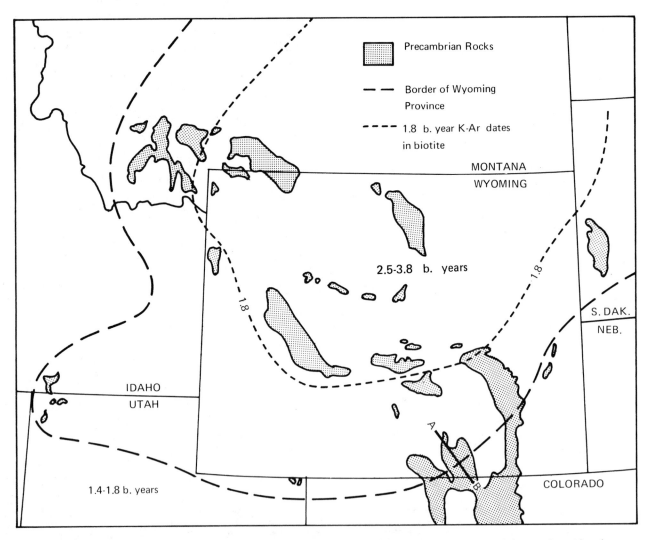

5.8. Map of the Wyoming Province showing the effect of later thermal heating on the outer parts of the province (data in
      southeast Wyoming after Hills and Armstrong, 1974).

Limpopo belt immediately north of Pikwe (fig. 5.11), which has up to 200 km of right lateral displacement.

Boundaries of structural provinces may or may not coincide with age-province boundaries. Age-province boundaries are typically gradual, extending for tens to hundreds of kilometers beyond structural boundaries. Such gradational boundaries are produced by a gradual change in either metamorphic grade or in degree of granitization. As was pointed out above, the Grenville Front marks the western boundary of the Grenville Structural Province, but Archean dates extend many tens of kilometers east of the contact into the Grenville Province (fig. 5.12). Another example is illustrated in fig. 5.8, where a wide metamorphic "overprint" is superimposed on the shear zone that separates the Wyoming Province from the Proterozoic terrane to the south. Although Precambrian rocks crop out at only a few places in this part of the Rocky Mountains, existing mineral dates indicate the presence of a zone several hundred kilometers in width in which daughter isotopes in mineral systems in the Archean rocks have been partially mobilized during the Hudsonian orogeny at 1.8–1.9 b.y. K-Ar mineral dates in this zone are generally $\lesssim 1.8$ b.y. The 1.8 contour in fig. 5.8 represents the maximum distance in from the contact where biotite began to retain argon at 1.8 billion years.

Examples of Orogenic Provinces

Archean Provinces. Archean provinces are roughly equadimensional in plan view and range from less than 500 km to about 2000 km in maximum dimension. They are composed of rocks ranging in age from about 2.5 to >3.5 b.y. Rocks 3.5 to 3.8 b. years in age have been described from many locations (Plate I).

Major Archean provinces of the world are shown in fig. 5.5 and Plate I. The largest preserved province is the Superior Province in eastern Canada (fig. 5.3). North America (including Greenland) has, in addition, four more provinces: the Slave, Godthaab, Nain, and Wyoming provinces. Eight small provinces are preserved in Africa (Clifford, 1970), at least three in South America, three in Australia, seven in Eurasia, and one or more in Antarctica. That the original extent of these provinces was probably considerably greater is evidenced by the abundance of relict Archean dates in younger Precambrian terranes. For example, relict dates from the Churchill Province (shown in fig. 5.3) suggest that the five North American Archean provinces may have origi-

nally been part of the same crustal province which was, in large part, reworked by orogeny at 1.8–1.9 b.y.

Archean provinces are composed of either a granite-greenstone or a high-grade metamorphic association, or both (Condie, 1981a). Statistical sampling in Canadian provinces indicates that granitic rocks (mostly granodiorites and tonalites) and gneiss-migmatite terranes dominate in both associations (Eade and Fahrig, 1971). Structural trends in Archean provinces range from rather continuous to complex and irregular, even within the same geographic area. An example of the complexity of Archean structural trends is shown for the Superior, Slave, and Nain Provinces in North America in fig. 5.3. In a few instances, extrapolation of structural trends from one province to another seems to be justified (as for instance between central-west Africa and northeastern Brazil [fig. 5.5]).

Granite-greenstone terranes are characterized by discontinuous greenstone belts engulfed in a "sea" of granite and gneiss. The major granite-greenstone terranes of the world occur in the Superior and Slave provinces in North America (fig. 5.3), in the Rhodesian and Kaapvaal provinces in southern Africa (Figure 5.4), and in the Yilgarn and Pilbara provinces of western Australia (Plate I). *Greenstone belts* are linear-to-irregular-shaped volcanic terranes which average 20–100 km wide and extend for distances of several hundred kilometers (Anhaeusser et al., 1969). They contain exposed thicknesses ranging up to about 15 km of predominantly mafic volcanic rocks (Glikson, 1970; Goodwin, 1968). An idealized map of a typical Archean granite-greenstone terrane is shown in fig. 5.9. Most greenstone belts exhibit a synformal structure with open-to-isoclinal folds, with major faults paralleling fold axes. The keel-shaped outcrop pattern so characteristic of some belts results from diapiric granitic intrusions around the edges. The overall deformation patterns reflect dominantly vertical stresses. Rocks within greenstone belts are commonly metamorphosed only to the greenschist facies, and primary textures and structures are well preserved. Three basic types of granitic rocks are recognized in association with greenstone belts: gneissic complexes, diapiric granitic plutons, and discordant granitic plutons. All three types are shown in fig. 5.9. The gneissic complexes and diapiric plutons range from granodiorite to (more commonly) tonalite or trondhjemite in composition, and are gradational with each other. They contain inclusions of volcanic rocks, graywackes, and other rock types found in greenstone belts. Diapiric granites have circular-to-elliptical

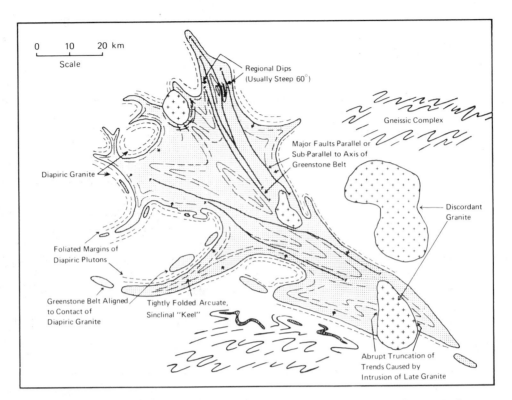

5.9. Idealized map of an Archean greenstone belt showing characteristic features (after Anhaeusser *et al.*, 1969).

shapes and are commonly a few tens of kilometers in diameter. Texturally they are more homogeneous than gneissic terranes, yet still commonly exhibit a gneissosity that follows the outline of greenstone belts. A steep metamorphic gradient may occur in greenstone belts adjacent to diapiric granites, indicating a shallow level of emplacement. Discordant granites are the youngest granites in greenstone-granite terranes, cross-cutting all older rock types with strongly discordant relationships. They are usually nonfoliated K-rich granites, and are commonly porphyritic.

Studies in the southern part of the Superior Province show that subprovinces with abundant greenstone belts alternate with subprovinces in which metasedimentary rocks (chiefly metagraywacke) dominate. Granitic rocks are more abundant than volcanic and sedimentary rocks in both subprovinces, with gneisses and migmatites most abundant in the metasedimentary types. The distribution of some of these alternating volcanic-granitic and sedimentary-granitic subprovinces is shown in fig. 5.10. Existing data indicate that the volcanism and sedimentation in the subprovinces occurred, in part, simultaneously (Ayres, 1969).

Field evidence indicates that some gneissic complexes and most diapiric granites are intrusive

into greenstone terranes. However, there are some areas in which field relationships indicate that the greenstone volcanic rocks were extruded onto granite-gneiss basement (Condie, 1981a). The fact that gneissosity in Archean gneissic granites subparallels schistosity in greenstone belts makes the identification of intrusive contacts or unconformities in many areas difficult. It is possible that the gneissic and diapiric granites represent rejuvenated pre-greenstone sialic basement in which daughter isotopes were partially (or completely) remobilized during a later thermal event. Careful U-Pb dating of these granites is needed to evaluate this problem. At the present time, however, field data seem to indicate that at least some Archean greenstone belts were erupted onto older sialic basement.

Archean high-grade metamorphic terranes are most abundant in southwest Greenland and in the Archean portions of the Baltic, Ukrainian, Anabar, Aldan, and Indian shields (Plate I). Three rock associations are recognized in Archean high-grade metamorphic terranes: gneiss-migmatite-granulite complexes, amphibolite-facies assemblages, and layered mafic and ultramafic intrusions (Windley and Bridgewater, 1971; Salop and Travin, 1972). Most Archean high-grade metamorphic terranes have

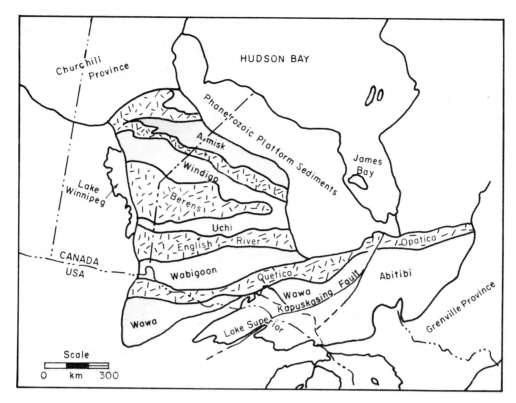

5.10. Archean volcanic-granitic (gray) and sedimentary-granitic subprovinces in the southern Superior province, Canada (after Ayres, 1969; Goodwin, 1971).

undergone at least one period of granulite-facies metamorphism (650–800°C, 20–35 km burial depth) which appears to represent the culmination of a long sequence of igneous, metamorphic, and deformational events (Windley, 1973). The Archean high-grade terranes are composed chiefly of quartzofeldspathic, gneiss-migmatite complexes that appear to have formed several hundred-million years before the major granulite-facies metamorphism. Infolded into the gneiss-migmatite terranes are synclinal keels of amphibolites, ultramafic rocks, and minor metasediments (see Chapter 10). Although primary structures are rarely preserved in these rocks, they are interpreted by some workers as infolded greenstone belts that have undergone higher grades of regional metamorphism. Layered igneous intrusions compose a small but important part of the high-grade metamorphic terranes. Individual intrusions may range up to ≥ 10 km in thickness and 200 km in strike length, and may preserve compositional layering and cumulus textures (i.e., textures indicative of crystal-liquid separation). Most evidence suggests that these bodies were intruded into the gneiss-migmatite terranes (or their predecessors) prior to granulite-facies metamorphism. Layered anorthosite bodies are also found and are of the particular interest in that some are similar

to lunar anorthosites in the high An content of their plagioclase (An_{80-100}) and their association with chromite.

The structural style in Archean high-grade metamorphic terranes is exceedingly complex. It is characterized by fold interference patterns that lack linear regional trends. The close association of folding with the formation of granitic plutons and migmatites suggests that partial melting accompanied deformation. Overall, a high degree of plasticity characterizes the metamorphism and deformation.

The Limpopo Belt. The Limpopo mobile belt is a region of highly deformed and metamorphosed rocks (chiefly gneisses and granitic rocks) lying between the Kaapvaal and Rhodesian provinces (both > 2.5 b.y.) (fig. 5.4). It is approximately 200 km wide and can be traced along strike for 900 km. The belt has been subdivided into three structural zones (fig. 5.11): a central zone (B), in which gneisses and other rocks are irregularly deformed with most structures trending oblique to the strike of the belt, and two marginal zones (A,C), in which structural grain is generally parallel to the trend of the belt. Major ductile shear zones separate the marginal areas from the central zone. Gravity studies suggest that the shear zones dip

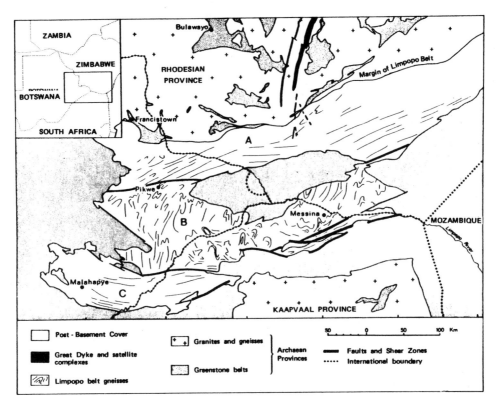

5.11. Limpopo mobile belt, southern Africa (after Cox *et al.*, 1965).

to the south and reflect thrusting of the Kaapvaal Province over the Rhodesian Province. Supracrustal rocks in the central zone of the Limpopo belt, which are highly metamorphosed, appear to represent assemblages of shale, quartzite, and carbonate. In the Messina area, they have been intruded with a thick anorthosite-gabbro complex. Archean greenstone belts extend into the marginal zones of the Limpopo belt both on the north and south, and boundaries of the Limpopo belt are gradational with the Archean granite-greenstone terranes.

Detailed structural and petrographic studies indicate that the Limpopo belt has undergone several periods of deformation and regional metamorphism (Mason, 1973; Coward et al., 1976). Metamorphic grade increases gradually from zeolite facies in central Zimbabwe to granulite facies in the Limpopo belt (Condie, 1981a); a similar increase in grade occurs approaching the belt from the south. Most data suggest that such increasing grades of metamorphism reflect increasing geothermal gradients as the Limpopo belt is approached. Structural studies indicate four periods of major deformation: D_1, an early deformation involving thrusting and nappe formation; D_2, deformation produced by emplacement of

syntectonic plutons; D_3, development of a widespread cleavage; and D_4, late isoclinal folding.

Recent detailed geochronology studies allow rather detailed reconstruction of the geologic history of the Limpopo belt and the adjacent Kaapvaal and Rhodesian provinces (table 5.1). The oldest rocks recognized are highly deformed tonalitic gneisses about 3.8 b.y. in age. Perhaps most striking is the long, complex geologic history of the belt ranging from at least 3.8 b.y. to about 2.0 b.y., when late diabase dikes were emplaced. Also, it is important to note that the timing of events in the Limpopo belt is similar to that of events in adjacent granite-greenstone provinces. This feature, together with the increasing metamorphic grade as the belt is approached from the north and south, suggests that the Limpopo belt and adjacent granite-greenstone provinces have evolved as part of the same plate, a constraint which is important in considering the origin of mobile belts (see Chapter 10).

The Grenville Province. The Grenville Province is exposed in eastern Canada (fig. 5.3) and probably underlies much of the western Appalachian Mountains and adjacent cratonic sedimentary basins, as evidenced by relict-age terranes and basement samples retrieved from drillcores. How far south the province extends and what its relationship is to similar-age rocks in Texas and Mexico are intriguing questions that are currently unanswered. In a pre-drift reconstruction of the North Atlantic, the Grenville Province extends into western Europe, exposed as the Gothics Province in Scandinavia (fig. 5.5). In eastern Canada, the Grenville Province has been divided into seven subprovinces, based chiefly on structural and metamorphic styles (Wynne-Edwards, 1972). It is notable that the Grenville Province is thrust against the Superior Province on the west along the Grenville Front (fig. 5.6) and that the crust beneath the Grenville Province is thicker than that beneath the Superior Province.

The Grenville Province is composed chiefly of gray tonalitic gneisses (~ 60 percent) that represent, in large part, reactivated Archean gneisses. Minor amounts of gabbro, diabase, and other igneous rock are intrusive into the gneisses. Anorthosite bodies and associated granites are also intrusive into the gneissic complex (fig. 5.12). These large masses of anorthosite, which comprise about 75 percent of world's known anorthosite, are typically layered and deformed. Most are characterized by positive gravity anomalies which reflect mafic roots. Other than the granitic plutons associated with the anorthosites, granites are rather minor in the Grenville Province.

Supracrustal rocks are known chiefly from three areas (fig. 5.12). These successions, which appear to have been deposited between 1.2 and 1.3 b.y. are chiefly made up of paragneisses, quartzites, and carbonate, and attain thicknesses of up to 15 km. In the southern part of the Grenville Supergroup, a thick succession of calc-alkaline volcanic rocks, carbonate, and graywacke is exposed. Existing geologic data suggest that most of the supracrustal rocks in the Grenville Province were deposited unconformably on the tonalitic gneiss complex. The volcanic rocks and graywackes may have been deposited on oceanic crust.

Structural trends in the Grenville Province are extremely complex and, in general, lack consistency except in small areas (fig. 5.3). Folds in the gneiss complex have wave lengths of the order of 15 km and small amplitudes, and commonly define broad basins and domes. Folds in the supracrustal successions, on the other hand, are of smaller wave length (~ 5 km) and dip steeply. At least three major periods of folding have been defined in the Grenville Province. A dominantly east-trending fold direction occurs in much of the western part of the province and appears to represent Archean folding, since the folds can be traced into the Archean Superior Province on the west. The folds are refolded about northwest-trending axes that probably developed during the Hudsonian orogeny at 1.8–1.9 b.y.. A third set of tight isoclinal folds trending northeast characterize the Grenville orogeny at about 1.2 b.y.

Metamorphic grade in the Grenville Province is typically amphibolite facies, although it ranges from greenschist to granulite facies. Granulite-facies terranes characterize the central part of the province and are closely associated with anorthosite massifs, which may have provided the additional heat for these terranes to form. Medium-pressure facies series characterize most of the province, and the distribution of metamorphic grades suggests an average exposure level of most of the province of 10 to 25 km.

Extensive radiometric dating in the Grenville Province indicates that it records $\gtrsim 1500$ m.y. of geologic history. Much of the western and central parts of the province and an unknown amount of the eastern part of the province represent reworked Archean crust. At 1.8–1.9 b.y. much of the province was deformed and intruded with granitic rocks at the time of the Hudsonian orogeny, which affects large portions of the Canadian Shield. The Grenville Front fault zone was at least in part initiated at this time. Between 1.4 and 1.5 b.y. most of the anorthosite and related granite bodies were emplaced, and granulite-facies metamorphism occurred in the vicinity of the anorthosite massifs. Between 1.2 and about 1.4 b.y. thick successions of chiefly quartzite, shale, and carbonate were deposited unconformably on reworked Archean gneisses. Some arc-type volcanism and sedimentation occurred in the southern part of the province at about 1.3 b.y. At 1.1 to 1.2 b.y., the Grenville orogeny, the major and most widespread period of deformation, occurred throughout the province. This event is characterized by isoclinal folding, regional metamorphism, and minor granitic plutonism. K-Ar ages indicate that the Grenville Province was uplifted and eroded to approximately its present level by 800 m.y. ago.

The Pan-African Province. The Pan-African Province is a vast array of interconnected mobile belts that cover much of Africa and eastern South America (fig. 5.5). Although in some areas, such as northeastern Africa, extensive granitic and volcanic activity is recorded, most of the Pan-African system is represented by one or several closely spaced thermal-

Table 5.1. Summary of Major Events in the Kaapvaal and Rhodesian Provinces, and the Limpopo Belt, Southern Africa.

Age (b.y.)	Kaapvaal Province	Limpopo Belt	Rhodesian Province
	South		North
—		Deposition of clastic sediments ↑↓	↑
—			
—		Mafic dike intrusion (1.9 b.y.) ↑↓	↑
2.0—		Localized shearing	
—			
—		Mafic dike intrusion (2.2 b.y.) ↑↓	↑
—			
—		Intrusion of satellites of the Great Dyke (2.46 b.y.)	Intrusion of the Great Dyke (2.46 b.y.)
2.5—	Intrusion of post-tectonic granites (2.6–2.7 b.y.)	Deposition shallow-water sediments; intrusion of Bulai ($\sim$2.7 b.y.) and Singelele granites (2.6 b.y.); high-grade metamorphism (2.6–2.7 b.y.); deformation	Intrusion of post-tectonic granites ($\sim$2.5 b.y.)
—	Cratonic sedimentation ($\pm$ volcanism) (2.4–2.9 b.y.)		Formation of Bulawayan (upper and lower) and

90

Emplacement of Lochiel- and Dalmein-type granites (2.8–2.9 b.y.)	D$_4$, deformation D$_2$, deformation D$_3$ High-grade metamorphism and plutonism (2.8–2.9 b.y.)	Shamvaian greenstone belts; granitic plutonism; regional metamorphism (2.6–2.7 b.y.) Intrusion of Mashaba, Chingenzi, and related plutons (2.8–2.9 b.y.)
	Intrusion of Messina layered complex and mafic dikes (3.15 b.y.)	
	Deposition of Messina and related formations (3.2–3.3 b.y.)	
Formation of tonalite-trondhjemite gneisses (≳3.3 b.y.)	Intrusion of Zanzibar tonalite (3.3 b.y.)	Intrusion of Mushandike and Mont D'Or granites (3.3–3.4 b.y.)
	Deformation D$_1$	Thrusting from southwest and regional metamorphism
Formation of Barberton and related greenstone belts (~3.5 b.y.)	Deposition of the Beitridge Formation (3.4–3.5 b.y.)	Formation of Sebakwian-type greenstone belts (≳3.5 b.y.)
		Formation of tonalite gneisses in the Shabani area (≳3.5 b.y.)
	Mafic dike intrusion (3.6 b.y.)	
	Formation of Sand River tonalitic gneisses (~3.8 b.y.)	

Time scale: 3.0 — 3.5 — (b.y.)

91

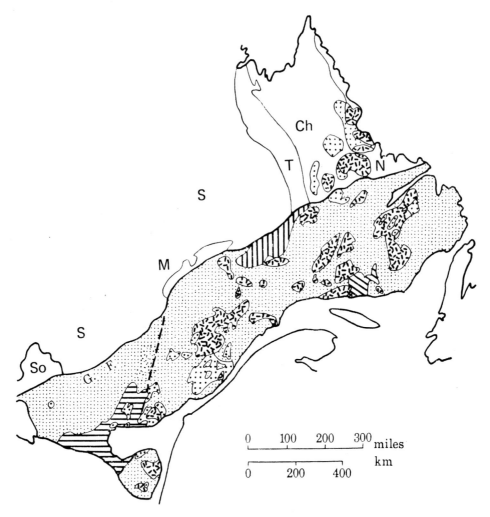

5.12. Sketch-map of the Grenville Province. Lined patterns—supracrustal successions; large dots—granites; short dashes—anorthosites. So, Southern Province; S, Superior Province; M, Mistassini and Otish basin; T, Labrador Trough; Ch, Churchill Province; N, Nain Province; GF, Grenville Front. From Baer (1976).

deformational events in the interval of 500–700 m.y. ago (Shackleton, 1976). In general, the Pan-African province in Africa can be subdivided into three regions (fig. 5.13): (1) a large region comprised chiefly of rejuvenated older sialic crust; (2) a region of island-arc and oceanic volcanic terrane in northeastern Africa and Arabia; and (3) sedimentary basins in western and southern Africa, where thick successions of sediments and volcanic rocks accumulated. Islands of older Precambrian crust isolated in Pan-African belts, which may cross-cut earlier mobile belts without any offset. This is illustrated by the Zambezi belt in southern Africa where it crosses the older Irumide belt (fig. 5.4). Field relations and Sr isotope studies indicate that much of the Pan-African system in Africa and South America developed on and by the

reworking of older sialic crust. The sediments deposited in such areas as the West Congo, Damara, and Katanga belts (fig. 5.13) are dominantly quartzites, shales, and carbonates and were deposited on older sialic crust. Only in the Pharusian-Dahomeyan belt in West Africa (and its continuation in Brazil) and in extreme northeast Africa and Arabia were volcanics and sediments deposited in part on oceanic crust (see Chapter 10). Ophiolites and arc volcanics are found in these areas.

Structurally the Pan-African belts are extremely complex, containing many phases of deformation. Strain is heterogeneous, and it appears that shortening in one place is accompanied by extension in another. Total shortening across the Zambezi belt caused by folding is estimated to be less than 25

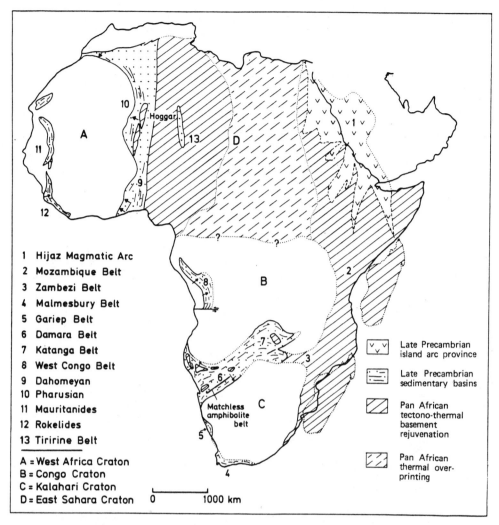

5.13. Map of Africa showing areas affected by the Pan-African orogenic event (after Kröner, 1979). Madagascar in two possible positions.

percent (Barr, 1976). Recumbent folds and nappes occur in some areas, although most data suggest they are superficial and that structures are steeply dipping at depth. Shear zones accompanied by mylonite are common along the margins of many Pan-African belts, such as along the western margin of the Mozambique belt in western Kenya (fig. 5.4). Transcurrent faults are also characteristic of many belts.

Metamorphic grade changes rapidly, as can be observed for example in the Pharusian belt in northwestern Africa. This suggests significant vertical motions between blocks within the belts. Granulite-facies terranes are particularly abundant in some belts, indicating substantial uplift and erosion. Medium-pressure facies series are typical of Pan-African belts, with blueschist facies found at only one minor location in Namibia.

The Hoggar area in northwest Africa is representative of the complex geologic history recorded in Pan-African systems (fig. 5.14). Relict-age subprovinces record dates in excess of 3.0 b.y., and major deformational and plutonic events are recorded at about 2.0 and 1.0 b.y. Both ophiolites and arc volcanics formed in the western part of the Hoggar region between 65 and 800 m.y. and major deformation in this area, which appears to be related to a continental collision, is recorded at 550–650 m.y. This was followed by rapid uplift and erosion.

The Cordilleran Province. The Cordilleran Province lies at the perimeter of and almost completely surrounds the Permian supercontinent (fig. 5.5). The most extensive information about this province comes from the North American, New Zealand and Japanese

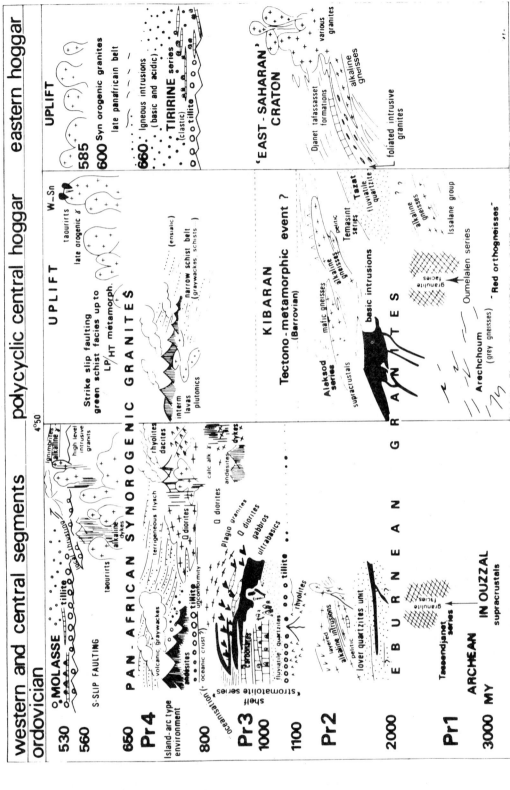

5.14. Summary of the main geologic events in the Hoggar region, northwest Africa (from Bertrand and Caby, 1978).

Table 5.2. Major Phanerozoic North American Orogenies in the Cordilleran, Appalachian, and Innuitian-Caledonian Provinces*

Cordilleran		Appalachian		Innuitian-Caledonian	
Laramide	40–80			Eurekian	10–60
Sevier-Columbian	80–130				
Nevadan	130–160				
Sonoma	200–280				
		Alleghenian	250–300		
Antler	300–375			Ellesmerian	300–375
		Acadian	350–400	Caledonian	350–400
		Taconic	400–450		

*Durations of orogenies given in millions of years; see Plate I for locations of provinces

segments. In North America, the Cordilleran Province has been divided into numerous subprovinces which are characterized by differing geologic histories. It also contains numerous relict-age terranes, some of which date back to the Archean (for example, the Wyoming Province, fig. 5.8). Relict Precambrian terranes have been described in North and South America and in Siberia. Some subprovinces have rather unique features and geologic histories, such as the Colorado Plateau and the Basin and Range Province in the western United States (see Chapter 9). Other provinces, such as the Columbia River Plateau (in the northwestern United States) and the Sierra Madre Occidental (in Mexico) are comprised of large volumes of volcanics.

Supracrustal associations in the Cordilleran Province fall chiefly into one of two categories. The *eugeocline* association is composed chiefly of calc-alkaline volcanic rocks and graywackes and can readily be interpreted in terms of an arc system (Chapters 7 and 8). The *miogeocline* association is comprised chiefly of quartzites, shales, and carbonates, and reflects a stable continental margin or cratonic basin environment (Chapter 8). Other, less widespread associations include ophiolites (mafic and ultramafic rocks, Chapter 8), plateau basalts, large ash-flow tuff fields, and continental-rift successions of arkose, conglomerate, and bimodal volcanics (Chapters 8 and 9). Plutons and batholiths, such as the Coast Range complex in British Columbia, the Southern California batholith in California and Baja California, and the Andes batholith, are intruded into supracrustal successions. Batholiths contain numerous small plutons, which range in composition from gabbro to granite, with granodiorite and tonalite dominating.

Structural trends in the Cordilleran Province are exceedingly complex due to many, often superimposed, periods of deformation. Major areas in North and South America record middle to upper Paleozoic folding. The major orogenies in the Cordilleran system, however, are during the Cretaceous and Tertiary (table 5.2). During these times, major thrusting directed toward the continent developed in western North America. In the late Tertiary and Quaternary the stress regime changed to dominantly extensional, and the Basin and Range Province developed (Chapter 9).

Depth of erosion in the Cordilleran Province is not as great as that in Precambrian crustal provinces, and hence metamorphic grade, on the average, is rather low. Most supracrustal rocks are either unmetamorphosed or range only up to greenschist facies. Amphibolite-facies terranes are of only local significance, and granulite-facies terranes are generally not exposed. It is notable that blueschist-facies terranes occur in many parts of the Cordilleran system and are significant in terms of plate-tectonic evolution (Chapters 8 and 9).

The geologic history of the Cordilleran province during the last 500 m.y., and especially during the last 200 m.y., can be readily interpreted in terms of plate tectonics as reviewed in Chapter 9.

ANOROGENIC PROVINCES

Anorogenic provinces are comprised of sedimentary and/or volcanic successions that have not undergone significant compressive deformation, although they may be associated with extensional faulting or warped into broad basin and domal structures. Metamorphic grade in even the oldest successions, which date back to ~ 3.0 b.y. in South Africa, is less than or equal to lower greenschist facies. Anorogenic successions may be traced into orogenic provinces, and many sedimentary successions that were originally deposited in stable basins are now represented by highly deformed remnants in orogenic provinces. Because anorogenic successions are reworked during later orogenies, their preservation as undeformed successions decreases with increasing age. Anorogenic successions include plateau and continental-rift volcanics and cratonic sediments.

Plateau and Rift Volcanics

Plateau and rift volcanics are erupted at times of extensional faulting and, in some instances, record times of continental breakup (Chapter 8). The largest preserved plateau basalt field is the Siberian Traps (Plate I), which covers an area of approximately 1.5×10^6 km^2 and averages 500 m thick. Most fields cover areas of $1–5 \times 10^5$ km^2 and range from about 300 to 7000 m thick (Waters, 1961). Individual lava flows range up to 50 m thick. Most plateau and rift volcanics are tholeiitic basalts, although some fields, such as the Ethiopian field, are comprised chiefly of alkali basalts and associated volcanics. Some successions are bimodal, as will be discussed in Chapters 7 and 8. Radiometric ages suggest that plateau and rift basalt fields are erupted over short periods of geologic time. Data from the Columbia River Plateau basalts indicate that most of the succession was erupted over a 3 m.y. period between 13 and 16 m.y. ago. The oldest well-preserved rift basalts occur in the Keweenawan and Coppermine River fields in North America (Plate I) and were erupted between 1.1 and 1.2 b.y. ago.

Associated with plateau and rift basalt fields are diabase dike swarms. These may have served as feeder dikes for the basalts, since field data indicate that fissure rather than central-conduit-type eruptions dominate. Swarms of diabase dikes can be followed for many hundreds of kilometers in the Canadian Shield, and may have been feeder dikes for basalts subsequently removed by erosion (fig. 10.14). Also, geochemical studies suggest that large stratiform intrusions may represent magma chambers from which plateau basalts are erupted. Examples are the Muskox intrusion in northern Canada, associated with the Coppermine River basalts, and the Duluth complex in Minnesota, associated with the Keweenawan basalts.

Cratonic Sediments

Cratonic sedimentary rock successions unconformably overlie much of the continents (Chapter 4, Plate I). Deformation is characterized by broad basins and domes and, locally, by block faulting. Cratonic successions are composed chiefly of well-sorted, commonly cross-bedded quartzites and sandstones, thick monotonous sections of shale, and variable amounts of generally fossiliferous limestone and dolomite. In addition, some successions contain iron formation, subgraywacke, conglomerate, tillite, arkose, and variable amounts of volcanic rocks. Stratigraphic thicknesses range from a few kilometers to greater than 15 km. Cratonic sediments occur in circular to linear basins less than 500 km across, in large interconnecting platform basins, and as thick prisms along stable continental margins. The oldest preserved cratonic sediment succession (2–3 b.y. old) is in the Kaapvaal basin in South Africa (Chapter 10). Remnants of many cratonic basins 1.7–2.0 b.y. in age occur in Canada and northern Australia.

EPISODICITY OF OROGENY

One of the first problems encountered in identifying episodicity of orogeny is that of defining an orogeny, as was discussed earlier. Existing data indicate that Archean "orogenies" may have been quite different from later Precambrian and Phanerozoic orogenies in respect to scale, shape of orogenic belts, and sequence and magnitude of various events. The large amount of granitic rock that engulfs Archean greenstones, for instance, records a voluminous magmatic stage that does not occur in Phanerozoic or in most other Precambrian provinces. Another problem is encountered in identifying orogenies that are accompanied by little or no magmatic activity, and especially in belts where more than one orogeny has occurred. It is not always clear from field relationships whether a particular igneous body is orogenic or anorogenic. For the purposes of this discussion, only those orogenies involving substantial amounts of plutonism and regional metamorphism are included. As more data become available from Precambrian orogenic belts, it should be possible to define various types of orogenies and to relate these to mechanisms of crustal evolution.

Another problem in identifying episodicity of orogeny is related to the fact (already mentioned) that various daughter isotopes become immobilized at different times in an orogenic cycle. Hence, extreme care must be used in comparing radiometric dates obtained by different methods. The histogram approach of identifying episodicity must compare only dates of one method and of one mineral or rock type. As previously discussed, U-Pb zircon dates and

Rb-Sr whole-rock isochron dates appear to be of most value in dating orogenies. The fact that these kinds of dates are still not available in large numbers necessitates caution in interpreting radiometric dates in terms of worldwide orogeny.

Closely spaced orogenies become increasingly difficult to resolve as we go back in geologic time. The wide ranges of dates from many Precambrian mobile belts (100–200 m.y.) may record several orogenies. The fact that numerous orogenies are recorded on the continents in the last 300–400 m.y. gives further support to this possibility. To avoid this problem, Precambrian orogenies will be discussed in terms of orogenic periods as previously defined (i.e., periods of time recording several closely spaced orogenies).

Orogenic Periods

Major Precambrian orogenic periods from various continental shield areas are summarized in fig. 5.15. This figure is based on many published radiometric dates emphasizing, when possible, Rb-Sr and Sm-Nd whole-rock isochron and U-Pb zircon dates. Local names of the orogenic periods are also included in the figure.

Results indicate a definite worldwide episodicity of Precambrian orogenic periods, with periods generally lasting 100 to 200 m.y. Spacings between periods range from about 300 m.y. to 800 m.y. with the largest gaps occurring between the Kenoran (2.6–2.7 b.y.) and Hudsonian (1.7–1.9) events and between the Grenville (1.0–1.2) and Hudsonian events. The oldest orogenic event, at 3.7–3.8 b.y., has been identified in southwest Greenland, in the northeastern Baltic Shield, and in South Africa. It is noteworthy that some mare basalts from the moon correspond in age to these rocks. The oldest lunar crust, as represented by the lunar highlands ($\gtrsim 4.5$ b.y.), however, is older than any rocks yet dated from the Earth. The first important period of deformation and granitic plutonism is recorded on the continents at 2.8–3.0 b.y. This was followed by widespread greenstone volcanism, deformation, and emplacement of voluminous tonalite at 2.6–2.7 b.y. At least one and perhaps two orogenic periods occur in the interval 1.7–2.0 b.y.; this period (or periods) is widely distributed on the continents (fig. 5.5). In terms of total areal coverage, the Grenville event (1.0–1.2 b.y.) is of only moderate importance, and the Pan-African event (500–700 m.y.) is of importance only in Africa and South America.

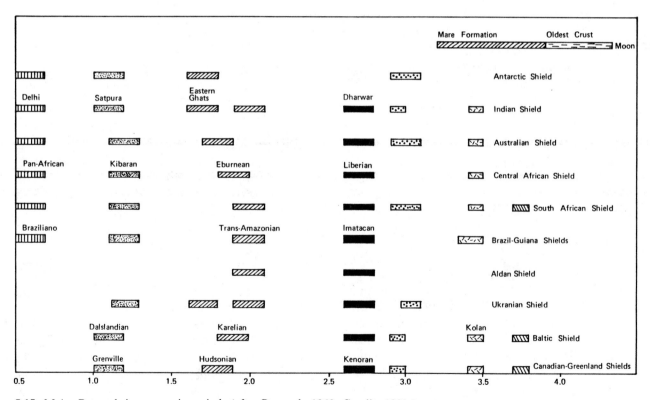

5.15. Major Precambrian orogenic periods (after Burwash, 1969; Condie, 1981a).

Phanerozoic Orogenies

It is possible in the Phanerozoic to resolve individual orogenies to better than 10 m.y. although most of the radiometric dates that are of sufficient accuracy for this purpose are from North America. An alternate approach that has been used in identifying Phanerozoic orogenies is to relate the percent of regression of shallow continental seas (as deduced from paleogeographic maps) to maximum intensities of orogeny. However, recent studies using the most accurate paleogeographic and radiometric data available seem to indicate that orogenies occur at times of maximum transgression of shallow seas in cratonic areas (Johnson, 1971; Rona, 1973). Also supporting this interpretation are data from the sea floor indicating that during the Cenozoic, increased spreading rates and relatively high oceanic ridges (displacing sea water on to the continents) correlate well with maximum transgression (Chapter 8) and probably with orogeny in continental areas. Histograms of radiometric ages from orogenic igneous and metamorphic rocks and ocean transgression-regression curves suggest a rough episodicity of major Phanerozoic orogenies. As an example, major Phanerozoic orogenies in North America are listed in table 5.2 with their corresponding ages. These orogenies affected large parts of the three North American Phanerozoic provinces. The data suggest that major orogenies last about 50 m.y. and occur with irregular spacings. Orogenies of local geographic extent are also recognized in all three provinces.

Causes of Episodicity

Causes that have been proposed for orogenic episodicity fall into one of four categories: (1) changing convection patterns in the Earth with time; (2) secular changes in the Earth's spin axis; (3) episodic variations in shear melting of the mantle produced by tidal power; and (4) episodic magmatism and deformation related to plate tectonics.

Runcorn (1965) has pointed out that convection patterns in the Earth are unstable and changes in physical conditions in the Earth can result in rather sudden changes in convection pattern. A growing core, for instance, could lead to increasing numbers of convection cells in the Earth with time. Runcorn proposes that major orogenies can be related to times of rapidly changing convection-cell sizes and distributions. The model has two serious disadvantages, however: (1) recent data indicate that the Earth's core formed within a few hundred million years after

accretion, and (2) the model does not account for orogenies before 2.7 b.y. or for minor orogenies of other ages. Sutton (1963) proposes a model based on four major continent-forming cycles (2.7–3.6, 1.9–2.7, 1.1–1.9, and < 0.6 b.y.). Each cycle lasts 750–1250 m.y. and involves an evolving convection system beginning with many small cells which increase in size and eventually fuse, giving rise to one Earth-wide cell. Widespread orogeny occurs at the beginning of each cycle and is caused by the existence of many small convection cells. Why such a cycle should repeat itself four times during geologic history is not specified in the model. Lambert (1980) has recently proposed that the 3.5 and 2.7 b.y. events may be related to the onset of shallow and deep convection in the Earth, respectively (see Chapter 11).

Williams (1973) suggests that the Earth's spin axis changes attitude with respect to the ecliptic with a period of about 2.5 b.y. while the moon stays in an orbit close to the ecliptic plane at all times. The result is two independent lunar torques applied to the outer parts of the Earth which alternately peak in intensity about every 620 m.y. A tidal torque which is caused by the pull of the moon on the Earth's tidal bulge peaks when the moon is in an equatorial orbit, and a precessional torque due to the pull of the moon on the Earth's equatorial bulge peaks when the moon is in a polar orbit. The alternate peaks in torque intensity in this model occur at 3.85, 3.25, 2.6, 2.0, 1.35, 0.75, and 0.13 b.y. thus corresponding closely to some orogenic peaks.

Another mechanism proposed by Shaw, Kistler, and Everden (1971) involves concentration of solid Earth tidal energy along mid-ocean ridge systems and in the asthenosphere by mechanisms of viscous dissipation, caused primarily by shear melting. This energy enters the continents as magmatic heat. Episodes of continental magmatism and orogeny with spacings of about 30 m.y. are explained in terms of periodic thermal instabilities related to shear melting in the mantle, which results in episodic losses of magma from the mantle. Epeirogenic oscillations with periods of the order of 200 m.y. and related orogeny and magmatism are caused by variations in the distribution of tidal energy dissipation between the solid Earth and shallow epicontinental seas.

Among the plate-tectonic models advocated to explain orogenic episodicity are the following (after Raymond and Swanson, 1980): (1) successive subduction of spreading centers which, because of their increased heat output, results in episodic increases in magma production and deformation in the continent overlying the subduction zone; (2) frictional heating along a subduction zone, accumulation of magma above the descending slab, and episodic release of magmas into the crust with accompanying deformation; and (3) plutonism and orogeny are related to episodic accretion of hydrous sediments and volcanics to the continent in the arc-trench gap. Although the plate-tectonic models may account for some episodicity of Phanerozoic convergent-plate margin orogeny, they do not seem capable of explaining most Precambrian orogenies (see Chapter 10).

SUMMARY STATEMENTS

1. Radiometric dates record times when rocks or minerals cool through immobilization temperatures of particular daughter isotopes. Incomplete remobilization of such isotopes by later thermal events produces relict dates. Whole-rock Sm-Nd and Rb-Sr isochron dates and U-Pb zircon dates are least likely to be reset, and hence most accurately record original formation dates of igneous rocks.

2. Geologic events that may be dated by radiometric methods are magmatism (plutonic or volcanic), metamorphism, regional uplift, diagenesis, and in some cases, orogeny.

3. A crustal province is a segment of the continental crust that records a similar range of radiometric dates and exhibits similar deformational styles. Relict-age terranes are islands of older crust within crustal provinces.

4. Crustal province boundaries are faults or shear zones, unconformities, or rapid changes in metamorphic grade.

5. Orogenic crustal provinces record complex and variable polyphase deformation, magmatism, and metamorphism. Such provinces are long-lived, commonly persisting for more than 500 m.y. before uplift and stabilization. Mobile belts are linear to curvilinear segments of orogenic provinces.

6. Archean provinces are composed of either or both a granite-greenstone association or a high-grade metamorphic association. The granite-greenstone association includes greenstone belts (comprised chiefly of volcanic rocks) engulfed with granitic rocks (chiefly tonalites). High-grade terranes are composed dominantly of tonalitic gneisses, granulites, and amphibolites with minor metasediments.

7. Anorogenic provinces include plateau and rift volcanics and cratonic sediments. These provinces are commonly associated with extensional faulting and may be warped into broad basin and domal structures. Volcanic provinces are comprised chiefly of basalts; cratonic sediments are dominantly quartzites, shales, and carbonates.

8. Precambrian orogenic periods are episodic, lasting 100–200 m.y. and occurring every 300–800 m.y. The most widespread orogenic periods are at 2.6–2.7, 1.7–1.9, and 1.0–1.2 b.y. Less widespread events are recorded at 0.5–0.7, 2.9–3.0, 3.4–3.5, and 3.7–3.8 b.y. Phanerozoic orogenies occur with irregular spacings and last about 50 million years.

9. Suggested causes for orogenic episodicity are: (1) changing convection patterns in the Earth with time, (2) secular changes in the Earth's spin axis, (3) episodic variations in shear melting of the mantle produced by tidal power, and (4) episodic magmatism and deformation related to plate tectonics.

SUGGESTIONS FOR FURTHER READING

Baer, A.J., editor (1970) Symposium on basins and geosynclines of the Canadian Shield. *Geol. Survey Canada*, Paper 70–40. 265 pp.

Baragar, W.R.A., Coleman, L.C., and Hall, J.M., editors (1977) *Volcanic Regimes in Canada.* Geol. Assoc. Canada, Spec. Paper 16. 476 pp.

Condie, K.C. (1981) *Archean Greenstone Belts.* Amsterdam: Elsevier. 434 pp.

Faure, G. (1977) *Principles of Isotope Geology.* New York: Wiley. 464 pp.

Geological Survey of Western Australia (1974) *Geology of Western Australia.* Perth: West. Australia Geol. Survey Memoir 2. 541 pp.

Price, R.A., and Douglas, R.J. W., editors (1972) Variations in tectonic styles in Canada. *Geol. Assoc. Canada, Spec. Paper* No. 11. 688 pp.

Windley, B.F. (1977) *The Evolving Continents.* New York: Wiley. 385 pp.

Chapter 6

Sea-Floor Spreading

THE THEORY OF SEA-FLOOR SPREADING

Sea-floor spreading is the process by which the oceanic lithosphere splits at oceanic ridges and moves away from ridge axes with a motion like that of a conveyor belt as new lithosphere is created and fills the resulting crack (see fig. 1.1). Oceanic lithosphere is consumed in the asthenosphere at subduction zones to accommodate the newly created lithosphere such that the surface area of the Earth remains constant. The alternate possibility—that the Earth is expanding to accommodate the growth of the lithosphere— seems unlikely, as will be discussed in Chapter 8. Hess (1962) is credited with proposing the theory of sea-floor spreading, although the name was suggested by Dietz (1961).

The most definitive evidence for sea-floor spreading comes from the study of the linear magnetic anomalies that characterize the sea floor (Chapter 4). Vine and Matthews (1963) first suggested that the alternate stripes of positive and negative magnetic anomalies were caused by bands of basaltic rock in layer 2 of the oceanic crust that were alternatively magnetized in normal and reversed directions of the Earth's magnetic field. They propose that the new lithosphere is formed over convective upcurrents beneath oceanic ridges by magmatic processes. As magmas emplaced in the axial rift zones cool through the Curie temperature of magnetic minerals, they acquire a magnetization in the direction of the existing magnetic field of the Earth. The newly magnetized crust then splits and is forced apart to make room for fresh injections of magma.

Many seemingly unrelated facts from the fields of geology, paleontology, geophysics, and geochemistry are consistent with and find a unified explanation in the sea-floor spreading model. For instance, the observed increase in age and thickness of deep-sea sediments in going away from oceanic ridges is predicted by the model, as are the compositional changes found in basalts dredged from the oceanic crustal layer 2 (see Chapter 7). First-motion studies of earthquakes, discussed later in this chapter, are also consistent with the model. The absence of deep-sea sediments older than Jurassic is explained by the fact that current sea-floor spreading rates of a few centimeters per year indicate that the ocean floors should be completely renewed every 200–300 m.y. Sea-floor spreading is also appealing in that it provides a mechanism for continental drift for which a great deal of evidence exists (Chapter 8), and which prior to this time had not been accepted by many geoscientists because of the lack of a plausible mechanism.

Rock Magnetism and the Geomagnetic Time Scale

In order to fully comprehend the magnetic evidence for sea-floor spreading, it is first necessary to discuss how rocks become magnetized. Permanent magnetization acquired by rocks is known as *remanent magnetization* and is of two general types. *Hard magnetization* generally reflects the direction of the Earth's magnetic field at the time a rock formed, and it is stable for long periods of geologic time. *Soft magnetization*, on the other hand, is acquired in weak, often transient fields (such as lightning) and may be partly or completely lost or its direction changed. Only hard magnetization is of importance in paleomagnetic studies and, because it can be partly obscured by soft magnetization, the latter must be removed by demagnetization techniques prior to measuring the hard magnetization. Three types of hard magnetization are known in rocks. *Thermal remanent magnetization* (TRM) is acquired by a magma when it cools through the Curie point of the major magnetic minerals (500–600°C). The direction of magnetization acquired by the rock is almost always that of the existing field although rarely it is in the opposite direction. *Detrital remanent magnetization* (DRM) occurs in some clastic sediments and results from alignment of magnetic grains in the direction of the magnetic field, although often with some rotation caused by depositional or diagenetic processes. *Chemical remanent magnetization* (CRM) is produced in rocks by the formation of new magnetic minerals during diagenesis, alteration, or metamorphism. Some rocks have acquired hard magnetization in a direction opposite to that of the Earth's present magnetic field. Such magnetization is known as *reverse magnetization*. Although some reverse magnetization in igneous rocks appears to have been produced by *self-reversal mechanisms*, such as the co-crystallization of two magnetic minerals with different Curie points, it is now clear that most reverse magnetization is acquired during periods of reversed polarity in the Earth's magnetic field.

One of the major discoveries in paleomagnetism is that stratigraphic successions of volcanic rocks and deep-sea sediment cores can be divided into sections that show dominantly reversed and normal magnetizations. From such data, *polarity intervals* are defined as segments of time in which the magnetic field was dominantly reversed or normal. From the combined magnetic data from volcanic rocks and deep-sea sediments, a *geomagnetic time scale* was formulated (Cox, 1969), extending back to about 4.5 m.y. (fig. 6.1). Although polarity intervals of short duration ($\leq 50,000$ years) cannot be resolved with K-Ar dating of volcanic rocks, they can be detected and commonly dated by other methods in deep-sea sediments, which contain a continuous (or nearly continuous) record of the Earth's magnetic history for the last 100–200 m.y. Paleomagnetic studies of sediment cores also indicate that polarity reversals occur over rather short periods of time of the order of 1000–2000 years, and that a decrease in intensity of the field by 60–80 percent over about 10,000 years precedes reversals, followed by a buildup of intensity for the next 10,000 years (Cox, 1969). The last reversal in the magnetic field occurred about 20,000 years ago.

Two types of polarity intervals are defined on the basis of their average duration: a *polarity event* (10^4–10^5 years) and a *polarity epoch* (10^5–10^6 years).

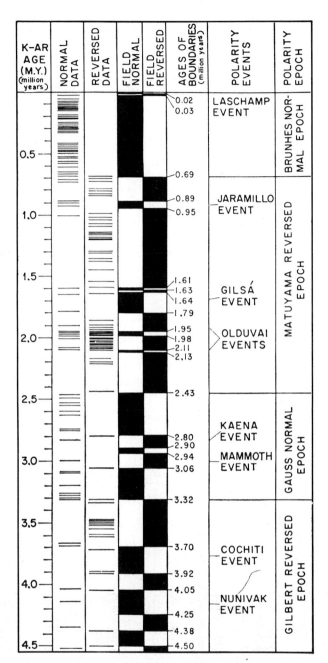

6.1. The geomagnetic time scale to 4.5 m. years (from Cox, 1969). Each short horizontal line represents the magnetic polarity and K-Ar date of one volcanic cooling unit. The duration of events is based in part on paleomagnetic data from deep-sea sediments.

A polarity epoch may contain several-to-many polarity events and can be dominantly reversed (e.g., the Brunhes), dominantly normal (e.g., the Matuyama), or mixed (figure 6.1). Based on the distribution of oceanic magnetic anomalies, it is possible to extrapolate the geomagnetic time scale back to more than 80 m.y. (Heirtzler et al., 1968). Independent testing of this extrapolation from dated basalts in Iceland indicates the predicted time scale is correct to within a few percent back to at least 10 m.y. (McDougall, 1979). Results suggest that over the last 80 m.y. the average length of polarity events has decreased with time. Reversals in the Earth's magnetic field are documented throughout the Phanerozoic, although the geomagnetic time scale cannot be extrapolated beyond about 200 m.y., the age of the oldest oceanic crust. Just how far back in time reversals occur is unknown at present, due chiefly to poor resolution of Precambrian pole positions. Currently, reversals have been documented back to about 1.5 b.y.

The percentage of normal and reversed magnetization for any increment of time has also varied with time. During the Phanerozoic, four intervals can be defined (fig. 6.2). The Cretaceous, Jurassic and late Ordovician-Silurian are characterized chiefly by normal polarities, and the early and late Paleozoic dominantly by reversed polarities. Spectral analyses indicate the presence of three polarity periodicities at 300, 110, and 60 m.y. (Irving and Puelaiah, 1976).

Sea-Floor Spreading Rates

One of the most profound discoveries in paleomagnetism, in terms of its impact on the geological sciences, as reported by Vine and Matthews (1963), is that linear magnetic anomaly patterns on the ocean floors correlate with reversed and normal polarity intervals in the geomagnetic time scale. An example of such correlations for the East Pacific Rise is given in fig. 6.3. The magnetic profile is shown reversed to accentuate the symmetry about the ridge axis, and the suggested correlations with polarity intervals are indicated on the bottom of the figure. A model profile for a half spreading rate of 4.4 cm/yr is also

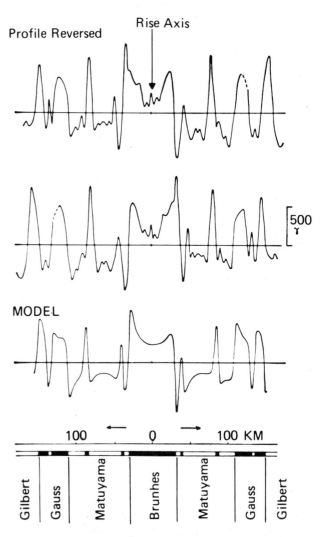

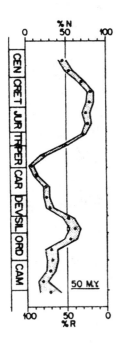

6.2. Distribution of magnetic polarities during the Phanerozoic averaged over 50 m.y. intervals (from Irving and Pullaiah, 1976). Percent of normal (N) and percent of reversed (R) poles are shown on horizontal axes.

6.3. Observed and model magnetic profiles across the East Pacific Rise at 51°S latitude and corresponding correlations with magnetic polarity intervals assuming a spreading rate of 4.4 cm/yr (Vine, 1966).

shown in fig. 6.3 and approximately reproduces the observed magnetic profile. Both the Jaramillo and Gilsa events produce sizable anomalies in the Matuyama epoch. The Kaena and Mammoth events in the Gauss epoch are not resolved, however. The lower limit of resolution of magnetic events in anomaly profiles with current methods is about 30,000 years. Although the distribution of magnetic anomalies seems to correlate well with polarity intervals, the amplitudes of these anomalies can vary significantly between individual profiles. Such variation appears to reflect, in part, inhomogeneous distribution of magnetite in basalts of the oceanic crust.

Magnetic anomalies can be correlated from one ocean basin to another as shown in fig. 6.4. For convenience, geophysicists number anomalies beginning with one at ridge axes. It is clear from the data in fig. 6.4 that the same anomalies do not occur

at the same distances from different ridges, and hence spreading rates must have varied from area to area. If the spreading rate has been constant in one ocean basin, it is possible to extend the geomagnetic time scale back to more than 4.5 m.y. using the magnetic anomaly patterns. Although data indicate that a constant spreading rate is unlikely in any ocean basin, the South Atlantic most closely approaches constancy (~ 1.9 cm/yr) and is commonly chosen as a reference to extrapolate the time scale (fig. 6.4). Subsequent paleontologic dates from sediment cores retrieved by the Deep Sea Drilling Project and a few radiometric dates of basalts dredged from the ocean floor substantiate, on the whole, an approximately constant spreading rate in the South Atlantic and allow the extension of the geomagnetic time scale to about 80 million years. The correlations of magnetic anomalies with distance from ridge axes (fig. 6.4) indicate that

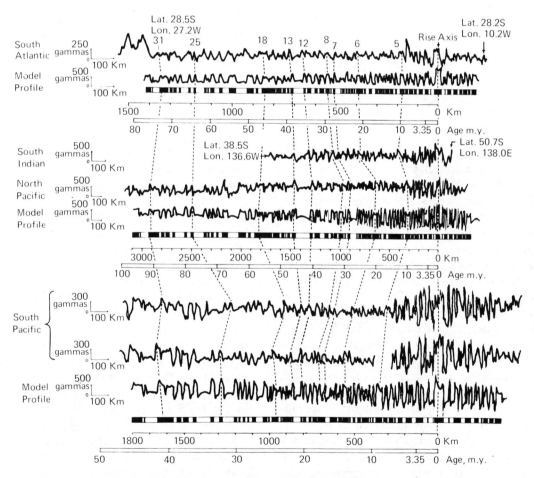

6.4. Magnetic profiles from the Atlantic, Indian, and Pacific Ocean basins (from Heirtzler *et al.*, 1968). Model profiles are given for the South Atlantic, South Pacific, and North Pacific based on the normal (black) and reversely (white) magnetized bands beneath each model profile. Proposed correlations of anomalies are shown with dashed lines. Each time scale is constructed assuming an age of 3.35 m. years for the end of the Gauss epoch.

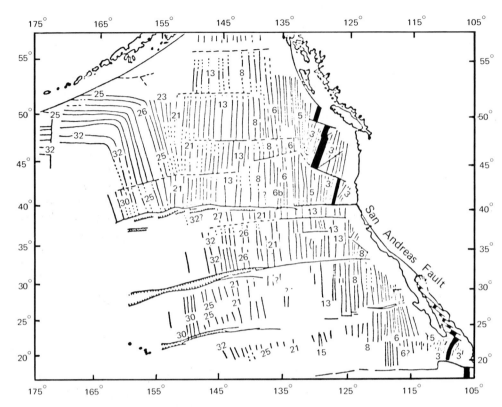

6.5. Magnetic anomaly map of the northeast Pacific basin (after Menard, 1969). Oceanic ridges shown in solid black.

spreading rates in the South Indian and North Pacific basins have been more variable and, on the average, faster than the spreading rate of the South Atlantic. Recent correlations of Mesozoic magnetic anomalies in the Western Pacific have allowed extension of the geomagnetic time scale in this region back to about 160 m.y. and indicate the presence of two unusually long magnetic periods of normal polarity, one in the middle Cretaceous (85–100 m.y.) and one in the early Jurassic (Larson and Pitman, 1972).

Spreading rates also can be estimated from dislocation theory, using the data derived from first-motion studies of earthquakes and from the observed dip-lengths of subduction zones (as defined by hypocenter distributions), if these lengths can be assumed to be a measure of the amount of underthrusting during the last 10 million years. Estimates of spreading rates, which are usually expressed as half-rates (the rate of one limb of an oceanic rise), commonly range within a factor of two of one another using all three methods.* Results indicate that spreading rates range from 1 to 20 cm/yr, averaging a few centime-

ters per year, and that variation can occur from one segment of a ridge to another and as a function of distance (time) from a ridge.

From estimated spreading rates, it is possible to contour the age of the sea floor. Such contours are known as *isochrons*. An example of an isochron map of the northeastern Pacific basin is given in fig. 6.5. Magnetic anomalies are numbered as shown in fig. 6.4 (no. $30 \simeq 87$ million years, no. $7 \simeq 30$ million years, etc.). It is obvious from the figure that the rate of spreading has varied between crustal segments bounded by east-west trending transform faults for the last 90 million years. It is possible to map the age of the ocean floor using magnetic isochrons. One such map for the Pacific Basin is given in fig. 6.6. The oldest crust in this area occurs immediately adjacent to the Izu-Bonin subduction zone south of Japan, and fossil evidence from cored deep-sea sediments suggests that it is Jurassic in age. Oceanic crust of upper Jurassic age has been also identified in the north Atlantic Ocean. Since the rate at which oceanic crust has been produced at ridges during the past several hundred million years is of the order of a few centimeters per year, it is unlikely that crust older than Jurassic will be found on the ocean floors today.

---

*Unless otherwise noted, all spreading rates referred to in this book are half-rates.

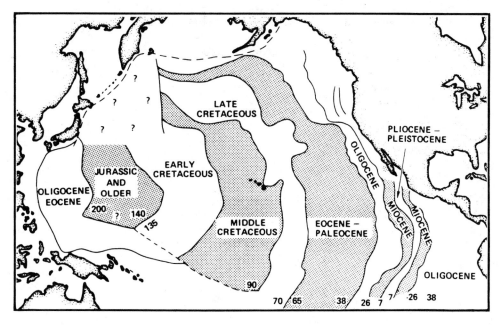

6.6. Age provinces in the Pacific basin as determined from magnetic anomaly distributions (after Douglas and Moullade, 1972).

Existing data suggest that the average age of the present oceanic crust is about 60 m.y. which is almost insignificant when compared to the average age of the continental crust of about 1.5 b.y.

Fragments of oceanic crust older than Jurassic are found in continental orogenic belts where they were tectonically emplaced during orogeny. Recent studies suggest that ophiolite complexes (see Chapter 8) found in Phanerozoic orogenic belts represent disrupted fragments of oceanic crust emplaced and preserved in the continental crust.

Earthquake Distributions and First-Motion Studies

Provided a sufficient number of seismic recording stations with proper azimuthal locations are available, it is possible to determine the directions of first motion at a site of earthquake generation. Such investigations are known as *first-motion studies* or *fault-plane solutions*. The results of such studies have been quite definitive in substantiating and enhancing our knowledge of sea-floor spreading.

Earthquakes occur along rather narrow belts on the Earth today (fig. 6.7), with these belts marking the boundaries between lithospheric plates that are largely devoid of earthquake activity (Isacks, Oliver, and Sykes, 1968). In general, there are four types of seismic boundaries, distinguished by their hypocenter distributions and geologic characteristics: (a) ocean ridges, (b) subduction zones, (c) transform faults, and (d) intracontinental boundaries.

Ocean ridges are characterized by shallow earthquake hypocenters ($\lesssim 70$ km) limited to axial rift zones. Earthquakes are generally small in magnitude, commonly occur in swarms, and appear to be associated with the intrusion and extrusion of basaltic magmas in the rifts. First-motion studies indicate that rift earthquakes are produced by dominantly vertical faulting, as is expected if new lithosphere is being injected upwards.

The most widespread and intense earthquake activity occurs along subduction zones such as the Peru-Chile, Aleutian, and Japan subduction zones (see Plate I). Hypocenters range from shallow (~ 20 km) to 700 km deep and define a *seismic zone* (also known as a *Benioff zone*), which dips at moderate to steep angles beneath arc systems and some continental borders. In terms of plate tectonics, the seismic zone is interpreted as representing a brittle region in the upper 10–20 km of a descending lithospheric plate (the latter of which may range from 50 to 100 km thick) (Isacks and Barazangi, 1977). Active volcanic chains overlie modern subduction zones. First-motion studies of earthquakes in subduction zones indicate variation as a function both of lateral distance along a subduction zone and of depth. Seaward from the trench in the upper part of the lithosphere where the plate begins to bend, shallow extensional mechanisms predominate (fig. 6.8). Be-

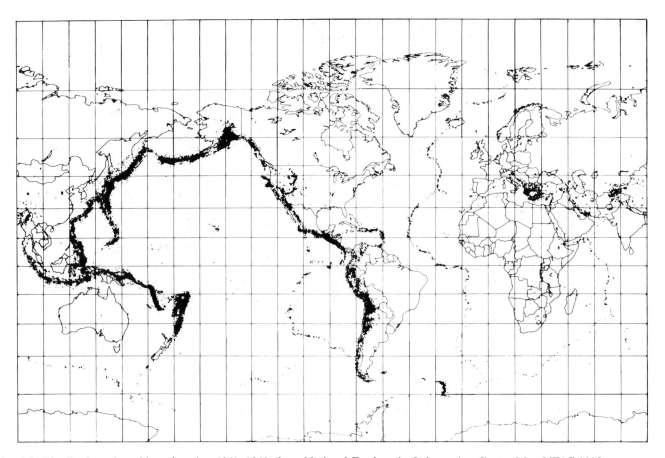

6.7. Distribution of world earthquakes 1961–1969 (from National Earthquake Information Center Map NE1C-3005).

cause of their low strength, sediments in oceanic trenches cannot transmit stresses, and hence they are usually flat-lying and undeformed. Seismic reflection profiles indicate, however, that sediments (and volcanics) on the landward side of trenches are intensely folded and faulted. As indicated in fig. 6.8, underthrusting mechanisms dominate at shallow depths in subduction zones (20–100 km). At intermediate depths (100–300 km), compressional or extensional motions dominate; at depths > 300 km, compressional motions are characteristic (Isacks et al., 1968). Earthquake data also suggest that descending slabs are broken into tongues that are absorbed independently in the asthenosphere.

Transform faults, like oceanic ridges, are characterized by shallow hypocenters (< 50 km deep), but they do not generally have associated volcanic activity. They cross continental or oceanic crust and may show apparent lateral displacements of hundreds of kilometers. First-motion studies along transform faults indicate horizontal motion in a direction away from oceanic ridges. Also, as predicted by sea-floor

spreading, earthquakes are restricted to areas between offset ridge axes. Transform faults may produce large structural discontinuities, and in some cases topographic breaks are left on the ocean floor.

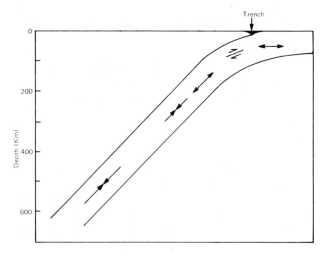

6.8. Summary of earthquake first-motions in subduction zones (from Isacks et al., 1968).

Such discontinuities are known as *fracture zones* and are useful in reconstructing ancient plate motions.

Intracontinental earthquake zones, as typified by the rather broad and irregular band of epicenters extending from southeast Asia through the Himalayas and Caucasus into southern Europe (fig. 6.7), are characterized by hypocenters chiefly $\lesssim$ 100 km deep and are associated with high, geologically young mountain ranges that have formed chiefly by compressive forces. Earthquake first-motions are quite variable, although commonly compressive. Some of these zones are interpreted as continent-continent collision boundaries (Chapter 8). Data on compressive stress distribution in continental areas as deduced from the combined results of first-motion studies and *in situ* stress measurements support the idea that stresses in continents are produced by the same mechanism that drives large lithospheric plates, which will be discussed later in the chapter.

PLATE CHARACTERISTICS

As was briefly mentioned in Chapter 1, the Earth's surface can be described as being composed of a mosaic of lithospheric plates that exhibit some combination of oceanic ridges or *divergent plate boundaries*, subduction zones or *convergent plate boundaries*, *transform faults*, and *intracontinental compressional zones*. At ridge axes, plates separate and lithosphere is generated; at transform faults, plates slide by each other and surface area is neither created nor destroyed; and at subduction zones, plates are consumed in the mantle beneath the leading edge of another plate. *Plate tectonics* is the study of deformation within plates and of the interactions of plates around their margins. Rayleigh-wave dispersion studies in the Pacific Basin indicate that the oceanic lithosphere increases in thickness as a function of increasing distance (and age) from the mid-ocean ridge (Leeds et al., 1974). It appears to reach a thickness of about 60 km in 50 m.y., 85 km in 100 m.y., and about 100 km by the time it arrives at the circum-Pacific subduction zones.

Plates range in size from less than 10^4 km² to about 10^8 km². Plate boundaries do not, however, always coincide with continental margins, as is shown in Plate I. Continents can be considered passengers on the plates, and continental drift is controlled by the plate motions. There are seven major plates on the Earth's surface today (Plate I): the Eurasian,

Antarctic, North American, South American, Pacific, African, and Australian plates. Intermediate-sized plates (10^6–10^7 km²) are the China, Philippine, Arabian, Iran, Nasca, Cocos, Caribbean, and Scotia plates. There are probably more than 20 plates with areas of 10^5–10^6 km², many of which have not as yet been precisely defined.

Plates may diminish or grow in area depending on the distribution of convergent and divergent boundaries. The African Plate, for instance, with the exception of the northern boundary, is entirely surrounded by active spreading centers and is growing. If the surface area of the Earth is to be conserved, other plates must be diminishing in area as the African Plate grows. Indeed, the plates in the Pacific area appear to be decreasing in area to accommodate the growth of the African Plate and other growing plates. If the current rate of shrinkage of the Pacific Basin continues, the American continents may collide with Asia in about 200 m.y.

Convergent Plate Boundaries

Convergent plate boundaries are defined by the dipping seismic or Benioff zones beneath arc-trench systems. Data indicate that earthquake hypocenters are highly concentrated in such seismic zones. The error in locating hypocenters is variable depending on depth and seismic-station control, but is commonly of the order of ±25 km. To illustrate the variation in seismic zones, eight cross sections of various arc-trench systems are shown in fig. 6.9. Many seismic zones, as exemplified by the Middle America, Central Aleutian Islands, and Bismarck arc areas (fig. 6.9a–c), extend to depths of less than 300 km. Others extend to depths as great as 700 km (fig. 6.9d). Some seismic zones exhibit large gaps (100–350 km wide)—for instance, the New Hebrides and central Peru zones (fig. 69g,h). Dips range from 30 to 90 degrees, averaging about 45 degrees. Considerable variation may occur along strike in a given subduction zone, exemplified by the Izu-Bonin arc system in the Western Pacific (Katsumata and Sykes, 1969). Hypocenters are linear and rather continuous on the northern end of this arc system (fig. 6.9e), becoming progressively more discontinuous toward the south. Near the southern end of the arc the seismic zone exhibits a pronounced gap between 150 and 400 km depth (fig. 6.9f). The magnitudes of earthquakes in the deeper parts of descending slabs are commonly greater than those at shallower depth, although they are less frequent.

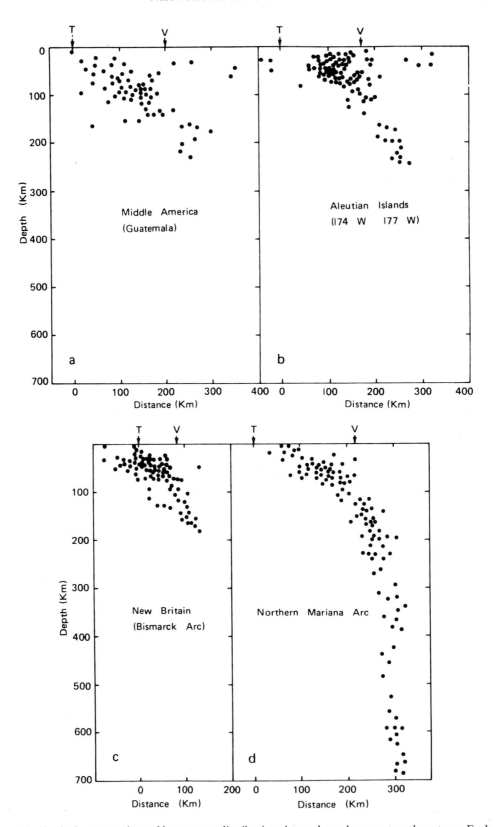

6.9. Vertical cross sections of hypocenter distributions beneath modern arc-trench systems. Each diagram shows earthquakes for 7–10 year periods between 1954 and 1969. T = trench axis; V = recently active volcanic chain. Distance is measured horizontally from each trench axis in kilometers. Hypocenter data from many sources and principally from National Earthquake Information Center, U.S. Coast and Geodetic Survey.

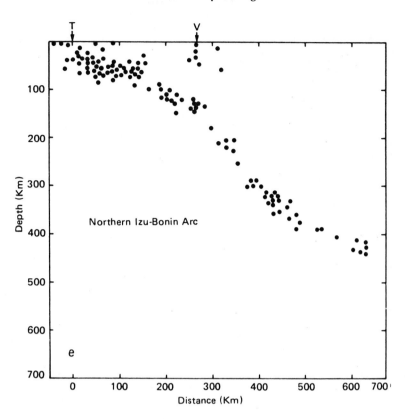

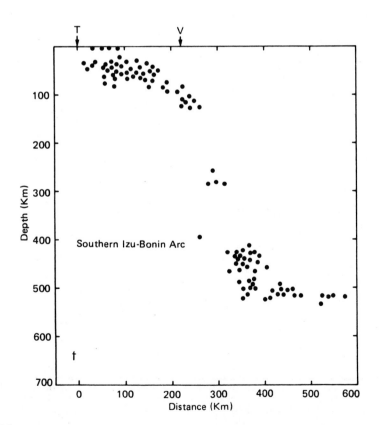

6.9. cont.

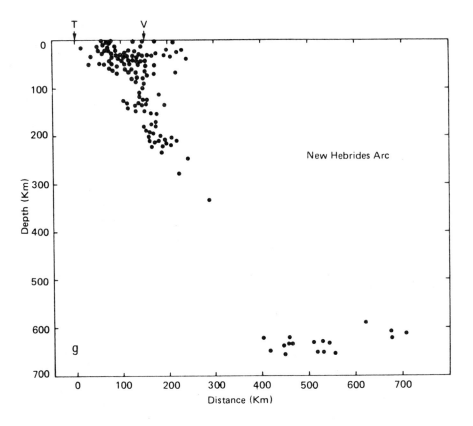

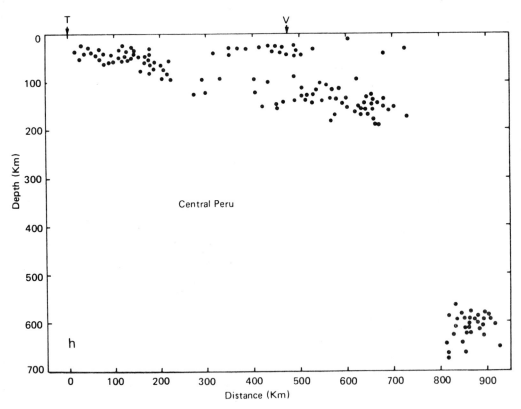

6.9. cont.

Possible stress distributions in subduction zones are illustrated in fig. 6.10. The stresses in a newly subducting slab result from forces applied to the slab from above as it enters the low-strength asthenosphere (a). As the slab enters stronger asthenosphere and part of the load is supported from below, a section in the center with only small or negligible stresses is produced (b). Deeper penetration results in the entire load of the slab being supported from below where it encounters the mesosphere and, hence, the slab is under compression throughout (c). A portion of the slab may break off before it encounters the mesosphere (d). A gap in seismicity as a function of depth in the slab is expected in (b) and (d). Shear-wave velocity distributions in descending slabs indicate that both cases occur in nature. Phase changes in descending slabs may also contribute to the production of earthquakes, as will be discussed later in the chapter. The distribution of hypocenters in descending slabs also suggests the existence of major discontinuities striking transverse to arc-trench systems (Carr et al., 1973). Such discontinuities seem to correlate well with changes in surface geology such as offsets or strike changes in volcanic chains or trench axes.

Relatively buoyant bathymetric features on the ocean floor, such as aseismic ridges, islands, and seamounts, appear to resist subduction (Kelleher and McCann, 1977). At locations where these features intersect active trenches, earthquakes are small and infrequent and active volcanism is diminishing or nonexistent. Under such conditions, the leading edge of the subducted slab may detach and a new subduction zone may develop on the seaward side of the buoyant crust which is only partially subducted. Seismic studies of earthquake motion, furthermore, indicate variations in the degree of coupling between

normal descending oceanic plates and overriding lithosphere (Kanamori, 1977). Coupling is very strong in the Peru-Chile and Alaskan subduction zones, resulting in large earthquakes, a low subduction-zone dip at shallow depths, and breakoff of the descending slab at shallow depths (fig. 6.9h). In such areas as the Kuril and northern Mariana arc systems (fig. 6.9d), descending slabs are largely decoupled from overriding plates. In these areas earthquakes are smaller and less frequent, and the oceanic slab extends to great depths in the asthenosphere. Another possible consequence of decoupling is a gradual retreat of subduction zones toward the ocean, which may be an important mechanism in the formation of some back-arc basins.

Subduction-zone dips are inversely related to sea-floor spreading rates, as the data in fig. 6.11 indicate. A model has been suggested by Luyendyk (1970) to explain this relationship, which assumes a constant sinking rate for all plates in the asthenosphere of 4 to 6 cm/yr. In such a model, the dip of the plate is a function only of the convergence rate. The leading edges of plates with slower convergence rates sink more in a given time interval than those with faster convergence rates, thus producing steeper dips in the former.

The lithosphere is a zone of high Q(>500); indeed, the discovery of high-Q zones dipping into the mantle at convergent plate boundaries gave con-

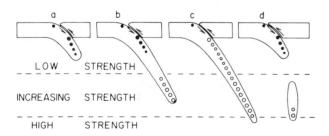

6.10. A model for stress distributions in descending lithosphere (after Oliver *et al.*, 1973). Key: ● = extensional stresses down-dip; o = compressional stresses down-dip. Size of circles qualitatively indicate relative amounts of seismic acitivity.

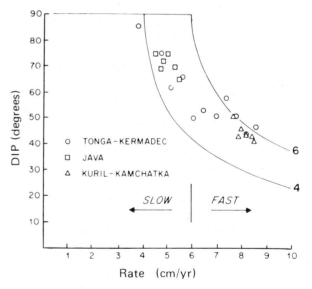

6.11. Subduction-zone dip versus convergence rate (cm/yr) (from Luyendyk, 1970). Curves refer to sinking rates of the descending plate into the asthenosphere at 4 and 6 cm/yr.

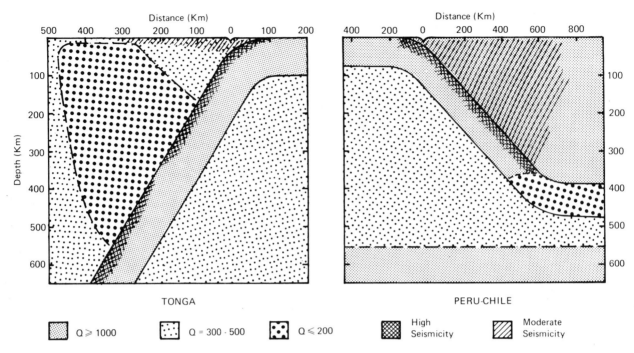

6.12. Schematic cross section of the Tonga and the Peru-Chile subduction zones showing the Q and seismicity distribution (from Barazangi and Isacks, 1971; Sacks and Okada, 1973). Horizontal distances measured from trench axes.

siderable impetus to the sea-floor spreading theory. Studies of the attenuation of body waves near convergent plate boundaries reveal a complex distribution of Q. A zone of extremely low seismicity and $Q(\leq 150)$, for instance, exists above the Tonga subduction zone (Barazangi and Isacks, 1971) (fig. 6.12), and correlates with the high surface heat flow observed in the overlying active marginal basin (the Lau-Havre trough—see fig. 4.2). A similar region occurs in the upper mantle beneath the Japan Sea. Unlike the Tonga region, the upper mantle above the Peru-Chile subduction zone (fig. 6.12) exhibits high to moderate seismicity and high $Q(1000-3000)$ (James, 1971; Sacks and Okada, 1973), suggesting a thickened lithosphere. It is possible, as will be discussed later, that partially molten mantle diapirs account for the very low-Q regions above some descending slabs. The relatively thick lithosphere beneath the central Andes (~ 350 km), on the other hand, may characterize regions in which plates carrying continental crust are overriding plates carrying oceanic crust.

Plate Motions

Because all plates on the Earth are moving, the motion of one plate is described relative to that of another which is arbitrarily fixed. The fact that plates

are moving on a sphere indicates that the motion of one plate relative to another along their mutual interface can be described in terms of a *pole of rotation* passing through the center of the Earth. Transform faults should lie on small circles about the pole of rotation for two adjoining plates, as for instance plates 1 and 2 in fig. 6.13. This characteristic is

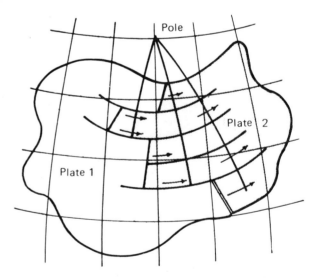

6.13. Pole of rotation for two plates on a sphere (after Morgan, 1968).

closely approached for most transform faults. Relative plate motions are best measured by their angular velocity, which increases from zero at the pole to a maximum at the equator of rotation. The arrows in fig. 6.13 are proportional to the angular velocity of plate 2 relative to plate 1. An increasing rate of separation of plates about a ridge axis with increasing distance from the pole is recorded by progressively increasing distances between specific magnetic anomalies. Similarly, the rate of plate convergence at subduction zones increases in going away from the pole of rotation. This is illustrated by the New Zealand-Kermadec-Tonga subduction system, which shows progressively deeper earthquake hypocenters going north, suggesting a greater depth of penetration by the Pacific plate. Because subduction zones do not bear any particular geometric relationship to poles of rotation, the motion of plates being consumed may vary from a maximum (convergence direction normal to trench) to purely transform motion, which is illustrated, for example, by the western part of the boundary between the Eurasian and African plates (fig. 6.14).

Poles of rotation for large segments of oceanic crust can be estimated from spreading-rate data and strikes of fracture systems. The poles for the North Atlantic and North Pacific are located in near proximity of each other—the northern poles located off the southern tip of Greenland and the southern poles south of Australia. The South Atlantic and South Pacific poles are also close or coincident, east of Baffin Island (northern) and in Antarctica (southern).

Making a steep angle with both of these sets of poles are the Arctic and Indian-Ocean poles, which extend from north of central Siberia to western Antarctica and from North Africa to south of New Zealand, respectively.

Small plates ($\lesssim 10^6$ km^2) seem to occur most frequently near continent-continent or arc-continent collision boundaries (see Chapter 8), and are characterized by rapid, complex motions. Examples are the Turkish-Aegean, Adriatic, Arabian, and Iran plates located along the Eurasian-African continent-continent collision boundary (Fig. 6.14), and the several small plates along the continent-arc collision border of the Australian-Pacific plates (fig. 6.15). The motions of the small plates appear to be controlled largely by the compressive forces of the larger plates. The Mediterranean, Black, and Caspian seas are remnants of the Tethys Ocean, which closed as Africa and India approached and finally collided with Eurasia (see Chapter 9). Only two minor subduction zones are active in this area today, one near the tip of Italy (the Calabria) and the other south of Greece and Turkey (the Hellenic). The interactions of the Australian and Pacific plates have been extremely complex during the Cenozoic. As fig. 6.15 shows, there appear to be at least five small plates in this area.

Not only plate sizes and shapes but poles of rotation (spreading directions) and spreading rates change with time. Some changes, such as changes in plate size and shape, may be gradual, while others, such as shifts in spreading centers, are abrupt and

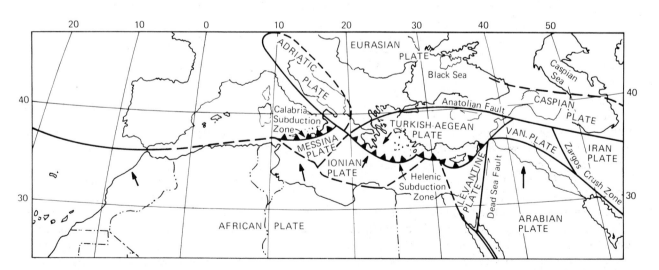

6.14. Approximate position of plate boundaries in the Mediterranean area (after Dewey *et al.*, 1973). Bold arrows refer to plate motions, relative to the Eurasian Plate. Key: ▲▲▲▲▲ = Subduction zones (triangles point in direction of dip); ══ = oceanic ridges ⟶ = transform faults.

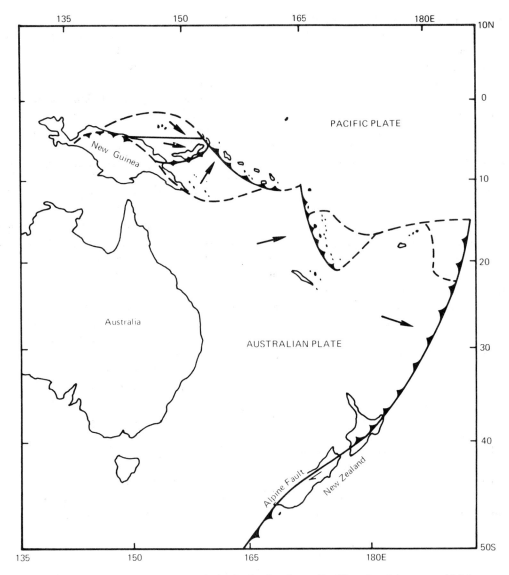

6.15. Summary of inferred plate boundaries in the Southwest Pacific (after Johnson and Molnar, 1972). Symbols as in Fig. 6-14; motions relative to the Pacific Plate.

may occur over periods as short at 10^5-10^6 years, as deduced from magnetic anomaly patterns. Continent-continent, continent-arc, and arc-arc collisions may result in the rapid cessation of subduction zones and the initiation of new subduction zones in other areas. An example of an area that has undergone a complex history of sea-floor spreading involving all of these types of changes is the Pacific Basin (fig. 6.16). By 150 m.y. ago, five spreading centers and two triple junctions existed in the Central Pacific (Larson and Pitman, 1972) (see next section). Subsequent spreading has resulted in growth of the Pacific Plate at the expense of the other three plates. Today the Kula Plate has been consumed (except for a remnant in the

Bering Sea), the Farallon Plate is represented by two minor remanants (the Juan De Fuca and Cocos plates), and the Phoenix (Nasca) Plate is greatly reduced in size (Plate I). A major shift in both spreading centers and directions of spreading occurred in the Eastern Pacific about 10 m.y. ago. Prior to this time spreading was from a now extinct ridge system, remnants of which occur in the Nasca Plate east of the East Pacific Rise and in the Pacific Plate north and west of the East Pacific Rise (Herron, 1972). The large E-W transform faults such as the Mendocino, Pioneer, and Clarion fracture zones (Plate I) were active during this time. The earlier ridge system was partly annihilated in a subduction zone

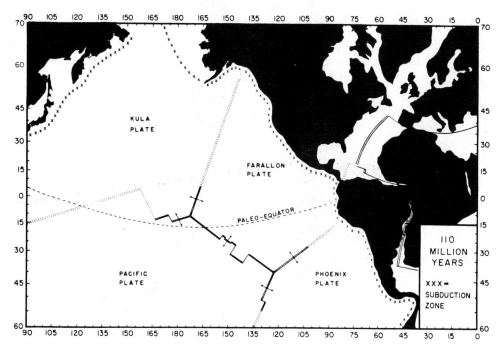

6.16. Pacific Ocean plates and plate boundaries at about 150 m. years (after Larson and Pitman, 1972). Key: xxx = subduction zones.

along the southwestern coast of North America (see Chapter 9).

Evidence suggests that oceanic ridges grow and die out by lateral propagation. One model for propagation of a medial rift zone is illustrated in fig. 6.17. The model assumes that the spreading rate of the dying rift decreases to zero while the rate on the growing rift increases from zero to full rate over approximately the same distance. The change from fast to slow spreading produces a zone of rough topography which forms a "V," with the apex at the tip of the growing rift. A zone of intense shearing is produced in the crust between the tip of the propagating and the tip of the dying rift. Offset magnetic anomalies and bathymetry consistent with propagating rifts, with and without transform faults, have been described along the Galapagos Ridge and in the Juan de Fuca Plate (Hey, 1977; Hey et al., 1980).

Transform Faults and Triple Junctions

Transform faults differ from classical transcurrent faults in that the sense of motion relative to offset along oceanic ridge axes is opposite to that predicted by transcurrent motion (Wilson, 1965) (fig. 6.18). It is clear that the motion observed along transform faults cannot have produced the offsets of ridge axes. These offsets may have developed at the time spreading

began and reflect inhomogeneous fracturing of the lithosphere, as was discussed above. There are three types of transform faults: ridge-ridge, ridge-trench, and trench-trench faults. Ridge-ridge transform faults are the most common type, and many are shown in Plate I. The Easter Island (southeast Pacific) and Romanche (Atlantic) fracture zones are examples of ridge-ridge transform faults. Most ridge-ridge transforms retain a constant length as a function of time (fig. 6.18), whereas ridge-trench and trench-trench transforms decrease or increase in length as they evolve. The Alpine transform fault in New Zealand serves as an example of how a trench-trench transform may increase in strike length (fig. 6.19). In this figure, the upper trench $a-b$ (destined to become the Kermadec Trench) consumes only plate Y (the Pacific Plate) and the lower trench $b-c$ (destined to become the Puysegur Trench) consumes only plate X (the Australian Plate). As Y is consumed between a and b, but not between b and c, $b-c$ must be steadily offset from $a-b$ to form two trenches joined by a transform fault.

Because of the time-dependent thermal behavior of the oceanic lithosphere (Chapter 4), transform faults must have dip-slip as well as strike-slip components. Studies of oceanic transform-fault topography and structure indicate that zones of maximum displacement are very localized (< 1 km wide) and that

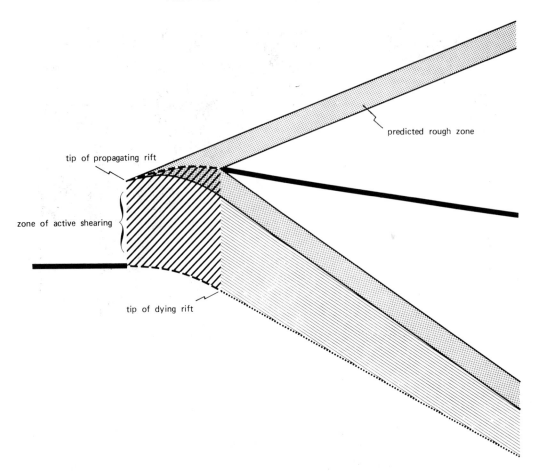

6.17. Crustal patterns produced by propagating and dying oceanic rifts (after Hey *et al.*, 1980). ▬▬ rift with maximum spreading rate; – – – rifts with transitional spreading rates; ·········· extinct rift axis.

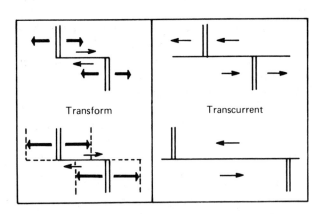

6.18. Motion on transform and transcurrent faults relative to an oceanic ridge axis (double vertical lines) (after Wilson, 1965). Note that offset increases with transcurrent motion while it remains constant with transform motion. Bold arrows refer to spreading directions, small arrows to plate motions.

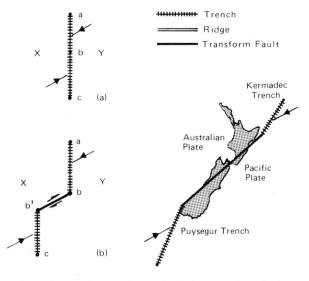

6.19. Origin of the trench-trench Alpine transform fault in New Zealand-Tonga (after McKenzie and Morgan, 1969).

they are characterized by an anastomozing network of faults (Burke and Sengör, 1979). Steep transform valley walls are comprised of inward-facing scarps associated with normal faulting. Large continental transform faults form where pieces of continental lithosphere are squeezed within intracontinental convergence zones such as the Anatolian Fault in Turkey. Large earthquakes ($M \geq 8$) separated by long periods of quiescence occur along "locked" segments of contental transforms whereas intermediate-magnitude earthquakes characterize fault segments in which episodic slippage releases stresses. Aseismic slip occurs along well-lubricated segments of transforms. Large earthquakes along continental transforms appear to have a period of about 150 years.

Triple junctions (or *triple points*) are points where three plates meet. Such junctions are a necessary consequence of rigid plates on a sphere, since this is the only way a plate boundary can end. There

are 16 possible combinations of ridge, trench, and transform-fault triple junctions (McKenzie and Morgan, 1969), of which only six are common. These are illustrated in fig. 6.20 with examples. Triple junctions are classified as stable or unstable, depending on whether or not they preserve their geometry as they evolve. The geometric conditions for stability are described with vector velocity triangles in the figure. Only RRR triple points are stable for all orientations of plate boundaries. It is important to understand evolutionary changes in triple junctions, because changes in their configuration can produce changes that superficially resemble changes in plate motions.

Temperature Distributions at Plate Boundaries

A model for the temperature distribution beneath oceanic ridges is given in fig. 6.21. The model is based on a boundary-layer solution for steady-state, two-dimensional cellular convection. It neglects the possible contribution of radioactive heat sources in the upwelling current and it assumes that the upper mantle behaves in a viscous manner and exhibits Newtonian viscosity which decreases exponentially with temperature and depth. Flow lines of particle motion in the convecting currents are shown by

Geometry	Velocity triangle	Stability	Example
RRR		All orientations stable	East Pacific Rise and Galapagos Rift Zone.
TTT		Stable if ab, ac form a straight line, or if bc is parallel to the slip vector CA	Central Japan.
TTF		Stable if ac, bc form a straight line, or if C lies on ab	Intersection of the Peru-Chile Trench and the West Chile Rise.
FFR		Stable if C lies on ab, or if ac, bc form a straight line	Owen fracture zone and the Carlsberg Ridge. West Chile Rise and the East Pacific Rise.
FFT		Stable if ab, bc form a straight line, or if ac, bc do so	San Andreas Fault and Mendocino Fracture Zone.
RTF		Stable if ab goes through C, or if ac, bc form a straight line	Mouth of the Gulf of California.

6.20. Geometry and stability requirements of six common triple junctions (after McKenzie and Morgan, 1969). Dashed lines ab, bc, and ac in the velocity triangles join points, the vector sum of which leave the geometry of AB, BC, and AC, respectively, unchanged. The junctions are stable only if ab, bc, and ac meet at a point. Key: — = trench; ═══ = ridge; ⇌ = transform fault.

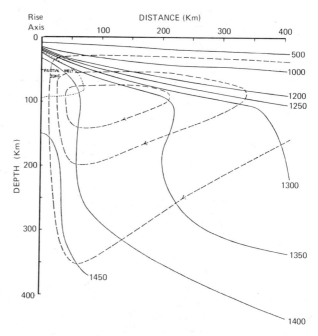

6.21. Model for temperature distribution beneath an oceanic ridge (after Turcotte and Oxburgh, 1972b). Temperature contours labeled in degrees C.

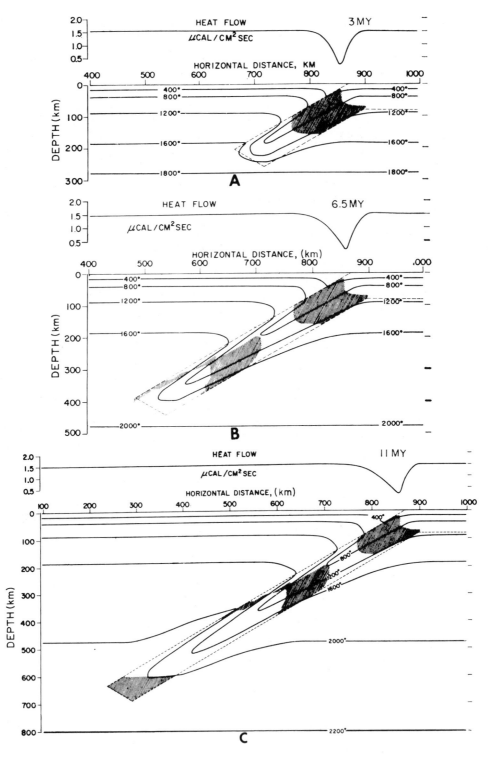

6.22. Temperature distribution (in degrees C) in a descending lithospheric slab as a function of time (after Toksöz *et al.*, 1971). Spreading rate is 8 cm/year. Shaded zones indicate regions of the gabbro-eclogite, olivine-β phase, and 650-km phase changes, in order of increasing depth.

dashed lines with arrows. A region of low viscosity, high heat flow, and high ascending velocities is associated with the relatively hot, rising currents in the model, while the downward flow is characterized by cooler, highly viscous, slow-moving material. The predicted region of partial melting is also shown.

The lithospheric plate cools as it moves laterally away from oceanic ridges; then as it descends into the asthenosphere at convergent plate boundaries, it begins to warm up again. Consideration of possible heat sources to warm the descending slab indicates that the following are most significant (Toksöz et al., 1971): (a) conductive and radioactive heat transfer from the surrounding mantle, (b) adiabatic compression of the slab, (c) frictional heating along slab faces (principally the upper face), and (d) phase changes within the slab. Heating caused by internal radioactive sources is negligible compared to the above sources. The temperature distribution within such a downgoing slab can be calculated by finite-difference methods for various convergence rates (Toksöz et al., 1971). An example for a convergence rate of 8 cm/yr at about 3, 6.5, and 11 million years after initiation of subduction is given in fig. 6.22. Note that isotherms are progressively bent downward by the cool, descending slab. The calculated heat-flow distribution above each stage in the descent of the slab is shown; as with observed heat-flow distributions over convergent boundaries, a distinct minimum occurs over the trench. It can be shown that this minimum results primarily from conductive heat transport from the surrounding mantle into the slab at shallow depths. The descending slab in this model reaches thermal equilibrium at about 650 km.

The relatively high heat flow observed above some marginal seas (Chapter 4) may be caused by frictional heating along the upper surface of descending slabs (Turcotte and Oxburgh, 1972b; Uyeda, 1977). The heat flow away from such a slab q is given by

$$q = v\tau \qquad (6.1)$$

where v is the velocity of plate descent and τ is the shear stress. For q = 4 HFU and v = 5 cm/yr, τ = 1 kbar, which is in agreement with values of shear stress deduced from seismic data. Because of the probable low thermal conductivity of upper-mantle rocks, it is necessary to call upon upward mass transfer to move this heat to the surface. Two models have been proposed for such transport (Karig, 1971; Sleep and Toksoz, 1971), illustrated in fig. 6.23. The first (a) involves an ascending mantle diapir of partially

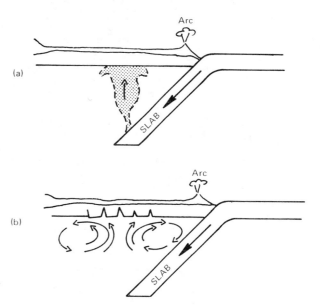

6.23. Models for material transport above descending lithospheric slabs (modified after Sleep and Töksoz, 1971): (a) rising mantle diapir; (b) secondary spreading center.

melted, shear-heated mantle material, which buoyantly rises to the base of the lithosphere. The second (*b*) involves the presence of a secondary spreading center induced by the drag along the upper face of a slab. Both models are consistent with the very low Q values and low seismicity above descending slabs beneath marginal seas. As was previously mentioned (Chapter 4), the magnetic anomaly distributions in some marginal-sea basins tend to support the secondary spreading-center model. It is possible that both mechanisms of mass transfer operate simultaneously.

Stress Distribution within Plates

Stress distributions in the lithosphere can be estimated from combined data from geological observations, first-motion studies of earthquakes, and direct measurement of *in situ* stress (Richardson et al., 1979; Zoback and Zoback, 1980). Results indicate the existence of *stress provinces* which show similar stress orientations and earthquake magnitudes and have linear dimensions ranging from a few hundred to 2000 km. Maximum compressive stresses for much of North America trend E–W to NE–SW (fig. 6.24) and E–W to NW–SE for cratonic regions of South America. Western Europe is characterized by dominantly NW–SE compressive stresses, while much of Asia is more nearly N–S. Within the Indian plate,

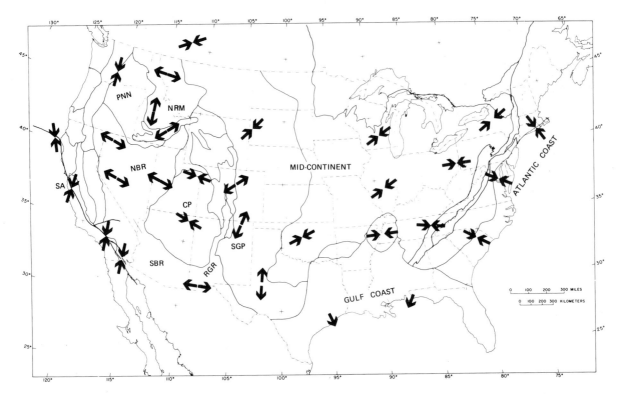

6.24. Stress provinces in the United States (from Zoback and Zoback, 1980). Arrows represent directions of least (outward-directed) or greatest (inward-directed) principal horizontal compression. Province Abbreviations: PNW, Pacific Northwest; SA, San Andreas; SN, Sierra Nevada; NBR, SBR, northern and southern Basin and Range Province; RGR, Rio Grande rift; CP, Colorado Plateau; SR, Snake River Plain; NRM, northern Rocky Mountains; SGP, southern Great Plains.

compressive stresses range from N–S in India to nearly E–W in Australia. Horizontal stresses are variable in Africa but tend to suggest a NW–SE trend for maximum compressive stresses in West Africa and an E–W trend for the minimum compressive stress in East Africa. Existing data from oceanic lithosphere away from plate boundaries is generally characterized by variable deviatoric compressive stresses. The overall pattern of stresses in both the continental and oceanic lithosphere is consistent with the present distribution of plate motions as deduced from magnetic anomaly distributions on the sea floor.

Examples of stress provinces in the United States are shown in fig. 6.24. Most of the central and eastern parts of the United States are characterized by compressive stresses, ranging from NW-SE in the Atlantic Coast to dominantly NE–SW in the midcontinent area. Much of the western United States is characterized by extensional and strike-slip stress patterns, although the Colorado Plateau, the Pacific Northwest, and the area of the San Andreas Fault are dominated by compressive deformation. The abrupt

transitions between stress provinces in the western United States imply a crustal or uppermost mantle source for the stresses. The correlation of stress distribution and heat-flow distribution in this area suggests that widespread rifting in the Basin and Range Province (and other rift areas) is linked to thermal processes in the mantle. The broad transitions between stress provinces in the central and eastern United States reflect deeper stresses at the base of the lithosphere related to drag resistance and compressive forces transmitted from the Mid-Atlantic Ridge spreading center.

SEA-FLOOR SPREADING IN THE WESTERN PACIFIC

The Western Pacific provides an example of a complex sea-floor spreading history. As is shown in fig. 6.16, the Mesozoic history of the Pacific Basin involves the interaction and evolution of several plates. The Pacific Plate may have grown from a triple junction of the Phoenix, Farallon, and Kula plates (Hilde, Uyeda, and Kronenke, 1977), with the Kula-Phoenix ridge originally extending into the Tethys Ocean. Between 200 and 150 m.y. the Kula plate was being subducted beneath eastern and southeastern Asia and decreased in size as the Pacific Plate grew. Data from magnetic anomalies and deep-sea drilling suggest a plate configuration by 135 m.y. as shown in fig. 6.25a. By 100 m.y., the Pacific Plate grew such that the western end of the Kula Plate was being consumed beneath Japan (b). Analysis of magnetic anomalies in the Japan Sea indicate that it began to open at this time. The subduction of the Kula-Pacific Ridge may have been responsible for its opening. Subduction of the Kula Plate continued, and by about 100 m.y. ago the crust of the Bering Sea, which may represent a remnant of the Kula Plate, had formed. The trapping of a fragment of the Kula Plate in the Bering Sea may have occurred as the Aleutian subduction zone began in late Cretaceous time, perhaps along a fracture zone in the Kula Plate (b). The Tethys Ridge was also subducted beneath Tibet at

about 100 m.y., and India began to be rifted away from Antarctica by 125 m.y. Also, the South China Sea began to open about this time and Borneo was separated from mainland China.

By 65 m.y., India approached the subduction zone beneath Tibet, and the Ninety-East Ridge was a transform fault between the India and Australia-Antarctic Plates (c). By about 50 m.y., the ridge on the southern margin of the Pacific Plate began to be subducted, Antarctica began to be rifted away from Australia, and the Ninety-East Ridge transform became extinct as the Australian and India Plates became part of the same plate. Between 80 and 60 m.y., the Tasman Sea opened off southeastern Australia. The Phoenix-Pacific Ridge migrated rapidly to the southeast and a new plate may have developed in the South Pacific (c). By 65 m.y. only a small amount of the Kula Plate remained in the northeast Pacific Basin. At about 45 m.y., spreading directions in the Pacific Plate changed from NNW to WNW, and this change is reflected by the bend in the Emperor-Hawaiian chain (fig. 8.18; Plate I). The ridge system

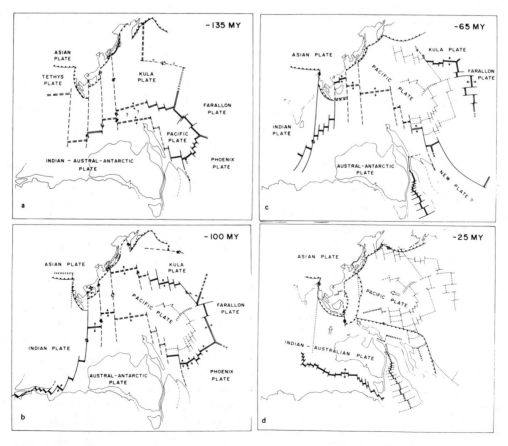

6.25. Schematic diagrams showing plate boundaries in the Western Pacific basin (after Hilde *et al.*, 1977). a) 135 m.y. ago; b) 100 m.y. ago; c) 65 m.y. ago; and d) 25 m.y. ago.

along the southern margin of the Pacific Plate was subducted along the northern margin of the Australian Plate at about 50 m.y. With the change in spreading direction of the Pacific Plate at 45 m.y., the transform faults south of Japan and Korea and south of Tonga became zones of compression, and subduction zones developed. The Philippine, Bonin-Mariana, and Tonga-Kermadec arc systems developed along these subduction zones, trapping older oceanic crust in back-arc basins (d). With increased subduction rates at about 25 m.y., back-arc spreading began in some or most of these basins (viz., the Philippine Sea) and remnant arc systems were isolated within the basins (Karig, 1971).

MECHANISMS OF SEA-FLOOR SPREADING

Convection Models

Two basic types of convection models have been proposed to drive plate motions. *Deep convection models* were proposed long before the theory of sea-floor spreading as a possible driving force for continental drift. They involve all or most of the mantle, as fig. 6.26 illustrates. In these models, a hot convective column rises beneath an oceanic ridge and flows outward, carrying the lithospheric plates with continents as passengers. *Shallow convection models* (fig. 6.27) involve only the asthenosphere, and because of the plastic behavior of the asthenosphere, these latter models have been widely accepted. Until recently, two major problems seemed to eliminate any convection models involving the lower mantle: the idea that the mantle is compositionally stratified, and data which suggested that the lower mantle is too viscous

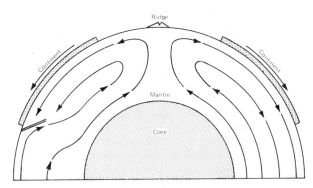

6.26. Model showing deep convection in the Earth (after Orowan, 1969).

to convect. Also pointed out as not being consistent with whole-mantle convection is the fact that earthquakes are not observed at depths greater than 700 km and the possibility that phase changes in the mantle cannot be preserved in a deep-convecting mantle. Recent investigations, however, have shown that all of these problems are not or may not be important (Davies, 1977; Phillips and Ivins, 1979). Available geophysical and geochemical data do not require the deep mantle to be chemically zoned, at least in terms of major elements, and recent estimates of the mantle's viscosity suggest that it is rather uniform at about 10^{22} poise throughout (except in the LVZ, where it is about 4×10^{20} poise). Available seismic evidence does not eliminate the possibility that descending slabs penetrate below 700 km, and thermodynamic calculations indicate that phase changes will not prevent whole-mantle convection. Tending to favor whole-mantle convection, although not unambiguously, are the long-wavelength gravity anomalies in the Earth (fig. 3.11) and the large size of the plates on the Earth's surface, implying convection cells extending to depths roughly equal to plate sizes (~ 3000 km).

Laboratory and theoretical models of convection are also not definitive. These studies indicate that there may be more than one scale of convection in the Earth (McKenzie et al., 1980). Laboratory studies of convection in materials with properties of the mantle (i.e., with large Rayleigh numbers) show that irregular and unsteady convection patterns may exist in the Earth. Although plate motions and long-wavelength gravity anomalies in the Earth clearly imply convection within the Earth, existing data and modeling techniques do not allow rigid constraints to be placed on the scale or scales of convection cells.

Forces Driving Lithospheric Plates

Two models have been proposed for the forces driving plates: the *viscous-drag model* and the *buoyancy model* (fig. 6.27) (Hargraves, 1978). The viscous-drag model involves upward convection beneath oceanic ridges and lateral spreading which, because of strong coupling between lithosphere and asthenosphere, drags plates along. As cells descend, they pull the cooling lithosphere with them into the asthenosphere. Return flow may occur either at the base of the asthenosphere or in the deep mantle. Most models for driving plates, however, recognize the base of the lithosphere as a major thermal boundary across which decoupling occurs.

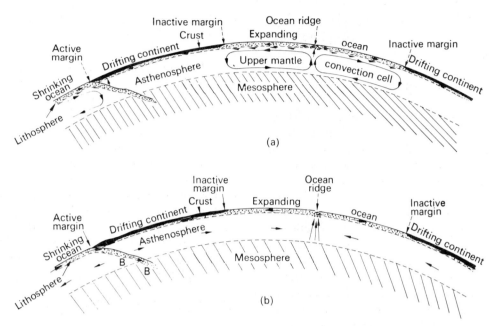

6.27. Viscous-drag (a) and buoyancy (b) plate tectonic models (after Bott, 1971; Hargraves, 1978).

The buoyancy model (fig. 6.27b) assumes an asthenosphere that is soft enough not to transmit significant horizontal shear stresses to the overlying lithosphere (Elsasser, 1971). Plate movement is caused by a combination of pushing apart of plates at oceanic ridges, gravity sliding of plates away from ridges, and pulling of plates by their leading edges as they descend at subduction zones. Pushing at ridges is accomplished by magma wedging along the axial zones where new lithosphere is formed. The pulling or negative buoyancy of the plate as it descends is caused by an increase in its density, resulting from the combined effect of cooling and phase changes. Calculations indicate that gravity sliding of a plate away from an oceanic ridge is alone sufficient to produce a sliding rate of 4 cm/yr for a slope of only 1/3000 on the surface of the asthenosphere. However, most data indicate that negative buoyancy is the major, if not the predominant, force in this model. Slow return flow occurs in the asthenosphere or lower mantle to offset the addition of lithosphere. The viscous-drag and buoyancy convection models are not mutually exclusive, and various aspects of both models may operate in the Earth today. For instance, stress distributions in the lithosphere indicate that although negative buoyancy forces in subduction zones dominate in driving plates, viscous-drag forces are required beneath lithosphere carrying cratonic continental crust (Richardson et al., 1979).

It is probable that convection styles and patterns in the Earth have changed with time. Whether or not convection currents have driven lithospheric plates in the Precambrian as they may have during the Phanerozoic is a topic of considerable interest, and will be discussed in Chapter 10. Just prior to the time that sea-floor spreading became widely accepted, Runcorn (1965) proposed that a pattern of deep convection cells in the Earth has changed as the Earth's core grew. Assuming Newtonian convection, his results suggest that the convection pattern would progressively change from a few large cells to many smaller cells as the mantle became thinner at the expense of the core. This hypothesis, however, seems unlikely in terms of our present knowledge, in that most data suggest that the core formed during or rapidly after accretion of the Earth (Chapt. 2).

A model proposed by Dickinson and Luth (1971) has some features similar to that proposed by Runcorn but calls upon progressive growth of the mesosphere instead of the core. This model suggests that the mesosphere has grown by the addition of sinking lithospheric slabs at the expense of the asthenosphere. This evolutionary pattern begins with deep convection and evolves to shallow, cellular convection as the asthenosphere thins. An increase in the number and changes in the distribution of convection cells would almost surely accompany this evolutionary pattern. It is of interest that this model also predicts an end to convection and plate motions within the next 10^9 years as the mesosphere completely consumes the asthenosphere!

The Mantle Plume Model

It has also been suggested that lithospheric plates are driven by forces associated with mantle plumes. *Mantle plumes* are diapirs of hot mantle material, which rise buoyantly, like salt domes in sediments, to the base of the lithosphere. Evidence for plumes comes from a variety of geophysical and geochemical data, some of which are discussed in Chapters 7 and 8. The distribution of 21 proposed major plumes active in the Earth today is shown in fig. 6.28. Some investigators have suggested that plumes are more numerous than those shown in the figure, with on the order of 20 plumes underlying Africa alone (Thiessen et al., 1979). Plumes are characterized by the following features (after Wilson, 1973): (1) a relative topographic high, including two cases where the oceanic crust is raised above sea level (Macquarie and St. Pauls); (2) the uplifts are capped by active (or recently active) volcanoes erupting magmas of the tholeiite or alkali series; (3) gravity highs over some plumes (Chapter 3); (4) one or two aseismic ridges or extinct volcanic chains lead away from oceanic plumes; and (5) high heat flow. Igneous-rock chains in both oceanic and

continental areas suggest that there may have been as many as 50 to 150 mantle plumes in the recent geologic past.

It is important, in terms of the possibility that plumes drive lithospheric plates, that most of the oceanic plumes occur on or near active oceanic ridges. Plume material may spread out both along and at right angles to oceanic ridges, and plates may be driven by drag at their base produced by plume motion or they may slide off the tops of plumes (triple junctions) in a manner similar to that suggested for the buoyancy convection model. Gravity and topographic distributions over proposed plumes are interpreted to indicate that plumes generate large stresses at the base of the lithosphere which, when considered collectively, are sufficient to drive lithospheric plates (Morgan, 1972a).

The Role of Phase Changes

Phase changes in descending plates which produce more dense mineral assemblages (as discussed in Chapter 3) may aid in pulling plates into the asthenosphere. Phase changes such as the gabbro-eclogite and olivine-β-phase transitions occur at depths 30–100 km shallower in descending lithospheric plates than in surrounding mantle because of depressed isotherms. The approximate locations of the gabbro-

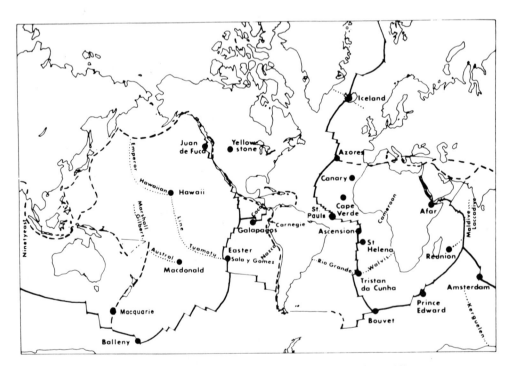

6.28. Distribution of 21 major mantle plumes (after Morgan, 1972b). Dotted lines represent linear volcanic chains that appear to represent hot-spot traces.

eclogite, olivine-β-phase, and of the 650 km discontinuity phase changes (Chapter 3) are indicated in fig. 6.22. The 650 km discontinuity is not displaced upward in the descending slab in the figure because in this model thermal equilibrium is reached at 650 km.

Phase changes in descending slabs may also contribute directly or indirectly to the production of earthquakes in the slabs. For instance, the tensional breaks characteristic of some plates at depths of 100–300 km (as previously discussed) may result from the pulling effect of a more dense portion of a slab at 300–400 km caused by the olivine-β-phase and pyroxene-garnet phase changes. Compressional earthquakes at 400–600 km in descending slabs may also reflect the excess density at 300–400 km and/or the resistance encountered by the slab as it strikes the base of the asthenosphere. Such intermediate and deep-focus earthquakes may also be produced directly by metastable mineral assemblages, carried outside of their stability fields in a descending slab and rapidly inverting (in $\lesssim 1$ sec) to stable assemblages (Ringwood, 1972b).

SUMMARY STATEMENTS

1. The theory of sea-floor spreading offers a unified explanation for a large amount of otherwise seemingly unrelated geological, geophysical, and geochemical data. The theory proposes that new lithosphere is generated at oceanic ridges, spreads laterally, and returns to the mantle in subduction zones.

2. Paleomagnetic studies indicate that the Earth's magnetic field has reversed its polarity many times in the geologic past. Combined results from volcanic rocks and deep-sea sediments allow the construction of a geomagnetic time scale back to about 4.5 m.y. Polarity intervals as short as 30,000 years can be resolved with existing methods.

3. The fact that the distribution of linear magnetic anomalies on the ocean floors correlates well with reversed and normal polarity intervals in the geomagnetic time scale gives major support to the sea-floor spreading theory. Magnetic anomalies appear to result from alternating bands of basalt injected at axial zones of oceanic ridges during normal and reversed magnetic intervals.

4. Sea-floor spreading rates (expressed as half-rates), as deduced from oceanic magnetic anomaly patterns, range from 1 to 20 cm/yr, averaging a few centimeters per year. Variations in rate occur from one segment of a ridge to another, as a function of distance (age) from a given ridge segment, and commonly on opposite limbs of the same segment.

5. Earthquakes occur chiefly along four rather narrow belts on the Earth which form the boundaries of lithospheric plates. These are axial zones of oceanic ridges, subduction zones, transform faults (where the plates slide by each other), and intracontinental compressional zones (remnants of continental collisions). First-motion studies of earthquakes along these boundaries are consistent with sea-floor spreading.

6. Hypocenter distributions in subduction zones range from continuous linear or curved patterns to poorly defined discontinuous patterns. A gap in seismicity occurs in some subduction zones, reflecting a piece of plate broken off or a region of low stress in the downgoing plate. Q and seismicity above descending slabs vary from very low (e.g., Tonga region) to high (e.g., Peru-Chile region).

7. Relative motions of lithospheric plates can be described in terms of a pole of rotation passing through the Earth's center. Transform faults commonly lie on small circles about the pole of rotation for two plates. Plate sizes, shapes, poles of rotation, and spreading rates have varied as a function of time.

8. Transform faults exhibit motion opposite to that needed to account for offset of oceanic ridges. Three types of transform faults exist, depending on which plate boundaries are offset: ridge-ridge (the most common), ridge-trench, and trench-trench. Plate boundaries terminate at triple junctions (points where three plates meet), which may preserve or change their geometry as they evolve.

9. Isotherms probably exhibit a mushroom shape beneath oceanic ridges, caused by upwelling mantle material spreading out at the base of the lithosphere. The relatively cool descending plate in subduction zones depresses isotherms. High

heat flow over some marginal seas requires material transport from the underlying subduction zone by rising diapirs, a secondary spreading center, or both.

10. Stress provinces in the lithosphere have dimensions of a few hundred to 2000 km. Overall stress patterns are consistent with plate motions as deduced from magnetic anomaly distributions on the sea floor.

11. Two models for the forces driving plates are the viscous-drag and buoyancy models. The viscous-drag model involves coupling between the lithosphere and asthenosphere as mantle convection currents spread laterally. In the buoyancy model, plate movement is produced by negative buoyancy in subduction zones, pushing at oceanic ridges, and gravity sliding of plates away from ridges.

SUGGESTIONS FOR FURTHER READING

Bird, J.M., editor (1980) *Plate Tectonics*, Second ed. Washington, D.C.: Amer. Geophys. Union. 992 pp.

Cox, A., editor (1973) *Plate Tectonics and Geomagnetic Reversals*. San Francisco: Freeman. 702 pp.

Jacobs, J.A., Russell, R.D., and Wilson, J.T. (1974) *Physics and Geology*, Second ed. New York: McGraw-Hill. 720 pp.

LePinchon, X., Francheteau, J., and Bonnin, J. (1973) *Plate Tectonics*. Amsterdam: Elsevier. 300 pp.

Toksöz, M.N., Uyeda, R.S., and Francheteau, J., editors (1980) *Oceanic Ridges and Arcs*. Amsterdam: Elsevier. 538 pp.

Chapter 7

Magma Associations

INTRODUCTION

Three principal *magma series* are recognized on the Earth, each composed of a group of closely related magma types that are emplaced in or on the Earth's crust. These are the *tholeiite, calc-alkaline,* and *alkali* series. If felsic and mafic end members predominate, the association is referred to as *bimodal.* Some major igneous-rock types are given in table 7.1 with their SiO_2 and K_2O contents and principal minerals. Various rock types in each magma series may be produced by varying degrees of melting in the upper mantle or lower crust and/or by other mechanisms (such as crystal-liquid separation) that will be discussed later in the chapter. The dominant rock type in the tholeiite series is *tholeiite,* a basalt containing little or no olivine. The calc-alkaline series is characterized by a predominance of *andesite* in volcanic terranes and of *granodiorite* in plutonic terranes. Smaller amounts of *tholeiite, rhyodacite, rhyolite,* and *shoshonite* (or their intrusive equivalents) also characterize this series. The alkali series, which is the least abundant of the three series, is characterized by *alkali basalt,* an olivine-bearing basalt that is relatively rich in alkali elements compared to tholeiite, Varying amounts of such alkali-rich rocks as *nephelinite, trachyte, latite,* and *phonolite* also characterize this series. *Picrite* is a rare olivine-rich tholeiite of varied occurrences.

The three magma series have quite distinct distributions when considered in terms of plate tectonics, as summarized in table 7.2 and fig. 7.1. Most magmas appear to be generated at plate margins, the exception being transform faults, where few if any

magmas are produced. The chief environments of magma production can be classified into plate-margin and intraplate environments, which, in turn, can be subdivided into seven plate-tectonic settings (table 7.2). Plate-margin settings are either subduction zones or oceanic ridges. The tholeiites produced beneath oceanic ridges and extruded into their axial zones are very low in alkali elements and hence are commonly referred to as *low-K tholeiites.* Large marginal-sea basins are also characterized by low-K tholeiites, whereas small back-arc and fore-arc basins are commonly characterized by calc-alkaline volcanics. Tholeiites and alkali basalts (and associated rocks) characterize oceanic and continental intraplate environments.

Basically, magma distribution seems to be related to tectonic stresses in the crust and upper mantle (table 7.2). Environments of extensional stress, such as oceanic ridges, marginal-sea basins, and continental rifts, are characterized by the tholeiite series or, in the case of continental rifts, by bimodal volcanism involving either or both the tholeiite or the alkali series. Subduction zones, which are associated with dominantly compressive stresses, are the sites for production of magmas of the calc-alkaline series. Regions of minor stress (compressional or extensional) such as ocean basins and continental cratonic areas are characterized by the alkali or tholeiite series.

Trace-element and isotope distributions in igneous rocks have been particularly revealing in terms of understanding the origin of magmas and in accounting for their distribution in terms of plate tectonics. Such distributions will be referred to in the description of magma associations that follow, and

Table 7.1. Some Chemical and Mineralogical Features of Igneous-Rock Types

	SiO$_2$%	K$_2$O%	Principal Minerals[†]
Low-K tholeiite	48–50	< 0.2	plg, pxs, ±ol
Tholeiite (tholeiitic gabbro)	48–52	0.2–1.0	plg, pxs, ±ol
High-Al tholeiite Al$_2$O$_3$ ≥ 16.5%	50–53	0.2–0.8	plg, pxs, ±ol
Andesite (diorite)	56–62	0.7–2.5	plg, pxs, amp
Dacite (tonalite)	62–68	1.2–2.5	plg, pxs, qtz, amp, biot
Rhyodacite (granodiorite)	65–71	2.5–4.0	plg, qtz, amp, biot
Rhyolite (granite)	71–75	4.0–5.5	K-feld, qtz, plg, biot
Shoshonite	54–56	2.8–4.0	plg, K-feld, pxs
Alkali Basalt (alkali gabbro)	44–46	0.7–1.0	plg, ol, pxs, ±foids
Nephelinite (ijolite)	38–40	1.0–2.0	neph, pxs
Latite (monzonite)	52–54	3.5–5.0	plg, K-feld, ±biot, ±amp
Trachyte (syenite)	58–61	5.0–7.5	K-feld, plg, ±biot, ±amp
Phonolite (nepheline syenite)	55–57	4.0–5.5	K-feld, neph, plg, ±amp, ±pxs
Picrite	43–46	< 0.4	ol, px, plg

Note: Extrusive rock names given with intrusive equivalents in parentheses.

[†]In order of decreasing abundance: plg = plagioclase, pxs = pyroxenes, amp = amphibole, neph = nepheline, biot = biotite, K-feld = K-feldspar, qtz = quartz, foids = principally nepheline or leucite, ol = olivine.

again in the section dealing with the origin of magmas. At the outset, however, a few preliminary remarks are necessary. Trace-element concentrations can vary significantly between rocks with approximately the same overall major-element composition. As an example, elements like Rb, Ba, and Cs vary by two orders of magnitude between low-K tholeiites and alkali basalts (table 7.3). Elements like these that are strongly concentrated in the liquid phase during partial melting or fractional crystallization are known as *incompatible elements*. Such elements are important

in the study of magma production, source composition, and evolutionary histories of magmas.

One particular group of elements that is important in evaluating magma histories is the rare-earth element group (abbreviated REE). Three types of variations can occur in the rare-earth elements during magmatic processes: (a) changes in the total REE abundance, (b) changes in the relative abundances of light and heavy rare-earth elements, and (c) fractionation of Eu from the other rare earths. Rare-earth distributions in rocks and minerals are commonly

Table 7.2. Plate Tectonic Classification of Magmas

	Plate Margin		Intraplate				
			Oceanic		Continental		
Plate Setting	Converging (subduction zone)	Diverging (oceanic ridge)	Marginal-sea basin	Large ocean basin (islands and seamounts)	Rift system	Cratonic area	Collision zone
Magma Series	Calc-alkaline (tholeiite)	Tholeiite (low-K)	Tholeiite (low-K) Calc-alkaline	Tholeiite Alkali	Bimodal Tholeiite Alkali	Alkali Bimodal	Bimodal Calc-alkaline (alkali)
Stress Regime	Compressive	Extension	Extension	Minor compressive	Extension	Minor compressive	Compressive

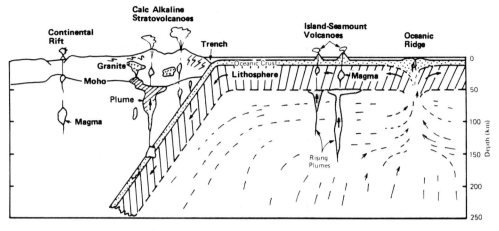

7.1. Diagrammatic cross section showing the relationship of magma generation to plate tectonics (modified after Ringwood, 1969b).

Table 7.3. Average Compositions of Basalts and Andesites

	Low Tholeitte		Continental Rift Tholeiite	Island Tholeiite	High-Al Tholeiite	oceanic Alkali Basalt	Cont. Rift Alkali Basalt	Arc Andesite	Low-K Andesite	High-K Andesite	Shoshonite
	Ridge	Arc									
SiO$_2$	49.8	51.1	50.3	49.4	51.7	47.4	47.8	57.3	59.5	60.8	52.9
TiO$_2$	1.5	0.83	2.2	2.5	1.0	2.9	2.2	0.58	0.70	0.77	0.85
Al$_2$O$_3$	16.0	16.1	14.3	13.9	16.9	18.0	15.3	17.4	17.2	16.8	17.2
Fe$_2$O$_3$*	10.0	11.8	13.5	12.4	11.6	10.6	12.4	8.1	6.8	5.7	8.4
MgO	7.5	5.1	5.9	8.4	6.5	4.8	7.0	3.5	3.4	2.2	3.6
CaO	11.2	10.8	9.7	10.3	11.0	8.7	9.0	8.7	7.0	5.6	6.4
Na$_2$O	2.75	1.96	2.50	2.13	3.10	3.99	2.85	2.63	3.68	4.10	3.50
K$_2$O	0.14	0.40	0.66	0.38	0.40	1.66	1.31	0.70	1.6	3.25	3.69
Cr	300	50	160	250	40	67	400	44	56	3	30
Ni	100	25	85	150	25	50	100	15	18	3	20
Co	32	20	38	30	50	25	60	20	24	13	~20
Rb	1	5	31	5	10	33	200	10	30	90	100
Cs	0.02	0.05	0.2	0.1	0.3	2	>3	~0.1	0.7	1.5	≳2
Sr	135	225	350	350	330	800	1500	215	385	620	850
Ba	11	50	170	100	115	500	700	100	270	400	850
Zr	85	60	200	125	100	330	800	90	110	170	150
La	3.9	3.3	33	7.2	10	17	54	3.0	12	13	15
Ce	12	6.7	98	26	19	50	95	7.0	24	23	32
Sm	3.9	2.2	8.2	4.6	4.0	5.5	9.7	2.6	2.9	4.5	3.2
Eu	1.4	0.76	2.3	1.6	1.3	1.9	3.0	1.0	1.0	1.4	0.95
Gd	5.8	4.0	8.1	5.0	4.0	6.0	8.2	4.0	3.3	4.9	4.2
Tb	1.2	0.40	1.1	0.82	0.80	0.81	2.3	1.0	0.68	1.1	0.50
Yb	4.0	1.9	4.4	1.7	2.7	1.5	1.7	2.7	1.9	3.2	1.7
U	0.10	0.15	0.4	0.18	0.2	0.75	0.5	0.4	0.7	2.2	1.3
Th	0.18	0.5	1.5	0.67	1.1	4.5	4.0	1.3	2.2	5.5	2.8
Th/U	1.8	3.3	3.8	3.7	5.9	6.0	8.0	3.2	3.1	2.5	2.2
K/Ba	105	66	32	32	12	28	16	58	49	68	36
K/Rb	1160	660	176	630	344	420	55	580	440	300	306
Rb/Sr	0.007	0.022	0.089	0.014	0.029	0.045	0.13	0.046	0.078	0.145	0.118
La/Yb	1.0	1.7	10	4.2	3.7	11	32	1.1	6.3	4.0	8.8

*Total Fe as Fe$_2$O$_3$.

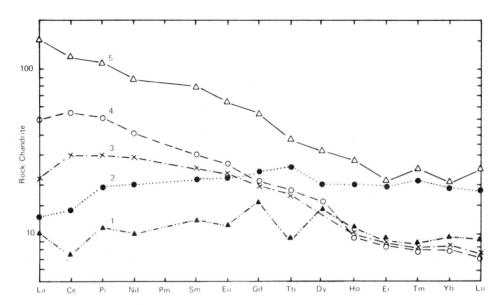

7.2. Chondrite-normalized REE patterns of average basalts (after Frey *et al.*, 1968; Schilling and Winchester, 1969, Jakes and Gill, 1970). (1) Arc tholeiite; (2) Ridge tholeiite; (3) Hawaiian tholeiite; (4) Hawaiian alkali basalt; (5) Oceanic-island alkali basalt.

presented graphically as chondrite-normalized values. Normalization is accomplished by dividing the concentration of each REE in a given rock by the corresponding concentration in chondritic meteorites. Such normalization is desirable in order to remove the odd-even effect (i.e., the fact that even-atomic-numbered REE are more abundant than odd ones). Typical REE distributions in volcanic rocks are shown in figs. 7.2, 7.4, and 7.8. When light REE (La to Sm) are concentrated relative to heavy REE (Tb to Lu), the patterns are referred to as light REE *enriched* (such as curves 2 and 3 in fig. 7.8); when heavy REE are concentrated relative to light REE, the patterns are referred to as light REE *depleted*. Patterns that show neither enrichment nor depletion are generally designated as *flat* or *chondritic patterns*. Eu can occur in two valence states ($+2$ and $+3$) and hence can become fractionated from other REE (which are $+3$ or possibly $+4$ in the case of Ce) by processes in which minerals that preferentially accept $+2$ ions are involved (Philpotts and Schnetzler, 1968). Feldspars are the most common examples of such minerals where Eu^{+2} closely follows Ca^{+2} and Sr^{+2}. When Eu is anomalously enriched (or depleted) in a rock, a sharp peak (or depression) is produced on REE diagrams known as a positive (or negative) *Eu anomaly*.

Separation of heavy isotopes (such as ^{206}Pb from ^{207}Pb and ^{87}Sr from ^{86}Sr) is not affected by some magmatic processes such as crystal-liquid sep-

aration. At the present time, a large number of $^{87}Sr/^{86}Sr$ values exist for volcanic rocks from many areas on the Earth. Pb and Nd isotope data are much fewer and Pb data are complex to interpret. A representative sampling of available Sr isotope results is given in fig. 7.3 in histogram form and will be referred to in subsequent discussions.

Numerous studies document the mobility of many elements during secondary processes such as deep-sea alteration and low-grade metamorphism (Condie, Viljoen, and Kable, 1977). Those elements particularly susceptible to remobilization during secondary processes are U, K, Rb, Cs, Ba, Si, Ca, Al, and volatile elements, while those least susceptible to change are Ti, Fe, Y, Zr, Nb, Ta, Hf, Co, Sc, and REE. Although light REE may be enriched somewhat during alteration, heavy REE are not affected. It is important in characterizing volcanic rocks and in evaluating their origin to place most weight on those elements that are relatively immobile during secondary processes, because even the youngest volcanics have undergone some alteration. Of the geochemical-tectonic classifications of modern basalts that have been proposed, those based on immobile elements have been most successful. Examples are the Ti-Zr-Y classification of Pearce and Cann (1973), the MgO-FeO_T-Al_2O_3 diagram of Pearce, Gorman, and Birket (1977), and the Th-Hf-Ta diagram of Wood, Jorou and Treul (1979). Employing these and other classifications, it is possible to group modern basalts into ocean-ridge, island-arc, and continental-interior categories.

PLATE-MARGIN ASSOCIATIONS

Oceanic Ridges

Ridge Volcanism. The most voluminous igneous rocks in oceanic areas are the low-K tholeiites produced beneath oceanic ridges. These tholeiites are commonly referred to as *ridge tholeiites* because they are extruded or intruded in the axial zones of oceanic ridges. Dredged samples of ridge tholeiite may be vesicular and exhibit glassy margins produced by rapid chilling during extrusion on the ocean bottom. They typically show pahoehoe (or ropy) surfaces that attest to their low viscosity during eruption. They also commonly occur as pillows on the ocean floor. *Pillows* are rounded, solidified masses of basalt (10–100 cm in diameter) that are thought to be produced by bulbous protrusions on submarine flows, which detach themselves from the parent flows and come to rest (while still hot and plastic) on the sea floor. Texturally, ridge tholeiites are fine-grained and often contain large, sparsely distributed, partially resorbed plagioclase crystals. The only modern example of ridge magmas being extruded above sea level is at Iceland. The Iceland tholeiites, however, have compositions more like continental tholeiites. Abyssal hills commonly parallel oceanic ridges and may represent older flows that have been slightly uplifted and moved away from modern ridges by sea-floor spreading. The widespread distribution of ridge tholeiites in oceanic areas is evidenced by studies of dredged samples and cores from the Deep Sea Drilling Project. Beside ridge tholeiites, tholeiitic diabases and gabbo, amphibolite, serpentinite, peridotite, and, rarely, diorite (and related siliceous rocks) have been recovered from the ocean floor. The recovery of these rock types lends support to the ophiolite model of the oceanic crust described in Chapter 8. Dredged ultramafic rocks are usually interpreted as fragments of mantle diapirs that have penetrated the crust.

Compositionally, ridge tholeiites differ from continental and island tholeiites by their relatively high contents of Al_2O_3 and Cr, and low contents of *large-ion lithophile (LIL) elements* such as K, Rb, Cs, Sr, Ba, Zr, U, Th, and REE (table 7.3). Fresh ridge tholeiites also exhibit high K/Ba and K/Rb ratios, low Fe^{+3}/Fe^{+2}, Rb/Sr, and La/Yb ratios, light-REE

depleted patterns (fig. 7.2), and low $^{87}Sr/^{86}Sr$ ratios ranging mostly from 0.702 to 0.704 (fig. 7.3) (Kay, Hubbard, and Gast, 1970; Peterman and Hedge, 1971). Geochemical studies of older ridge tholeiites from various locations and depths in the oceanic crust (retrieved in part by the Deep Sea Drilling Project) indicate that deep-sea, low-temperature alteration is common. In general, the degree of alteration of ridge tholeiites seems to increase with distance from active ridge systems. Such alteration is characterized by varying enrichments in Fe^{+3}, K, H_2O, B, Li, Rb, U, and Cs, and sometimes in Sr, Ba, and light REE, and by increases in the amounts of zeolites and iron oxides (Christensen et al., 1973). Altered ridge tholeiites typically exhibit $^{87}Sr/^{86}Sr$ ratios in the range 0.703 to 0.708 (Bass et al., 1973). Trace elements such as Sc, V, Co, Y, Zr, Ti, and heavy REE are least affected by alteration. Measurements of density and seismic-wave velocities in ridge tholeiite samples exhibiting varying degrees of alteration also suggest that observed decreases in these quantities in layer 2 of the oceanic crust as a function of increasing distance from ridges may result from increasing degrees of alteration (Christensen and Salisbury, 1972).

Hydrothermal Fields. Employing deep-sea photography and using small submersibles, it has been possible in recent years to carefully describe several hydrothermal fields on oceanic ridges (Corliss et al., 1979; Rona, 1980). The submersibles, in particular, enable scientists to make direct observations and measurements on the sea floor. On the Galapogos Rift at 86°W, several hydrothermal vents ranging in size from 400 to 1600 m^2 are spread out along a 2.5 km segment of a fissure system. The TAG hydrothermal field on the Mid-Atlantic Ridge (26°N) also occurs along a fault zone on the east wall of the medial rift valley. Chemical equilibria data suggest that water given off in these fields is up to 300°C in temperature. Animal communities living near the vents appear to be totally dependent upon energy derived from seawater-rock reactions and sulfur-oxidizing bacteria.

Geochemical data indicate that submarine hydrothermal waters are major sources of Mn and Li, minor sources for Ca, Ba, Si, and CO_2, and a sink for Mg. Fe, Ni, Cu, Cr, Co, and Cd are not present in the vent waters because of the presence of H_2S which leads to precipitation of these metals in mounds near the vents or, if trapped beneath the cap of relatively impermeable sediments, to massive sulfide deposits. As was previously mentioned in Chapter 4, a

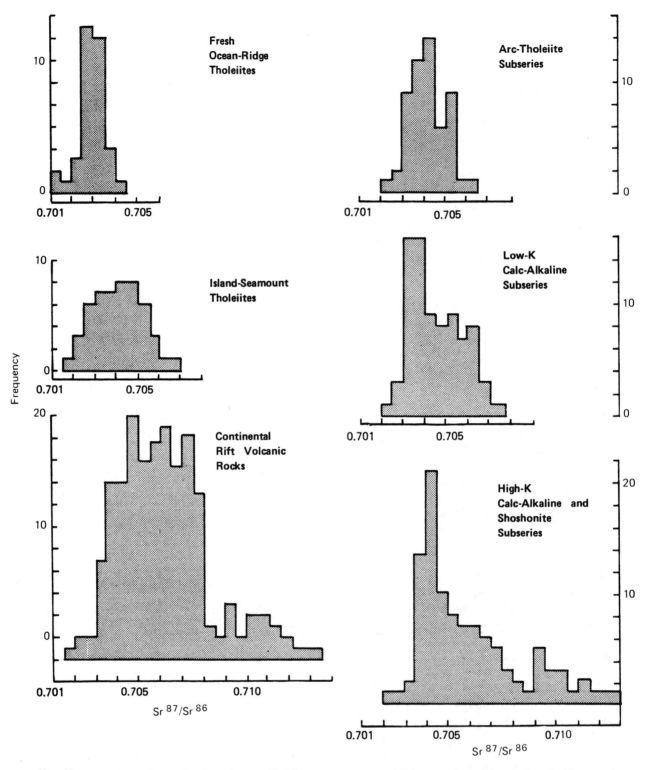

7.3. $^{87}Sr/^{86}Sr$ ratios in modern volcanic rocks compiled from many sources. All data recalculated such that the Eimer and Amend standard $^{87}Sr/^{86}Sr = 0.7080$. Frequency expressed in number of samples.

significant amount of the Earth's heat is lost through hydrothermal activity along oceanic ridges.

Subduction Zones

Modern volcanic rocks of the calc-alkaline series occur in arcs and along continental margins that overlie subduction zones. Subduction-zone-related volcanism commonly starts abruptly at a *volcanic front*, which roughly parallels oceanic trenches and begins 200–300 km inland from trench axes adjacent to the arc-trench gap (see fig. 8.16) (Sugimura et al., 1963). The volume of magma erupted decreases in the direction of dip of the subduction zone. The abrupt onset of volcanism at the volcanic front probably reflects the onset of melting in (or above) the subduction zone, and the decrease in volume of erupted magma behind the front may be caused by either a longer vertical distance for magmas to travel from the descending slab or a decrease in the water content of the slab as a function of depth, or both.

The most common volcanic rock type in arcs is andesite. Porphyritic textures are common in all volcanic rocks, and inclusions of crustal rocks are occasionally found. Calc-alkaline volcanoes are commonly large, steep-sided stratovolcanoes composed of varying proportions of lavas and fragmental materials (the latter commonly dominating). Their eruptions range from mildly explosive to violently explosive, and contrast strikingly to the eruptions of oceanic island and continental-rift volcanoes as indicated by the explosive indices (E) given in table 7.4. Large amounts of water are given off during eruptions. Rapid removal of magmas may result in structural collapse of the walls of stratovolcanoes, producing *calderas* such as Crater Lake in Oregon and the Valles Caldera in New Mexico. The final stages of eruption in some volcanic centers are characterized by the eruption of fluid, siliceous ash-flows that may travel great distances. Seismic shadow-zone studies indicate that modern magma reservoirs in subduction-zone areas are commonly 50–100 km deep. The migration of earthquake hypocenters from depths of greater than 200 km to shallower depths over periods of a few months prior to eruption seems to reflect the ascent of magmas at average rates of 1–2 km/day (Scheinmann, 1971).

Unlike calc-alkaline volcanic fields, calc-alkaline batholiths, which formed from magmas that cooled slowly within the crust, are composed chiefly of granodiorite and related granitic rocks. The difference in rock-type abundances reflects differences in viscosities and water contents of the magmas.

Table 7.4. Explosive Indices (E) of Modern Volcanic Provinces

	E*
Indonesia	99
Central America	99
Alaska-Aleutians	94
Andes	85
Japan	70
New Zealand-Tonga	67
Kamchatka-Kurils	55
Mediterranean Area	41
Atlantic Ocean	29
Indian Ocean-Africa	20
Central Pacific	12

$$*E = \frac{\text{quantity of fragmental material}}{\text{total erupted material}} \times 100.$$

After Rittman, 1962.

Andesite magmas, which are quite fluid, readily move to the Earth's surface, whereas siliceous magmas may be extremely viscous and crystallize before reaching the surface. Batholiths range in area from 10^4 to 10^5 km^2 and are composed of smaller constituent plutons that appear to have been emplaced in the crust by diapiric intrusion. They commonly contain fragments of crustal metamorphic rocks in varying degrees of digestion.

Compositional Polarity. Geochemical studies of modern subduction-zone related volcanic rocks suggest that the calc-alkaline series can be broadly subdivided into three completely gradational subseries: the arc-tholeiite, calc-alkaline (*sensu-stricto*), and *shoshonite* subseries (Jakes and White, 1972). Progressively more K_2O at a given SiO_2 level is observed in going from the arc-tholeiite to the shoshonite subseries. K_2O is rarely greater than 1.5 percent in the arc-tholeiite subseries, where it increases slowly with SiO_2. The shoshonite subseries, on the other hand, exhibits rapid increases (or, rarely, decreases) in K_2O with increasing SiO_2. The arc-tholeiite or calc-alkaline subseries characterizes volcanic fronts. REE element patterns vary from flat to highly enriched, and the range of $^{87}Sr/^{86}Sr$ ratios tends to increase in going from the arc-tholeiite to the shoshonite subseries (fig. 7.3). Existing geochemical data from Mesozoic-Cenozoic granitic batholiths in the circum-Pacific region also indicate similar systematic compositional changes in a direction normal to the strike of arc-trench systems of the same age. These changes in composition of igneous rocks as a function of increasing depth to subduction zone are known as *compositional polarity*. The distribution of Mesozoic-

Cenozoic economic mineral deposits in the Andes also suggests a polarity. Fe and Cu deposits are most abundant nearest the coast (and presumably formed above a rather shallow subduction zone), Ag, Pb, and Zn deposits are most frequent in a zone farther inland, and Sn and Mo deposits are most abundant farthest from the coast (Sillitoe, 1972). Compositional polarity does not occur above near vertical subduction zones. In these areas (such as in the Solomon Islands) there is a tendency for the three subseries to occur stratigraphically on top of each other, with the arc-tholeiites at the base of the section (Jakes and White, 1972).

The most abundant rock type in the arc-tholeiite subseries is *arc-tholeiite* followed by *arc-andesite*. Arc-tholeiites are similar in composition to ridge or altered ridge tholeiites (table 7.3), as exemplified by their high Al_2O_3 content, Fe enrichment, low alkali and related element concentrations, and rather high K/Rb ratios. They also exhibit relatively flat REE distribution patterns (fig. 7.2) and low $^{87}Sr/^{86}Sr$ ratios (0.702–0.704). They differ from fresh ridge tholeiites in their lower Ni and Cr and higher Ba contents. Arc-andesites (and more siliceous volcanics) differ from their counterparts in the calc-alkaline and shoshonite subseries by their low concentrations of K_2O, Rb, Ba, Sr, U, Th, and low Rb/Sr ratios, by their high CaO and K/Rb values (table 7.3), by their light-REE depleted patterns, and by their smaller range of $^{87}Sr/^{86}Sr$ ratios (fig. 7.3).

Andesite and dacite characterize the calc-alkaline subseries. Rhyolites (chiefly as ashflows) and high-Al_2O_3 tholeiites are of lesser importance. This subseries can be further broken down into a *high-K* and *low-K* group, which again appear to be completely gradational. The low-K group is characterized by elongate belts of relatively large stratovolcanoes composed of approximately equal amounts of flows and pyroclastics (chiefly ash and cinder.) Systematic changes in composition from the base to the top of a volcanic section are not found and andesites and dacites are intimately interbedded (Condie and Swenson, 1973; Condie and Hayslip, 1975). The Cascade, Japan, and Middle America volcanic chains are typical examples of the low-K group. Volcanoes of the high-K type range from large stratovolcanoes to small vents more or less randomly distributed in broad equidimensional volcanic fields. Andesites are mostly laharic breccias with siliceous ash flows becoming more important near the end of eruptive cycles. The abundance of breccias in most high-K volcanic fields may reflect more viscous (dry?) magmas than those characterizing low-K fields. The stratovolcanoes of the Andes and the Tertiary volcanic fields of the western United States are examples of high-K calc-alkaline volcanic areas. In plutonic regimes, tonalite is the most abundant rock type in the low-K group and granodiorite or quartz monzonite in the high-K group. At the same SiO_2 level, the high-K group is usually enriched in K_2O, Rb, Cs, Ba, Zr, U, and Th, and exhibits higher Fe^{+3}/Fe^{+2}, Fe/Mg, and Rb/Sr ratios (table 7.3). REE patterns in both groups are similar, exhibiting variable enrichment in light REE (fig. 7.4). The range in $^{87}Sr/^{86}Sr$ ratios in the low-K group is smaller (0.703–0.707) than that observed in the high-K group (0.704–0.710) (fig. 7.3).

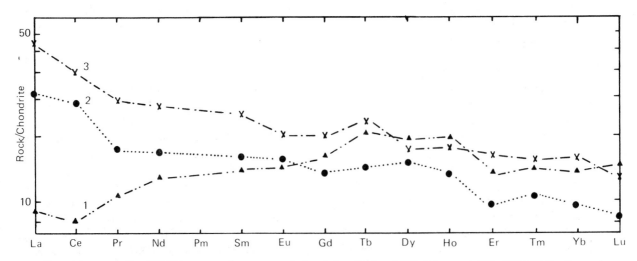

7.4. Chondrite-normalized REE patterns of average andesites (after Taylor, 1969; Jakes and Gill, 1970; Philpotts *et al.*, 1971). (1) Arc andesite; (2) Low-K, calc-alkaline andesite; (3) High-K, calc-alkaline andesite and shoshonite.

Volcanic rocks of the shoshonite subseries have a very limited distribution. Shoshonites, which are alkali-rich andesites, are the most abundant members of the subseries. The overall eruptive features are much like those described above for the high-K calc-alkaline group into which the shoshonite subseries is gradational. Major and trace-element concentrations and $^{87}Sr/^{86}Sr$ ratios are similar to those of the high-K calc-alkaline group, the chief difference being a greater enrichment in alkali and related elements at the same SiO_2 level (table 7.3).

Geochemical Polarity Indices. Data suggest that a linear correlation exists between silica-normalized K_2O contents of subduction-zone related volcanic rocks (K_2O contents at specific SiO_2 levels) and depth to subduction zone (SZ) (Hatherton and Dickinson, 1969; Dickinson, 1970) and with crustal thickness (C) (Condie, 1973). Silica-normalized K_2O values (60% SiO_2) for young volcanics for which

either or both crustal thickness C and depth to subduction zone SZ are known are plotted in fig. 7.5. Linear regression lines and correlation coefficients (significant at the 95 percent confidence level) are shown for each parameter, and the following are regression equations at 60 percent SiO_2 (after Condie, 1973):

$$C(km) = 18.2 (K_2O) + 0.45 \qquad (7.1)$$

$$SZ(km) = 89.3 (K_2O) - 14.3 \qquad (7.2)$$

A comparison of K_2O-versus-SZ relationship for specific arc systems in the circum-Pacific area, however, indicates that different K_2O-SZ curves exist (at given SiO_2 levels) for different arcs (Nielson and Stoiber, 1973). For instance, at 60 percent SiO_2 and 1.5 percent K_2O, the observed depth to subduction zone beneath modern arc volcanic terranes can range as follows: Central America, 110 km; Kurile-Kamchatka, 140 km; Honshu-Hokkaido, 160 km.

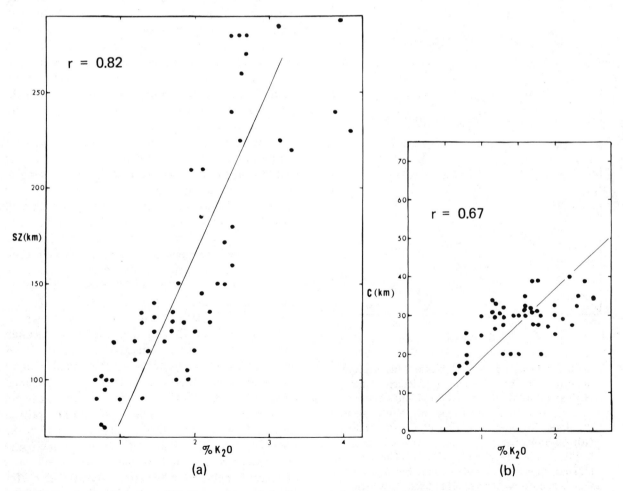

7.5. K_2O values at 60 percent SiO_2 for modern subduction-zone related volcanic rocks plotted as a function in (a) of depth to subduction zone (SZ) (modified after Dickinson, 1970) and in (b) of crustal thickness (C).

Equation 7.2 gives a depth of 120 km for these values, suggesting an inherent error of up to about 30 percent in the estimate of SZ. Allowing for this source of error, the K_2O indices appear to provide a means of roughly estimating depth to subduction zone and crustal thickness in the geologic past. In general, the modern arc-tholeiite subseries occurs on thin crust ($\lesssim 20$ km) and over shallow subduction zones ($\lesssim 150$ km); the calc-alkaline subseries occurs on crust of intermediate thickness (20 to 30 km) and over subduction zones of intermediate depth (100 to 200 km); and the shoshonite subseries occurs on intermediate to thick crust ($\gtrsim 25$ km) and over deep subduction zones ($\gtrsim 200$ km).

The distribution of Rb and Sr in young volcanic rocks related to subduction zones is also sensitive to crustal thickness (Condie, 1973). Rb-Sr variation curves for Tertiary and younger volcanic suites from the circum-Pacific area are plotted in fig. 7.6. Based

on the best estimates of crustal thickness available, the dashed lines divide the diagram into four regions of crustal thickness ($\lesssim 15$, 15–20, 20–30, and > 30 km). On the island of Vita Levu in the Fiji Islands, three volcanic suites occur. The Rb-Sr distributions of the three suites are quite different. From oldest to youngest, the suites are as follows (Gill, 1970): Fiji 1 (arc-tholeiite subseries); Fiji 2 (calc-alkaline subseries); and Fiji 3 (shoshonite subseries). Only the calc-alkaline subseries records the present crustal thickness of Vita Levu (25—30 km). The Fiji 1 suite may have recorded the crustal thickness in late Eocene or Miocene time. The young shoshonite group, however, records too great a crustal thickness. Tertiary rocks of both the calc-alkaline and shoshonite subseries occur in the Absaroka field in northwestern Wyoming, and the Rb-Sr distribution in both subseries records a crustal thickness > 30 km (observed thickness = 40 to 45 km). Although considerably more data are needed from regions that contain either all three or two of the three volcanic subseries near each other, existing information suggests that the calc-alkaline subseries, when present, records the most accurate crustal thickness in island arcs. If an arc has not evolved to the calc-alkaline stage, the arc-tholeiite subseries accurately records the crustal thickness (for example, Saipan). On continental margins, both the calc-alkaline and shoshonite subseries may record accurate crustal thicknesses.

Because trace-element distributions in modern volcanic rocks indicate that depth to subduction zone and crustal thickness are related, it is interesting to compare these two quantities in areas where both are known. Cross sections showing the relationship of crustal thickness C to depth to subduction zone SZ of four areas in the circum-Pacific region are shown in fig. 7.7. These sections are compiled from areas in which contours of equal crustal thickness and equal depth to subduction zone roughly parallel each other. Such contours are parallel or subparallel to each other around most portions of continental margins and of island arcs. Exceptions occur in the Aleutian and Kurile arcs, where crustal thickness decreases along the island chains yet no obvious shallowing of the subduction zone occurs. The Java and Japan curves in figure 7.7 exhibit an increase in C with SZ in the range of 20–35 km and 100–250 km, respectively. All but Saipan show a maximum in C in the range of SZ = 200–300 km. Also shown on the figure are points representing locations of young volcanoes or volcanic fields from the circum-Pacific region for which C and SZ are known. With the exception of the four central Andes volcanic fields, most of the

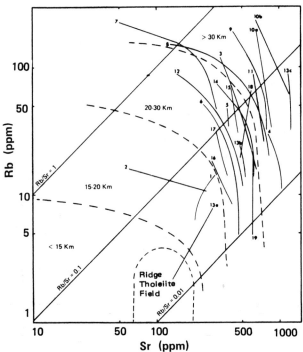

7.6. Rb-Sr crustal thickness grid (after Condie, 1973). Dashed lines separate crustal thickness fields (given in kms). (1) Tonga; (2) Saipan; (3) Medicine Lake Highlands; (4) Mount Shasta; (5) Mount Lassen; (6) Talasea, New Britain; (7) Taupo area, New Zealand; (8) Mount Taylor, New Mexico; (9) Central Andes; (10) Absaroka field, Wyoming (a = calc-alkaline, b = shoshonite); (11) Bougainville; (12) Guatemala; (13) Fiji (a = Fiji 1, b = Fiji 2, c = Fiji 3); (14) Guadalajara area, Mexico; (15) Mexico City area; (16) Izu Peninsula, Japan; (17) Asama, Japan; (18) Mount Rainier; (19) Mount Jefferson.

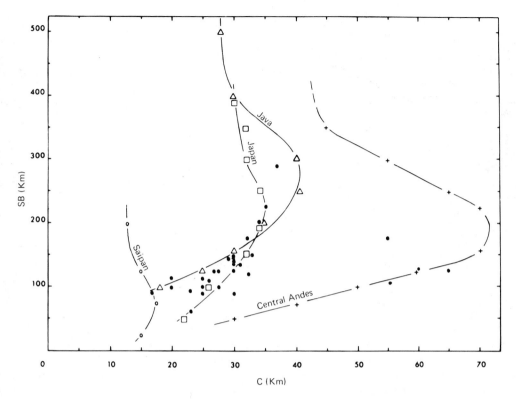

7.7. Depth to subduction zone (SB) versus crustal thickness (C) in the circum-Pacific region. Key: — = C-SB cross sections; ● = young volcanoes.

points define a broad band of increasing C and SZ roughly coincident with the positively correlated parts of the Java and Japan curves. Very few young volcanoes occur over subduction zones $\gtrsim 250$ km deep. These relationships indicate that in most areas a positive correlation does exist between crustal thickness and depth to subduction zone up to SZ $\cong 250$ km, and that silica-normalized K_2O values in volcanics from such areas should record both C and SZ as the geochemical data suggest. The central Andes trend is a major exception to the more normal trend defined by the Java-Japan curves.

The cause of a positive correlation between C and SZ up to subduction zone depths of 250 km is not clearly understood but probably results from crustal growth being controlled by magmatic processes in or near subduction zones. It is probable that one of the mechanisms by which the continental crust has grown throughout geologic time is by the over- and under-plating of magmas derived from subduction zones (see Chapter 11). Island arcs may evolve along paths on figure 7.7 leading roughly from the maximum in the Saipan curve (or slightly above) toward the maxima in the Java-Japan curves as growth proceeds. The anomalous crustal thickness beneath

the central Andes may result from special conditions where a continental plate overrides an oceanic plate. It is possible that continental or arc growth is controlled in part by the motion of plate boundaries at subduction zones. Such boundaries may migrate away from a continental or arc margin, with the crust above the subduction zone growing laterally as migration proceeds. Crustal thinning as subduction zone shallows could result from such a process. Crustal growth by arc-continent and arc-arc collisions (Chapter 11) may also be consistent with such a correlation.

INTRAPLATE ASSOCIATIONS

Marginal-Sea Basins

Low-K tholeiites, compositionally indistinguishable from ridge tholeiites, characterize the basins of large marginal seas (Hart et al., 1972). This striking similarity supports the recent magnetic and heat-flow data (Chapters 4 and 6) which suggest that marginal-sea basins result from secondary spreading centers and that ridge tholeiites are erupted at these spreading centers. In relatively small back-arc basins, tholeiites may be mixed with calc-alkaline volcanics from the arc proper.

Large Ocean Basins

Magma eruption within the oceanic crust is minor when compared to the vast volumes of magma extruded at spreading centers. It is manifest by volcanic islands and seamounts on the ocean floor, of which there appear to be two basic occurrences: volcanic chains and isolated volcanoes. Typical volcanic chains are the Hawaiian-Emperor, Tuamotu-Line, and Austral-Marshall-Gilbert chains in the Pacific basin (see fig. 8.18). Volcanic chains may result from rising mantle plumes (hotspots) as the oceanic lithosphere moves over them (Morgan, 1972b). Magmas produced in or above plumes are erupted and form seamount and island chains (see Chapter 8). The fact that such volcanoes continue to grow after they have crossed mantle plumes may result from the progressive tapping of a large lens-shaped magma chamber centered over the plumes, or from shallower magma chambers in the lithosphere that move with the volcanoes. Isolated volcanoes may be produced from small plumes that do not endure long enough to produce a volcanic chain or from occasional tapping of minor magma pockets in the upper mantle.

Magmas erupted from ocean-basin volcanoes are tholeiites or alkali basalts (or both). Many volcanic islands, such as the Hawaiian Islands, appear to be composed almost entirely of tholeiites with only minor alkali basalts, which are erupted only at the final stages of activity at specific volcanic centers. Volcanic centers such as Easter Island and Cobb seamount may be composed dominantly of alkali basalts. However, since the interior of these volcanoes have not been sampled, it is also possible that they are composed dominantly of tholeiites, with only cappings of alkali basalt. Dredging of samplings from seamounts and from beneath sea level on islands suggests that volcanic rocks from oceanic volcanoes differ from ridge volcanic rocks in being composed dominantly of *hyaloclastites*—i.e., deposits of volcanic debris composed chiefly of broken volcanic glass and pillow fragments (Bonatti, 1967). Such differences suggest that island-seamount magmas are considerably more viscous than rise magmas and readily fragment upon eruption into sea water.

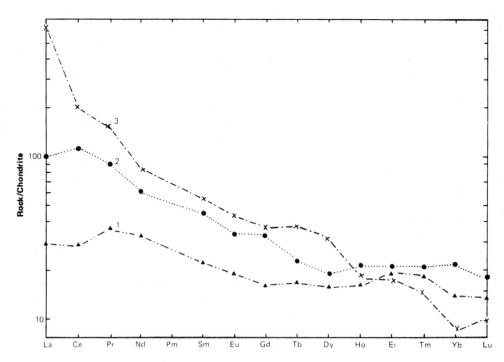

7.8. Chrondrite-normalized REE patterns of continental basalts (after Masuda, 1966; Frey *et al.*, 1968; Helmke and Haskin, (1973). (1) High-Al$_2$O$_2$ tholeiite; (2) cratonic rift tholeiite; (3) cratonic rift alkali basalt.

Tholeiites composing oceanic volcanoes are generally similar to continental tholeiites in composition (table 7.3). They are considerably more enriched in alkali and related elements and exhibit light REE enriched patterns when compared to ridge tholeiites (table 7.3, fig. 7.2). They also exhibit a larger range in $^{87}Sr/^{86}Sr$ ratios (0.702–0.706; fig. 7.3). oceanic alkali basalts have lower concentrations of most trace elements, lower Rb/Sr ratios, and higher K/Rb ratios than most continental alkali basalts. They exhibit enriched REE patterns, although not as enriched as continental alkali basalts (figs. 7.2 and 7.8) and their range of $^{87}Sr/^{86}Sr$ ratios is the same as oceanic island tholeiites (fig. 7.3).

Volcanic chains that appear to have formed as the oceanic lithosphere moved over mantle plumes (Chapter 8) commonly show a regular evolutionary history in terms of changing magma composition as exemplified by the Hawaiian Islands. In this area, the oldest rocks from a given volcano are olivine tholeiites, followed by an increasing abundance of Fe-rich quartz tholeiites, and terminating with minor volumes of alkali basalt (and its differentiates) (MacDonald and Katsura, 1964). Such a sequence of events is most readily interpreted in terms of a hotspot model, in which the early phases of magmatic activity reflect extensive melting in or above a mantle plume. As the volcano moves over the hotspot (still tapping the limbs of its magma chamber, however), less magma reaches the surface and shallow fractional crystallization becomes more prevalent, producing relatively more quartz tholeiites. As the volcano approaches the edge of the plume, isotherms have dropped and the depth and degree of melting decrease, resulting in the production of small volumes of alkali basalt (Morgan, 1972b). The fact that the $^{87}Sr/^{86}Sr$ ratios do not appreciably change over this evolutionary history in Hawaii indicates that the same mantle source has been tapped throughout (or sources of progressively greater depth yet with the same Rb/Sr ratios). Some island and seamount volcanoes may be composed entirely of alkali basalt and its derivatives, indicating a small and rather consistent degree of melting throughout their histories, which in turn may reflect rather small and/or deep mantle plumes.

Continental Rift Systems

Included in the category of continental rift systems are cratonic rifts that bisect stable cratonic areas such as the East African rift system, multiple rift systems such as the Basin and Range Province, and plateau (or flood) basalt regions such as the Columbia River Plateau and the Deccan Traps. All of the volcanic rocks found in these areas share a similar tectonic environment characterized by extensional stresses. Magmas are commonly erupted from small to intermediate-sized cinder cones or, in the case of plateau basalts, from extensive fissure systems. Only rarely are large stratovolcanoes developed (such as Kilimanjaro in Africa). Siliceous magmas usually erupt as domes, plugs, or minor flows, the exception being the voluminous flows of phonolite and trachyte in the East African rift system.

Rift magmas are characterized by the bimodal association or/and the tholeiite or alkali series (table 7.2). Because large volumes of magma may be erupted, large volcanic fields such as the Columbia River Plateau in the northwestern United States, the Ethiopia field in eastern Africa, and the Deccan Traps in India may be formed (Plate I). *Bimodal volcanic suites* are characterized by penecontemporary eruption of basaltic and felsic magmas with a sparsity of intermediate members. This contrasts to the calc-alkaline series, in which intermediate members commonly dominate, as is illustrated in fig. 7.9. In this figure, low values of the differentiation index are basalts and high values are felsic volcanic rocks. The volume of bimodal volcanic rocks can vary from minor, such as those found in the Basin and Range Province (≤ 20 m. years in age), to quite large, such as are found in Ethiopia and Iceland. Commonly the felsic members are less abundant than the mafic ones, a feature probably related to the greater viscosity of the former.

Compositionally, rift alkali basalts and tholeiites are highly enriched in LIL elements when compared to all other basalt types (table 7.3). Enrichments persist in volcanic members with higher SiO_2 contents (trachytes, phonolites, rhyolites) when compared to their counterparts in the calc-alkaline series. Very significant light REE enrichments characterize continental rift volcanic rocks (fig. 7.8) and $^{87}Sr/^{86}Sr$ ratios exhibit a large range extending from 0.703 to greater than 0.710 (fig. 7.3). There is a tendency in some rift systems for a zonation of volcanic types from the most SiO_2-saturated types in rift valleys to the least saturated on the limbs and flanks of the rift. Such a zonation, for example, is reported in the northern Rio Grande rift, where tholeiites dominate in the floor of the rift and alkali basalts around the edges (Lipman, 1969). Geochemical variation in the alkali series in going from alkali basalts (or nephelinites) to trachytes and phonolites is similar to that observed in the shoshonite subseries.

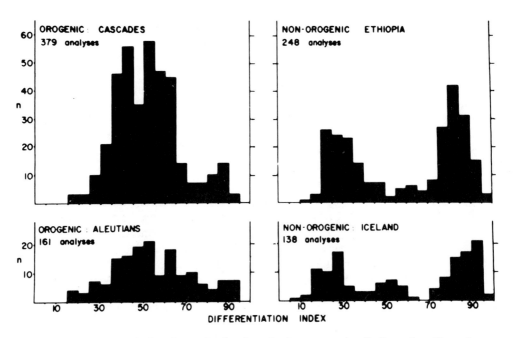

7.9. Histogram of chemical analyses of volcanic rocks from two calc-alkaline suites (Cascades, Aleutians) and two bimodal suites (Ethiopia, Iceland) (after Martin and Piwinskii, 1972). n = number of analyses.

Cratonic Areas

Igneous-rock occurrences in continental cratonic areas are rare. Where found, they occur as small intrusive complexes, dikes, sills, volcanic necks, pipes, or (rarely) as small volcanic fields. The rocks are typically highly alkalic even at low SiO_2 (< 50%) levels and may also be bimodal. They also show large enrichments in light REE and generally exhibit $^{87}Sr/^{86}Sr \gtrsim 0.705$. Kimberlites are of special significance in that they contain fragments of ultramafic rocks and eclogite from the upper mantle (Chapter 3) as well as diamonds and other high-pressure minerals that indicate depths of origin for kimberlitic magmas of ≥ 200 km.

Collision Zones

Collision zones are characterized by variable igneous rock associations with the relative abundances of rock types changing as a zone evolves (fig. 8.17). Just prior to collision, the overriding plate is characterized by calc-alkaline volcanism. As the crust thickens, calc-alkaline volcanism gradually gives way to bimodal volcanism and locally, alkali volcanism. Geochemical and isotopic features of bimodal volcanics associated with collisional zones are similar to continental rift volcanics and require the same constraints on their origin.

THE ORIGIN AND SOURCE OF MAGMAS

Seismic, heat flow, and experimental petrologic data indicate that mafic and intermediate magmas are generated in the upper mantle. Melting in magma source areas requires thermal energy of about 100 cal/gm of magma produced. This energy may be supplied directly from local concentrations of radioactive elements in the mantle, energy given off during earthquakes, or by upward convection currents or plumes from deeper in the mantle. In subduction zones, frictional heating along the seismic zone in the descending slab may be sufficient to cause melting. Since the melting point of silicates increases with load pressure, a decrease in pressure caused perhaps by faulting in a descending slab could reduce the effective melting temperature at a given depth, resulting in melting. Water depresses the melting point of most silicates and hence an increase in water content (caused, for instance, by the breakdown of hydrous minerals carried into the mantle in subduction zones) could cause melting. All of these mechanisms may operate in varying degrees in the generation of magmas. Heat brought up by convection beneath oceanic ridges is probably chiefly responsible for melting in these areas, whereas rising mantle plumes may be responsible for most volcanic activity within ocean basins. Melting in and/or above subduction zones is thought to result chiefly from water liberated by dehydration of descending oceanic crust, and perhaps from frictional heating along the top of the descending slab (McBirney, 1969b; Turcotte and Oxburgh, 1972b).

The composition of magmas produced in the mantle is dependent upon the composition of the source rock and the degree of melting. The degree of melting is, in turn, dependent upon the temperature, pressure, and water content of the source. Higher temperatures and water contents produce larger degrees of melting, whereas greater pressures result in smaller degrees of melting. Large degrees of melting of upper mantle rocks (ultramafic rocks, eclogites) produce basaltic (or ultramafic) magmas, and small degrees produce intermediate magmas. Melting can be described by two idealized processes that represent end-members of a continuum of melting conditions. *Equilibrium* (or *batch*) *melting* occurs when the melt and the residual solids remain in equilibrium until the melt is extracted from the source. *Fractional melting* occurs when magma is continually extracted from the source as melting proceeds such that equilibrium does not exist (except instantaneously) between melt and residual solids. More complicated melting models such as *dynamic melting* involve successive removal of melt increments as melting proceeds and trapping of small amounts of liquid in the source. The composition of magmas may be changed by such processes as fractional crystallization, wall-rock reaction, and crustal contamination. *Fractional crystallization* is the process by which crystals are removed from a crystallizing magma by gravitational sinking or floating (or some other process). The resulting liquids become increasingly SiO_2- and alkali-rich. *Wall-rock reaction* is the process by which incompatible elements (i.e., elements that do not readily substitute in mantle minerals), such as K, Rb, Ba, Sr, U, and Th, become preferentially enriched in a magma at or near the source by selective migration of these elements out of wall rocks. Wall-rock reaction could also occur at shallower depths provided the magmas were collected and remained in chambers at such depths for time periods long enough for significant wall-rock reaction to occur. *Crustal contamination* may occur when a magma passes through the crust and partially digests crustal rocks or when crustal rocks are involved in the generation of the magmas either in the upper parts of subduction zones (subducted sediments and oceanic crust) or in the deep roots of mountain chains.

Data from the fields of experimental petrology and geochemistry provide important boundary conditions for the origin and evolution of magmas. It is possible in the laboratory to reconstruct the P-T regimes in the Earth where magmas are produced. It is also possible to vary water content and oxygen fugacity and to use starting materials of compositions similar to those expected in the upper mantle and lower crust (Chapters 3 and 4). Experiments may begin with natural (or synthetic) upper-mantle mineral assemblages or with natural (or synthetic) igneous rocks. The upper and lower stabilities of minerals can be determined under a variety of conditions and the composition of melts coexisting with various minerals can be determined by analyzing glasses produced by rapid chilling. The results of experimental petrologic studies that represent closed systems must be carefully evaluated in terms of the open or partially open systems that exist in the Earth.

Results of existing experimental data for the origin of basaltic and calc-alkaline magmas by the partial melting of upper-mantle and lower-crustal rocks are summarized in fig. 7.10 and table 7.5. The composition of basaltic magmas produced by a given amount of melting (which corresponds to a specific amount of water in the melt) as a function of depth (at 0.1 percent H_2O) is given in fig. 7.10. Stability boundaries for residual phases are also indicated. The

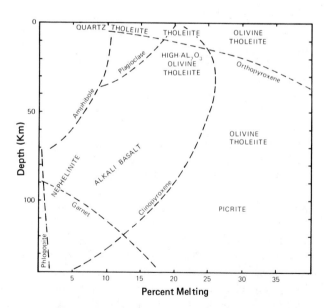

7.10. Generalized petrogenetic grid for basaltic magmas derived by melting of peridotite (of pyrolite composition) with 0.1 percent H_2O (modified after Green, 1972a). The dashed boundaries marked with a mineral name show that this mineral will occur as a residual phase after extraction of magma types to the left of the boundary. Olivine is present as a residual phase for all magma types in the figure.

Table 7.5. Summary of Experimental Data for the Production of Calc-Alkaline Magmas

Parent rock	Hydrous Conditions				Dry Conditions			
	Depth (km)	Degree of Melting			Depth (km)	Degree of Melting		
		10%	10–20%	>20%		10%	10–20%	>20%
Gabbro	≲75	Dacite	Andesite	Tholeiitic andesite or Tholeiite	15–35	Dacite	Andesite	Tholeiitic andesite or Tholeiite
Amphibolite	≲75					—	—	—	—
Eclogite	≳75				≳75	Andesite	Tholeiitic andesite	Tholeiite

*After Green and Ringwood (1967), Green and Ringwood (1968), Wyllie (1973).

composition of melts obtained by complete or partial melting of various source rocks for calc-alkaline magmas under wet and dry conditions is given in table 7.5.

Major advances in analytical, experimental, and theoretical geochemistry provide a basis for continually improving methods for studying the origin and source of magmas (Eggler, 1979; Frey 1979). It is now clear that the significant fractionation of trace elements observed in many igneous rocks cannot be the result of simple melting or fractional crystallization. It appears necessary to involve either or both partial melting of compositionally distinct sources or multistage melting and/or crystallization. REE distributions are particularly sensitive to certain mechanisms of magma evolution. Flat REE patterns, for instance, can be produced by complete melting of a rock with flat REE patterns or partial melting of a rock in which the phases that contribute substantially to the early melts have flat REE patterns.

Enriched light-REE patterns may be produced by such mechanisms as equilibrium melting, in which the melt equilibrates with residual phases that strongly reject light REE (such as garnet), or by fractional crystallization, in which such light-REE discriminating phases are removed from the magma. Negative Eu anomalies in igneous rocks may reflect removal of feldspar (which preferentially accepts Eu^{+2}) by fractional crystallization together perhaps with relatively reducing conditions such that the Eu^{+2}/Eu^{+3} ratio is high. Positive Eu anomalies would seem to be explained most readily by feldspar accumulation, and the absence of Eu anomalies by either the negligible role of feldspar in the origin of a particular magma or closed-system crystallization in which feldspar (en-

riched in Eu) and a phase depleted in Eu (perhaps pyroxene) crystallize in the correct ratios to cancel out any Eu anomaly. Existing data indicate that these examples of REE behavior are oversimplified and that complex models involving several minerals must be considered when evaluating the origin of magmas. Estimates of REE distribution coefficients have been instrumental in constructing such models.

Although it is possible to obtain $^{87}Sr/^{86}Sr$ ratios by directly analyzing modern volcanic rocks, initial $^{87}Sr/^{86}Sr$ ratios in older igneous rocks must be obtained from isochron intersections (described in Chapter 5). The initial $^{87}Sr/^{86}Sr$ ratio in an igneous rock is related to the Rb/Sr ratio in the source area (higher Rb/Sr ratios produce magmas with higher $^{87}Sr/^{86}Sr$ ratios), the degree of partial melting of the source if significant differences exist between the Rb/Sr ratios of residual minerals, and the amount of contamination with crustal materials which—because of their high Rb/Sr ratios—will enrich the magmas in radiogenic Sr (increasing the $^{87}Sr/^{86}Sr$ ratio) (Faure, 1977). Contamination with very young crust will not have this effect, since not enough time has passed for significant amounts of ^{87}Sr to accumulate. Each of the above variables must be considered in using the Sr isotope tracer. The distribution of Rb and Sr in mantle source areas is not well known, but analysis of ultramafic and mafic rocks and geochemical models of the Earth's composition allow some restraints to be placed on their distribution. Low and intermediate initial $^{87}Sr/^{86}Sr$ ratios in igneous rocks, for instance, are controlled by the Rb/Sr ratio in mantle source regions for moderate amounts of melting. High initial $^{87}Sr/^{86}Sr$ ratios ($\gtrsim .708$), on the other hand, are most readily accounted for by crustal contamination of mantle-derived magmas or a crustal origin. All members of a suite of igneous rocks that evolved from the same parent magma by fractional crystallization should have the same $^{87}Sr/^{86}Sr$ ratios since this process does not fractionate heavy isotopes. The $^{206}Pb/^{204}Pb$, $^{208}Pb/^{204}Pb$, and $^{143}Nd/^{144}Nd$ isotopic ratios can be used in a similar manner to monitor the U/Pb, Th/Pb, and Sm/Nd ratios of the source.

Oceanic Magmas

Experimental, geochemical, isotopic, and geophysical data are consistent with an origin for ridge tholeiites (and tholeiites formed at the axes of marginal basins) involving 10–30 percent melting of depleted mantle peridotite (Chapter 3) at depths of 60–70 km (Green, Hibberson, and Jaques, 1979). A residue composed of olivine and pyroxenes is left behind, forming a layer in the uppermost oceanic mantle of depleted peridotite as sea-floor spreading proceeds. Studies of magma temperatures and models of geothermal gradients beneath oceanic ridges indicate that most ridge tholeiites are probably generated at depths < 70 km. The low concentrations of alkali and related elements and rather high concentrations of Ni and Cr are accounted for in this model by calling upon a depleted mantle source area and by relatively large degrees of partial melting. The depleted light-REE patterns, high K/Rb ratios, and low $^{206}Pb/^{204}Pb$ and $^{87}Sr/^{86}Sr$ ratios (reflecting low Rb/Sr and U/Pb ratios in the source) are generally believed to support this concept of a depleted source (Gast, 1968; Kay, Hubbard, and Gast, 1970). This model suggests that source rocks have gone through one or more cycles of earlier melting in which light REE are preferentially extracted and Rb is removed relative to K and Sr (thus increasing the K/Rb and decreasing the Rb/Sr in the source). Many ridge tholeiites are Fe-rich quartz tholeiites that have probably been produced by shallow (≤ 15 km) fractional crystallization involving removal of Mg-rich olivine, pyroxenes, and plagioclase. A slight enrichment in LIL elements and light REE and the occurrence of small negative Eu anomalies in these rocks support such an origin.

Several mechanisms have been suggested for the origin of island-seamount tholeiites and alkali basalts: (a) fractional crystallization of rise tholeiite magma involving the upward movement of alkali elements by gaseous transfer, (b) wall-rock reaction of ridge tholeiite magma, (c) fractional crystallization of ridge tholeiite magma at depths ≥ 35 km, and (d) partial melting of an undepleted mantle source.

A fractional crystallization origin involving gaseous transfer is not supported by experimental petrologic data, which suggest that gaseous transport of elements in the system basalt-H_2O-CO_2 is unlikely (Holloway, 1971). The often high $^{87}Sr/^{86}Sr$ ratios in island basalts also argue against a common mantle source for ridge tholeiites and alkali basalts. Several lines of evidence do not favor (although neither do they completely eliminate) a wall-rock reaction origin for island tholeiites and alkali basalts. For instance, thermal model calculations indicate that there is no simple way to increase the temperature of large volumes of wall rock around a magma chamber, a requirement necessary to dissolve incompatible elements (Gast, 1968). Also, an adequate mechanism of mass transfer of elements from significant volumes of wall rock into magma chambers has not as yet been demonstrated. Deep fractional crystallization of ridge tholeiite magma is consistent with experimental data and may be consistent with Sr isotope data (assuming a less depleted source for alkali basalt magmas). It is, however, severely limited by trace-element distributions which necessitate the removal of unreasonably large amounts of olivine and pyroxenes by fractional crystallization to account for the very large enrichments of alkali and related elements in alkali basalts (Gast, 1968; Hart et al., 1970).

The fourth mechanism, which proposes that tholeiite and alkali basalts characterizing oceanic islands and seamounts are derived by partial melting of a peridotite mantle source less depleted than the mantle source from which ridge tholeiites are derived, appears to be consistent with most existing data (Hart and Brooks, 1981). A deepening of melting on the flanks of oceanic ridges is also compatible with probable temperature distributions in oceanic areas (see fig. 6.21), which indicate that geothermal gradients decrease away from ridges. This suggests that melting should occur at greater depths and smaller volumes of magma should reach the surface. The degree of melting would control the composition of magma produced (< 15% = alkali basalt; > 25% = island tholeiites). Smaller degrees of melting would explain the enrichment of LIL elements (including light REE) in alkali basalts, since these elements should strongly partition themselves into liquid phases at mantle depths. The absence of Eu anomalies in such basalts indicates that plagioclase did not play a major role in their production.

Examining further the composition of the sources of oceanic basalts, we see that Sr, Pb, and Nd isotopic ratios exhibit significant variation, indicating heterogeneous mantle sources (Hart and Brooks,

1981). Variations occur on a variety of scales ranging from intra-island variation to regional variations between islands and oceanic ridges. Ridge-type mantle, as mentioned above, is significantly depleted in LIL elements, whereas island-type mantle is not. Gradual transitions between these two types of mantle have been documented south of Iceland and near the Azores in the North Atlantic (fig. 7.11). These transitions have generally been interpreted as mixing of the two mantle sources (Hart et al., 1973; Schilling, 1973), although a simple two-component mixing model does not explain all geochemical data in detail. The distinction between these two sources is also exemplified by elements and element ratios such as K, Ti, P, REE, Ba, La/Sm, Ba/Sr, Rb/K, and Nb/Zr which progressively increase in basalts in going from oceanic ridge to islands. Such changes cannot readily be explained by differences in degree of melting or fractional crystallization. U/Pb and Rb/Sr isotopic studies indicate that the geochemical differences between ridge and island mantle have existed for 1 to 2 b.y. (Chapter 11)

Differences between Sr and Pb isotopic compositions of tholeiites and alkali basalts from the same islands cannot readily be explained in terms of source differences. Trace-element and isotopic results are best explained by varying degrees of melting of the same ultramafic source in which garnet or amphibole remain in the source. The geochemical differences between island and ridge basalts, however, seem to demand differences in mantle sources. Two general models have been suggested to explain the spatial arrangement of these sources in a convecting mantle as follows:

1. *The plume model* (Brooks, James, and Hart, 1976). In this model, ridge basalts are derived from shallow depleted asthenosphere, and island basalts from undepleted plumes which rise from depths below the asthenosphere (fig. 7.1). A major problem with this model is the fact that the ocean floor is covered with seamounts and islands, each of which would demand a plume (Hart and Brooks, 1981), thus straining the original plume model proposed by Morgan (1972b).

2. *The depleted lower-mantle model* (Tatsumoto, 1978). In this model, ridge basalts are derived from upwelling of depleted lower mantle and island basalts from an undepleted oceanic asthenosphere. This

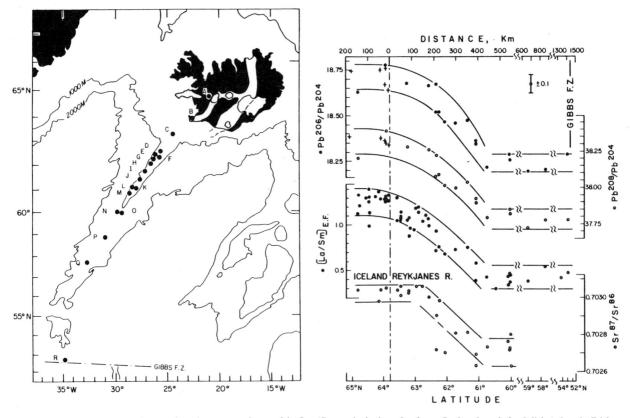

7.11. Variation in lead and strontium isotope ratios and in La/Sm ratio in basalts from Iceland and the Mid-Atlantic Ridge south of Iceland (from Sun *et al.*, 1975). [La/Sm]$_{E.F.}$ = chondrite-normalized La/Sm ratio.

model is faced with the problem that many (if not most) island basalts appear to be produced at hot-spots in the upper mantle which are fixed or move relatively slowly compared to plate motions (Chapter 8).

Subduction-Zone-Related Magmas

Several origins have been suggested for the production of subduction-zone related magmas, of which the most important are as follows:

1. Fractional crystallization of a high-Al_2O_3 tholeiite magma;
2. Partial or complete melting of sialic crustal rocks;
3. Partial melting of wet ultramafic rocks in the upper mantle, followed perhaps by fractional crystallization;
4. Mixing of mantle-derived basaltic magma with sialic crustal rocks or with felsic magmas; and
5. Partial melting of upper-mantle and crustal rocks in subduction zones.

Experimental, geochemical, and isotope data provide a quantitative basis to evaluate each of these origins (Green, 1980). Any origin for the calc-alkaline series must account for the compositional polarity exemplified by the gradation of the three magma subseries previously discussed.

Fractional Crystallization. Experimental data (Boettcher, 1973; Eggler and Burnham, 1973; Green and Ringwood, 1968) indicate that andesite and more siliceous members of the calc-alkaline series can be produced by fractional crystallization of high-Al_2O_3 thoeliite magma at depths of 35–100 km by removal of amphibole and clinopyroxene ($\pm$plagioclase, orthopyroxene) or at depths > 100 km by removal of garnet and clinopyroxene. Because of the instability of amphiboles at shallower depths, it is unlikely that basaltic magmas can fractionate to produce andesites at depths much less than 35 km. Geochemical variation of andesites erupted from Paricutin in Mexico from 1943 to 1952 show increases in SiO_2 and K_2O and decreases in MgO, FeO, Al_2O_3, and CaO (Wilcox, 1954). Such trends are broadly consistent with a progressive fractional crystallization model for the Paricutin magmas. Several significant arguments, however, have been presented against fractional crystallization as the major process by which calc-

alkaline magmas evolve:

1. With the exception of the arc-tholeiite sub-series, andesite is the most voluminous member of the calc-alkaline series, suggesting that it rather than tholeiite is the primary magma produced directly in the mantle.

2. Transition trace elements (Ni, Cr, Co, V) are not significantly different in concentration in andesites and many, if not most, high-Al_2O_3 tholeiites (Taylor et al., 1969). Since these elements are strongly enriched in the phases crystallizing from tholeiite (amphiboles, pyroxenes, olivine), they should be rapidly reduced in concentration in the derivative andesitic melts.

4. Fractional crystallization of tholeiite should produce enrichments in LIL elements in the remaining melts since these elements and especially Rb and Cs are strongly partitioned into the liquid phase (Taylor, 1969). Such enrichments and increases in element ratios are small when andesites are compared to high-Al_2O_3 tholeiites (table 7.3).

5. To produce the felsic members of the calc-alkaline series by continued fractional crystallization of andesite, plagioclase must be removed in significant amounts. The fact that Eu anomalies are negligible (and slightly positive when found) in calc-alkaline volcanic rocks greatly limits the amount plagioclase that can be removed during fractional crystallization.

Existing data seem to indicate that fractional crystallization is not the major process by which calc-alkaline magmas are produced. It probably does, however, play a minor role in the evolution of most calc-alkaline magmas.

Melting of Sialic Crust. Crustal melting models, which are consistent with the results of experimental studies, involve the production of felsic members of the calc-alkaline series by partial melting of continental (or arc) crust. Thermal models of the crust together with experimental data indicate that felsic magmas can be produced by hydrous melting in orogenic segments of the crust at depths $\gtrsim 20$ km (Wyllie, 1973). Temperatures in the crust, however, are probably not high enough to produce andesite magmas even under hydrous conditions. Models involving the melting of sialic crust do not seem adequate as a general explanation for the origin of all felsic members of the calc-alkaline series for the following reasons:

1. The models do not account for the production of felsic magmas in areas where little or no sialic crust exists (such as the Kuril Islands and outer part of the Aleutian chain).

2. It is commonly not possible to obtain even approximate agreement between the concentrations of many trace elements in modern calc-alkaline volcanic rocks and estimates of their concentrations in various models for the composition of the crust.

3. Sr and Pb isotope distributions in most calc-alkaline volcanic rocks indicate that they cannot represent remelted average sialic crust.

Perhaps the best candidates for a crustal melting origin are some of the large calc-alkaline batholiths and felsic ash-flow tuffs. The rather high initial Sr isotope ratios in the Sierra Nevada batholith ($Sr^{87}/Sr^{86} \cong 0.707$), for instance, seem to necessitate an origin involving either partial melting of the lower crust (with a rather low Rb/Sr ratio) or significant contamination of mafic magmas with sialic crustal material with typically high Rb/Sr ratios (Presnall and Bateman, 1973).

Partial Melting of Ultramafic Rocks. One possible origin for calc-alkaline magmas is by partial melting of peridotite either in a plunging lithospheric plate or above it. Experimental studies of small degrees of melting of peridotite under hydrous conditions has led to differing conclusions regarding whether or not calc-alkaline magmas can be produced in significant quantities by such a mechanism (Kushiro, 1972; Nicholls and Ringwood, 1972). Olivine must be a liquidus (or near-liquidus) phase for andesites to be produced by hydrous partial melting of peridotite. Experimental studies indicate that phase relationships are significantly affected by small changes in oxygen fugacity (Boettcher, 1973) and that olivine is a liquidus phase at oxygen fugacities measured in natural andesites and basalts ($f_{O_2} \cong 10^{-11}$ atm). Geothermal gradient and water content of the source, however, are also of critical importance in determining if significant amounts of calc-alkaline magmas can be produced at these oxygen fugacities. Provided the correct water content, f_{O_2}, and temperature exist, experimental data seem consistent with the formation of andesite magma by the partial melting of peridotite at depths < 60 km. This model is grossly consistent with some trace-element distributions and Sr isotope ratios (the relatively low $^{87}Sr/^{86}Sr$ values). One serious disadvantage of the model is the fact that it cannot explain the apparent continuum of magma types from the arc-tholeiite to the shoshonite subseries because experimental results seem to limit its role to depths $\lesssim 60$ km in the mantle.

Mixing Models. Mixing models for the origin of calc-alkaline magmas call upon the contamination of tholeiitic magmas produced in the mantle with varying amounts of sialic crustal materials or with felsic magmas (Eichelberger, 1975). If felsic magmas are called upon, they are derived by melting of crustal rocks or by fractional crystallization of tholeiitic magma. Some of the problems discussed in the sections dealing with the melting of sialic crust and fractional crystallization apply also to mixing models, and seem to limit their importance in most cases. For instance, when tholeiites are mixed with average sialic crustal rocks to produce andesites and dacites in terms of major-element composition, LIL trace elements are more abundant in the mixtures by a factor of two or more than observed in andesites or dacites (Taylor, 1969). Mixing of tholeiite with depleted siliceous granulites (which may exist in parts of the lower continental crust—see Chapter 4) can be manipulated to produce approximate balance for some trace elements. However, it seems doubtful that such high-grade metamorphic rocks exist in the lower crust in oceanic arc areas; they more likely characterize stable cratonic areas, where they represent the end result of craton-forming processes.

The rather high $^{87}Sr/^{86}Sr$ ratios in some high-K calc-alkaline rocks along continental margins seem to necessitate contamination with crustal materials (Whitford et al., 1977). Such contamination may occur in descending lithospheric slabs and involve continental sediments that have been subducted. Contamination may also occur in deep crustal roots of orogenic regions where melting temperatures of crustal rocks are exceeded, or above mantle diapirs arising from subduction zones that melt their way into the lower crust. Trace-element and isotopic distributions in Cenozoic arcs would seem to limit the contribution of continental sialic material in the magmas produced to less than 10 percent, however.

Melting in Descending Slabs. Experimental data indicate that calc-alkaline magmas can be produced by varying amounts of melting of amphibolite, gabbro (diabase, basalt), or eclogite under conditions ranging from dry to water-saturated (table 7.5) (Green and Ringwood, 1968; Wyllie, 1973). The only magmas that may not be able to be produced by this mechanism are the extremely felsic ones (viz., rhyolites). The mechanism involves two stages of magma generation, production of tholeiites at oceanic ridges and partial melting of these tholeiites in subduction zones (fig. 7.1). The ridge tholeiites, gabbros, and amphibolites in the oceanic crust descend into the mantle at subduction zones and invert to eclogite beginning at depths of 50–100 km. Varying but small amounts of deep-sea and continental sediments may be dragged down the subduction zone and contribute to the production of calc-alkaline magmas. This mechanism has the advantage of being able to account for most of the geochemical and isotopic distributions in calc-alkaline magmas as well as being consistent with experimental and geophysical data.

A Model for the Origin of Calc-Alkaline Magmas.

Some subduction-related volcanic rocks have Sr, Nd, and Pb isotopic ratios that lie within the range of ocean ridge or oceanic island basalts, while others have much higher ratios (fig. 7.3). Together with trace-element model studies, the isotopic data indicate a heterogeneous source for these magmas. Those island-arc basalts and andesites that exhibit consistently low isotopic ratios, such as the Tonga-Kermadec and Marianas arc systems, suggest depleted mantle sources (descending slab ± overlying mantle) similar to those of ocean-ridge basalts. Others clearly require subduction of continental sediments, illustrated by the Sunda-Banda arc northwest of Australia (fig. 7.12). In this arc, volcanic rocks west of Solor have typical ocean-ridge Sr isotope ratios, while those east of Solor show variably enriched ratios probably caused by the subduction (and contribution to melts) of continental sediments from Australia. Very large isotopic ratios characteristic of felsic volcanic rocks in some continental arc systems (like the Andes and western Japan) appear to reflect an undepleted continental lithosphere source or large amounts of crustal contamination either from subducted sediments or direct incorporation of continental crust into the magmas.

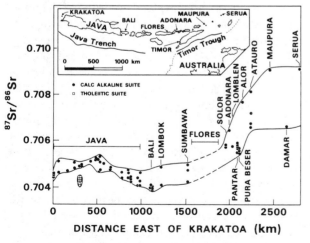

7.12. Strontium isotope variations along the Sunda-Banda Arc, Indonesia (after Whitford *et al.*, 1977 and S. R. Hart, 1981, personal commun.).

The most adequate model for the origin of calc-alkaline magmas is a complex model that allows some contribution from each of the models mentioned in previous sections. The model must have inherent depth-dependent processes to account for the compositional polarity of calc-alkaline magmas. Such a model is summarized below, as a function of increasing depth of magma generation in a subduction zone (after Fitton, 1971; Perfit et al., 1980; Whitford et al., 1979).

At depths ≤ 80 km, the distribution of geotherms in subduction zone areas suggests that temperatures are close to the wet liquidus of basalt and that rather large amounts of andesite (or tholeiite) melt are produced and erupted. This accounts for the large volumes of volcanic rocks at the volcanic front. Some of the magmas may also be produced by partial melting of peridotite above the descending plate caused by the upward escape of water from the plate. Fractional crystallization of the andesites (or tholeiites) and/or smaller degrees of melting of the basaltic parent may produce the less common felsic magmas of the arc-tholeiite subseries. The fact that altered ridge tholeiite is the dominant parent rock (in the form of basalt, gabbro, diabase, or amphibolite) being melted explains the similar trace-element (including REE patterns) and isotope distributions in arc and altered ridge tholeiites. REE and other LIL-element data, however, seem to necessitate some interaction of most melts with the overlying mantle wedge. Iron enrichment in arc tholeiites is best explained by shallow crystallization involving removal of Mg-rich olivine, pyroxenes, and plagioclase. Some young volcanic areas (as was mentioned above) allow more than 10 percent contribution to the magmas of subducted deep-sea and/or continental sediments based on trace element and Sr and Pb isotope distributions (Armstrong and Cooper, 1971; Whitford et al., 1977). Other areas (such as the Cascades and Tonga), however, allow at most only 1–2 percent contribution of such sediments (Church, 1973; Oversby and Ewart, 1972).

At depths of 80 km to about 150 km, amphibolite, basalt, and gabbro are partly or completely converted to eclogite, and much of the water is lost from the descending plate. The loss of water results in the production of less magma and of somewhat more alkali-rich magmas (the calc-alkaline subseries). Where local concentrations of water occur, dacite and andesite may be produced by partial melting of eclogite followed by interaction with garnet peridotite above the descending slab. The relative enrichment in LIL trace elements is accounted for by the smaller degree of melting and the slightly enriched REE patterns by equilibration of melts with residual garnet. The low K/Rb ratios may reflect an increasing importance of clinopyroxene ($\pm$ phlogopite) as the solidus phase whereas amphibole (with high K/Rb ratios) is the dominant solidus phase at shallower depths. The rather wide range in $^{87}Sr/^{86}Sr$ and other isotopic ratios seems to suggest a range in the amount of contamination of the magmas with sialic crustal materials. The highest $^{87}Sr/^{86}Sr$ values occur in members of the high-K group and may be produced by hot mantle diapirs rising from the subduction zone (fig. 7.1) and eventually melting their way to the base of the crust and partially melting the crust. The mixing of mafic magma in the diapir with siliceous rocks in the lower crust could produce, for instance, the voluminous high-K andesites found in the Tertiary fields of western North America. Felsic ash-flow (rhyolite) and granitic batholith magmas may also be produced by such diapiric melting, in which the upper parts of the diapirs contain more differentiated magmas and the degree of melting of crustal rocks is small.

At depths greater than 150 km most of the water in the descending plate has been lost, and consequently the amount of melting and eruption decrease. A small degree of melting now produces shoshonitic magmas very enriched in alkali and related elements. Perhaps such magmas acquire part of their trace-element enrichments from wall-rock reaction in the ultramafic wedge above the descending slab. The very enriched REE patterns suggest extensive equilibration of the magmas with residual garnet in the mantle. The range in $^{87}Sr/^{86}Sr$ ratios is still large and may reflect diapiric melting in the lower crust as described above for the high-K calc-alkaline group.

The compositional polarity so well exhibited in some modern subduction-zone-related volcanic rocks is most adequately explained by some combination of the following: (a) progressive changes in the dominant solidus (or near-solidus) phase from amphibole at shallow depths to clinopyroxene ($\pm$phologopite) at greater depths; (b) scavenging of LIL elements by aqueous fluids liberated by dehydration of descending slabs, with the scavenging path length increasing with slab depth; and (c) larger degrees of crustal contamination of magmas erupted over deeper parts of a descending plate.

Continental Rift and Craton Magmas

Many origins have been proposed for continental rift and craton magmas. Experimental petrologic, trace-element, and Sr isotope data indicate that no single origin is satisfactory for these magmas and that complex schemes must be devised to explain their origin and evolution, even at single volcanic centers. It seems clear from experimental data that the three most important mafic magmas in the alkali series— i.e., alkali basalt, picrite, and nephelinite—can be derived by partial melting of peridotite at depths > 35 km ($\geq$70 km for picrite and nephelinite) by various degrees of partial melting (up to 35 percent) (fig. 7.10). The origin of the great variety of magmas in the alkali series is generally explained by one or a combination of the following:

1. Fractional crystallization of a parent alkali basalt, picrite, or nephelinite magma;
2. Mixing of one or more of the above primary magmas with varying amounts and types of continental crustal rocks; and
3. Wall-rock reaction of one or more of the above three magma types in the upper mantle.

Experimental data indicate that fractional crystallization of alkali basalt, picrite, or nephelinite magmas can produce various members of the alkali series by removal of varying amounts of olivine, pyroxenes, and garnet at depths > 35 km (Bultitude and Green, 1967; Green and Ringwood, 1967). Such fractional crystallization should enrich the residual melts in elements as observed in the alkali series. The highly light-REE enriched patterns may reflect garnet removal. Model calculations of the degree of enrichment, however, indicate that fractionation is not adequate in all cases to explain the extreme degree of enrichment of alkali and related elements in some members of the alkali series. The variable, often high $^{87}Sr/^{86}Sr$ ratios ($\gtrsim 0.710$) also seem to indicate that fractional crystallization alone is not an adequate mechanism to account for many alkali volcanic rocks.

Complex mixing theories for the origin of the alkali series have been proposed involving partial digestion of average crustal rocks, granite, limestone, and other specific rock types by alkali basalt, picrite, or nephelinite parent magmas. Such models are appealing for alkali volcanic series in which the Sr concentration in the lavas is negatively correlated with $^{87}Sr/^{86}Sr$ ratio—such as suites in eastern Uganda (Bell and Powell, 1970). In other volcanic fields with high $^{87}Sr/^{86}Sr$ ratios, however, the Sr concentration is also high (> 1000 ppm), and it is unlikely that crustal rocks (with Sr = 300–400 ppm) have significantly contributed to the magmas.

Wall-rock reaction of picrite, alkali basalt, nephelinite, or ultramafic parent magmas may enrich these magmas sufficiently in LIL elements to produce the observed concentrations of these elements in members of the alkali series. This mechanism, if important, would appear, however, to operate in conjunction with varying degrees of melting and/or fractional crystallization in order to produce the variety of compositions typical of the alkali series. The very light-REE enriched patterns in alkali rocks are most adequately explained by equilibration of the magmas with residual garnet during (or before) wall-rock reaction. For a wall-rock reaction model, the significant range in $^{87}Sr/^{86}Sr$ ratios in alkali volcanic rocks may be explained by one or some combination of (a) lateral and vertical variations in the Rb/Sr ratio in the mantle source areas beneath continents, or (b) wall-rock reaction involving varying degrees of solution of phases enriched in Rb (like phlogopite and amphibole) and hence also in ^{87}Sr.

Basalts from cratonic and continental rift settings exhibit a wide range of Sr, Nd, and Pb isotopic values indicating a heterogeneous source. Isotopic ratios and LIL-element concentrations are too high in most mafic volcanics, however, to be derived from depleted oceanic ridge sources. Existing data favor one of two sources: (1) an undepleted subcontinental lithosphere or a depleted subcontinental lithosphere that has been enriched in LIL elements by metasomatic fluids, or (2) a depleted mantle source followed by varying amounts of crustal contamination.

SUMMARY STATEMENTS

1. Major sites of magma generation are at convergent (subduction zones) and divergent (oceanic ridges) plate boundaries and appear to be related to compressive and extensional stresses, respectively. Lesser quantities of magma are generated in intraplate environments such as ocean basins, continental rift systems, and cratons.

2. Volcanism at oceanic ridges and in marginal seas is characterized by low-K tholeiite (ridge tholeiite). Such magmas are injected along the axial zones of ridges as sea-floor spreading proceeds. Ridge tholeiites are characterized by low concentrations of LIL elements, depleted light-REE patterns, and low $^{87}Sr/^{86}Sr$ ratios.

3. Magmas associated with subduction zones belong to the calc-alkaline series. At a given SiO_2 level in subduction-zone-related igneous rocks, LIL elements increase as a function of depth to subduction zone and (in most areas) crustal thickness, giving rise to a compositional polarity. This polarity is also commonly manifest by a change from flat to light-REE enriched patterns, a decrease in the K/Rb ratio, an increase in the range of $^{87}Sr/^{86}Sr$ ratios in volcanic rocks, a decrease in the amount of magma reaching the surface, and, in some cases, a zonation of economic mineral deposits.

4. Three completely gradational subseries can be defined from compositional polarity above modern subduction zones. The arc-tholeiite subseries (closest to the trench) is characterized by large volumes of tholeiite similar in composition to ridge and altered ridge tholeiites; the calc-alkaline subseries (*sensu strictu*) is characterized by having andesite as the most abundant volcanic rock and granodiorite as the most abundant plutonic rock; and the shoshonite subseries (farthest from the trench) is characterized by small volumes of high-alkali igneous rocks.

5. Ocean-basin volcanism forms islands and seamounts either as volcanic chains or as isolated volcanoes and is characterized by either tholeiites and alkali basalts (and their derivatives) or both. Volcanic chains appear to form as the oceanic lithosphere moves over mantle plumes (hotspots). Seamount and island volcanic rocks are characterized by moderate to high concentrations of LIL trace elements, light REE enriched patterns, and variable but generally low Sr and Pb isotopic ratios.

6. Continental rift systems are characterized by bimodal volcanic associations. Mafic magmas are most commonly erupted from small volcanoes or fissure systems. Igneous rocks in cratonic areas are rare and when found generally occur as small intrusive complexes or volcanic necks. Rift and cratonic magmas are characterized by variable but significant enrichment in LIL elements, light-REE enriched patterns, and a wide range in Sr and Pb isotopic ratios.

7. Heat brought up by convection or mantle plumes is probably responsible for melting of upper-

mantle rocks in oceanic areas. Melting in subduction zones probably results chiefly from water liberated from the oceanic lithosphere as it descends into the mantle, and perhaps from frictional heating of the descending slab.

8. The diversity of magma compositions observed on the crust may be produced by some combination of the following: partial melting of ultramafic (or other) rocks in the upper mantle; fractional crystallization of basaltic or intermediate magmas; wall-rock reaction at or near the sites of magma production; crustal contamination during the ascent of mantle-derived magmas through the crust; and varying degrees of melting of crustal materials either in subduction zones or in the deeper levels of orogenic crustal areas.

9. Most data are consistent with the following model for the origin of oceanic magmas. Olivine tholeiites at oceanic ridges are produced at shallow depths (60–70 km) by large degrees (10–30 percent) of melting of depleted peridotite; shallow fractional crystallization of these magmas produces quartz tholeiites with Fe enrichment. Island-seamount tholeiites and alkali basalts are produced beneath ocean basins at depths > 35 km (over mantle hotspots) by variable degrees of melting of undepleted peridotite; fractional crystallization of such magmas produces more differentiated magmas.

10. The production of calc-alkaline magmas at convergent plate boundaries appears to chiefly involve varying degrees of hydrous melting of subducted rocks followed by interaction with ultramafic rocks in the overlying mantle wedge and with sialic crust.

11. Compositional polarity of subduction-zone-related magmas probably results from a combination of the following: (a) progressive changes in the solidus phase, from amphibole at shallow depths to clinopyroxene ($\pm$phlogopite) at great-

er depths, (b) scavenging of LIL elements by aqueous fluids liberated by dehydration of descending slabs, with the scavenging path length increasing with slab depth, and (c) larger degrees of crustal contamination of magmas erupted over deeper parts of the descending plate.

12. Continental-rift and cratonic magma generation involves varying degrees of melting of upper-mantle peridotites followed by some combination of fractional crystallization, crustal contamination, and wall-rock reaction.

13. Sr, Pb, and Nd isotopic data and LIL-element distributions require inhomogeneous mantle sources for mafic magmas. Ocean-ridge tholeiites and some island-arc basalts are derived from depleted mantle and island basalts from undepleted mantle. Most subduction-related magmas require contributions from the descending slab, the overlying mantle wedge, and, in some cases, the continent (either directly or as subducted sediments). Cratonic and continental rift basalts are produced from undepleted or enriched mantle lithosphere or/and crustal contamination.

SUGGESTIONS FOR FURTHER READING

Cox, K.G., Bell, J.D., and Pankhurst, R.J. (1979) *The Interpretation of Igneous Rocks*. London: George Allen & Unwin. 450 pp.

Fitton, J.G. (1971) The generation of magmas in island arcs. *Earth and Planet. Sci. letters*, **11**, 63–67.

Hart, S.R., and Brooks, C. (1981) Sources of terrestrial basalts: An isotopic point of view. In *Basaltic Volcanism of the Terrestrial Planets*, Ch. 7. Houston: Lunar and Planetary Inst.

Wyllie, P.J. (1971) *The Dynamic Earth*. New York: Wiley. 416 pp.

Chapter 8

Plate Tectonics and Continental Drift

Although continental drift was first suggested in the 17th century, it did not receive serious scientific investigation until the beginning of the 20th century. Wegener (1912) is usually considered the first one to have formulated the theory precisely. In particular, he pointed out the close match of opposite coastlines of continents and the regional extent of the Permo-Carboniferous glaciation in the Southern Hemisphere. DuToit (1937) was the first to propose an accurate fit for the continents based on geological evidence. Continental drift, however, did not receive wide acceptance among geoscientists, and especially among Northern-Hemisphere geoscientists, until the last two decades. The chief reasons, aptly summarized by MacDonald (1964), related to lack of an acceptable mechanism and to geophysical data that indicated that continents have roots in the mantle (see Chapters 3 and 4) and hence could not be moved at the Moho, as was commonly proposed. The breakthrough came in the early 1960s, when the sea-floor spreading theory became widely accepted. Sea-floor spreading offers an explanation for how the continents may drift and yet retain their deep roots, by drifting at the base of the lithosphere, not at the Moho.

It is possible, using a variety of geological and geophysical data, to reconstruct the positions of the continents prior to their last breakup about 200 million years ago. Matching of continental borders, stratigraphic sections, and fossil assemblages were some of the earliest methods used to reconstruct continental positions. Today, in addition to these methods, we have polar wandering paths, sea-floor spreading directions, petrotectonic assemblages, and hotspot traces that may be used to reconstruct continental

and, more generally, plate positions in the geologic past.

THE EXPANDING EARTH HYPOTHESIS

An alternate explanation for some of the evidence of continental drift is that the Earth has expanded, with new surface area being created at oceanic ridges (Carey, 1976). This hypothesis generally presumes that the area of the continents has remained about constant and that the ocean basins have grown as the Earth expands. An expanding Earth and sea-floor spreading are not mutually exclusive processes, and the overwhelming evidence for sea-floor spreading and continental drift seems to necessitate that, if the Earth expanded, these processes have gone on concurrently.

Evidences for an expanding Earth are either ambiguous or are based on tenuous and ad hoc assumptions. One of the earliest evidences cited is the idea that the continents have become progressively emergent with time necessitating an increase in the Earth's radius of about 0.5 mm/yr (Egyed, 1956). Egyed's results, based on data collected from paleogeographic maps, suggest that the percentage of the continents flooded by shallow seas has decreased with time. This is interpreted by some geoscientists to indicate that the Earth is expanding and causing sea water to move progressively off the continents into the growing ocean basins. Egyed's results, however, have several sources of inherent error which render their interpretation uncertain. First, Egyed ignores the fact that the continents are only 88 percent (and not 100 percent) emergent today, and that some sea water is tied up in polar ice caps. Also, determining the emergent area of continents by scaling data from paleogeographic maps has many sources of error, including inadequate maps for most continents. Recent estimates of the areas covered by sea water as a function of time in North America (for which the most accurate paleogeographic maps are available) indicate a rather constant relationship between average elevation of the continents and sea level since the Cambrian (Wise, 1973). It has also been shown, as will be discussed later in this chapter, that the total volume of oceanic rises, which has probably changed with time, can be the most important factor in controlling flooding of the continents. Existing data are not sufficient to document progressive emergence (or submergence) of the continents with time.

Carey (1958, 1976) has discussed a variety of geometric arguments which to him seem to necessitate an expanding Earth. For instance, large gaps are left on some reconstructions of the Permian supercontinent, which disappear on a globe of smaller radius. Such problems, however, may or may not be real, in that other investigators have managed to obtain acceptable reconstructions (fig. 9.7). Glikson (1980) has suggested that the lack of evidence for extensive oceanic crust in the Proterozoic (i.e., the near absence of ophiolites) can best be explained by having an Earth with a smaller radius and Proterozoic supercontinents covering nearly the entire surface.

Major obstacles to an expanding Earth, however, seem to outnumber evidences in favor of expansion (McElhinny et al., 1978; Schmidt and Clark, 1980). New estimates of the radius of the Earth during the last 400 m.y. from paleomagnetic results limit the amount of expansion over this time to <0.8 percent. Also, crater distributions and ages on the lunar surface rule out any significant expansion of the moon (<0.06 percent) during the last 4 b.y. and a similar conclusion appears to apply to Mars, Mercury, and Venus. By analogy, the Earth would not be expected to have undergone significant expansion during the same time period. Biogeographic distributions of invertebrates during the early Paleozoic are also incompatible with an expanding Earth, in that they demand large expanses of oceanic crust between continental blocks (Burrett and Richardson, 1980). It has also not as yet been possible to find a satisfactory mechanism for planetary expansion. Results from the moon seem to limit a decrease in the gravitational constant over the last 4 b.y. to $<2 \times 10^{-10}$ yr^{-1}, which eliminates one of the more popular causes

suggested for expansion. Considering all sources of data, it would appear that the Earth has not expanded more than one percent since its formation 4.6 b.y. ago.

METHODS OF PLATE RECONSTRUCTION

Geometric Matching of Continental Borders

The geometric matching of continental shorelines was one of the first methods used in reconstructing continental positions. Both Wegener (1912) and DuToit (1937) employed this method in their early investigations. Later studies have shown that matching of the continents at the edge of the continental shelf or continental slope results, as it should, in much better fits than the matching of shorelines, which reflect chiefly the flooded geometry of continental margins. The use of computers in matching continental borders has resulted in more accurate and objective fits. As an example, fig. 8.1 shows a computerized, least-square fit of continents on both sides of the Atlantic Ocean at 500 fathoms depth. As indicated, areas of overlap and gaps are minimal. Many areas of overlap can be explained by additions to the edges of the continents after their rupture. Examples are the Niger Delta and the carbonate reefs of the Bahama Bank and Blake Plateau. A computerized fit, also at 500 fathoms, of Australia, Antarctica, India, and Madagascar to Africa and South America is shown in fig. 8.2. Although the fit of Australia to Antarctica seems quite certain based on geologic evidence, the geometric fits of Antarctica, Africa, India, and Madagascar that are shown are not unique and are, at the present time, a subject of some discussion (see Chapter 9).

Matching of Stratigraphic Sections and Crustal Provinces

Similarities in age and lithology of stratigraphic sections on opposite continental margins is another feature used by early investigators to support continental drift. One of the most striking and earliest-described similarities is that of the late Paleozoic (Gondwana) sections on continents in the Southern Hemisphere. In a general way, the late Precambrian-Paleozoic terranes define a continuous belt from eastern Australia through Antarctica into southern Africa and Argentina when the continents in the Southern Hemisphere are considered in their predrift positions

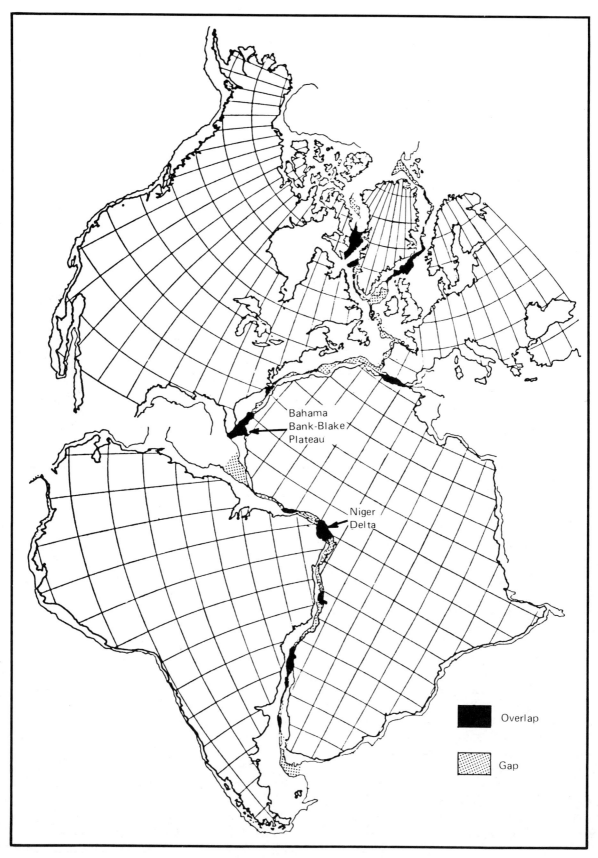

8.1. Computerized least square fit at 500 fathoms of the continents around the Atlantic Ocean (from Bullard *et al.*, 1965).

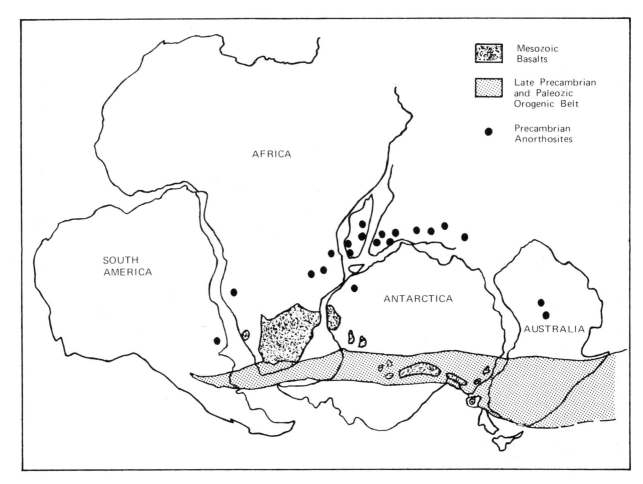

8.2. Computerized geometric fit at 500 fathoms of continents in the Southern Hemisphere (after Smith and Hallam, 1970).

(fig. 8.2). A belt of Triassic and Jurassic basalts and diabases of similar composition, which extends from South Africa through Antarctica into Tasmania, and a clustering of Precambrian anorthosites in East Africa, Madagascar, and India also support the overall fit of the continents shown in fig. 8.2.

Extensive radiometric dating of Precambrian terranes bordering the Atlantic Ocean have also provided strong evidence for continental drift. The fit of South America to Africa suggested by the matching of crustal provinces and structural trends (fig. 8.3) is very close to the 500 fathom, computerized fit of these continents shown in fig. 8.1. Most striking is the continuation of the Pan-African provinces from west-central Africa into eastern South America. Also, the > 2.0 b.-year-old provinces in western Africa and in the Congo area continue into Brazil. Precambrian provinces around the North Atlantic are considerably more complex and appear to involve extensive reworking of older terranes. A generalized distribution

of crustal provinces is shown on a predrift reconstruction of the continents in fig. 5.5; for the most part, it is consistent with the continental reconstruction shown.

Paleontologic Evidence

Modern ocean basins are effective barriers not only to terrestrial organisms (animal and plant) but also to most marine organisms. Larval forms of marine invertebrates, for instance, can only survive several weeks, which means they can travel only 2000–3000 km with modern ocean-current velocities. Hence, the geographic distribution of fossil organisms provides an important constraint on the sizes of oceans between continents in the geologic past. When continents break up and begin to drift apart, organisms that cannot readily cross the intervening oceans may evolve into many diverse and specialized groups on the various continental fragments. On the other hand,

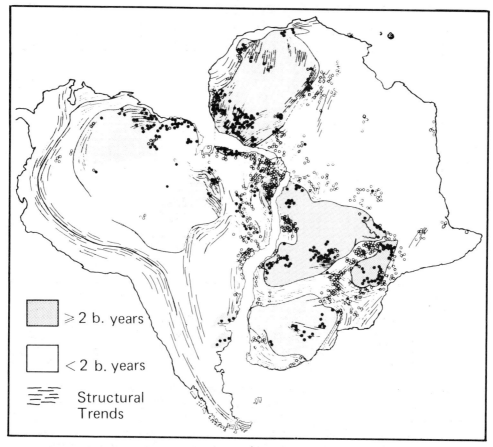

8.3. Predrift reconstruction of Africa and South America based on matching of crustal
provinces (Hurley, 1968c). Each point represents a radiometric age. Key: ● = ages > 2.0 b.
years; o = ages < 2.0 b. years.

when continents come together, groups compete for
survival, resulting in the extinction of some of the
least fit groups. It is possible to relate the origin of
some specialized groups and extinction of others to
the breakup and collision of continents (Hallam,
1974; Valentine, 1973). For example, the diversity of
invertebrates seems to correlate with continental fragmentation and assembly as deduced from other plate
reconstruction methods (Fallow, 1977; Valentine and
Moores, 1972). This is illustrated in fig. 8.4, where the
number of families of nine major invertebrate phyla
is plotted as a function of time. The breakup and
reassembly of the continents are diagrammatically
illustrated beneath the plot. Data suggest that a large
supercontinent (Pangaea—see Chapter 9) broke up
between the late Precambrian and the Ordovician,
giving rise to four or more continents with intervening oceans (labeled 1–4 in fig. 8.4). Between the
Ordovician and Permian these continents reassembled, only to fragment again in the Triassic. Con-

tinued fragmentation in the Cenozoic produced the
existing continents. An increase in the diversity of
invertebrates correlates well with increasing continental fragmentation and a decrease in diversity (caused
by extinctions) with reassembly of continental fragments.

The biogeographic distribution of Cambrian trilobites indicates the existence of several continents
separated by major ocean basins during the early
Paleozoic (fig. 8.5). Major faunal province boundaries
commonly correlate with suture zones between continental blocks brought together by later collisions.
Wide oceans are implied in the Cambrian between
North America and Europe, Siberia and Europe, and
China and Siberia. Minor faunal provincialism within
individual blocks probably reflects climatic differences. Studies of early Ordovician brachiopods
also indicate they belonged to at least five distinct
geographic provinces, which was reduced to three
provinces during the Europe–North American colli-

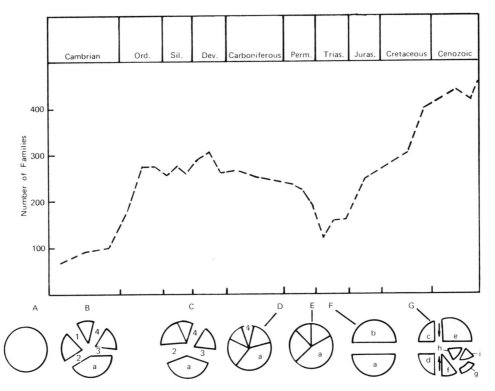

8.4. Correlation of diversity of invertebrates with time and with patterns of continental fragmentation and assembly (after Valentine and Moores, 1970). Events: (A) Pangaea; (B) fragmentation of Pangaea to produce pre-Caledonian (1), pre-Appalachian (2), pre-Hercynian (3), and pre-Uralian (4), oceans; (C) Suturing of Caledonian-Acadian orogenic systems; (D) Suturing of Appalachian and Hercynian systems; (E) Suturing of Urals—reassembly of Pangaea; (F) Opening of the Tethys Sea; (G) Closing of Tethys Sea, reopening of Atlantic, fragmentation of Gondwanaland. Continents: (a) = Gondwanaland; (b) = Laurasia; (c) = North America; (d) = South America; (e) = Eurasia; (f) = Africa; (g) = Antarctica; (h) = India; (i) = Australia.

sion in late Ordovician time. The opening of the North Atlantic basin in the Cretaceous resulted in the development of American and Eurasian invertebrate groups from an originally homogeneous Tethyan group. Also, ammonites from East Africa, Madagascar, and India indicate that only shallow seas existed between these areas during the Jurassic.

The similarity of mammals and reptiles in the Northern and Southern hemispheres prior to 200 m.y. ago demands land connections between the two hemispheres (Hallam, 1973). Early separation of Africa, India, and Australia (early to middle Mesozoic) led to the evolution of unique groups of mammals (e.g., the marsupials) in the Southern Hemisphere, while the fact that North America and Eurasia were not completely separated until early Tertiary accounts for the overall similarity of Northern Hemisphere mammals today. When Africa, India, and Australia collided with Eurasia in the mid-Ter-

tiary, mammalian and reptilian orders spread both ways, and competition for the same ecological niches was keen. Such competition led to the extinction of 13 orders of mammals.

Plant distributions are also sensitive to continental drift. The most famous are the *Glossoptera* and *Gangemoptera* flora in the Southern Hemisphere (Plumstead, 1973). These groups range from Carboniferous to Triassic in age and occur on all continents in the Southern Hemisphere and in northeastern China, suggesting that these continents (including at least part of China) were connected during this period of time (fig. 8.6). The complex speciation of these groups could not have evolved independently on separate land masses. The general coincidence of late Carboniferous and early Permian ice sheets and the *Glossopteris* flora (fig. 8.6) appears to reflect the fact that *Glossopteris* adapted to relatively temperate climates and rapidly spread over the high latitudes

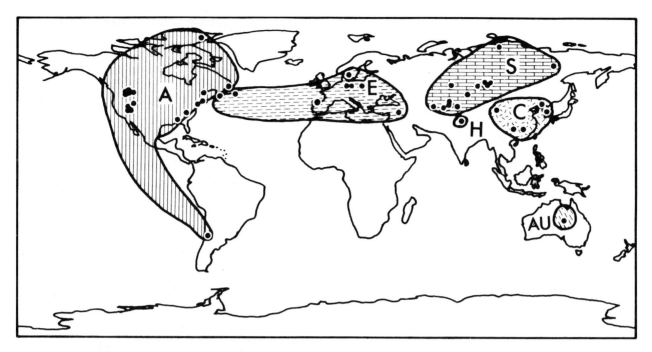

8.5. Trilobite biogeographic regions during the *Evinia-Elrathina* realms of the middle Cambrian (after Burrett and Richardson, 1980). A = American; E = European; S = Siberian; H = Himalayan; C = Chinese; AU = Australian.

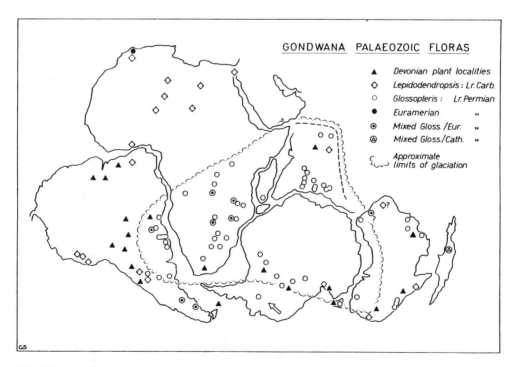

8.6. Paleozoic flora distribution in Gondwana (after Chaloner and Lacey, 1973). Gondwana reconstruction after Smith and Hallam (1970).

during the Permian. The breakup of Africa and South America is reflected by the present distributions of the rain forest tree *Symphonia globulifera* and a semiarid leguminous herb *Teramnus uncinatus*, which occur at approximately the same latitudes on both sides of the South Atlantic (Melville, 1973). Conifer distribution in the Southern Hemisphere reflects continental breakup, as is evidenced by the evolution of specialized groups on the various continents after continent dispersal in the early and middle Mesozoic.

Although a correlation between the rate of taxonomic change of organisms and plate tectonics seems to be well established, the actual causes of such changes are less well agreed-upon. At least three different factors have been suggested to explain rapid increases in the diversity of organisms during the Phanerozoic (Hallam, 1973):

1. The increase in the areal distribution of a particular environment results in a widespread distribution of various ecological niches in which a large number of organisms become established. For instance, an increasing diversification of marine invertebrates in the upper Cretaceous may reflect increasing transgression of continents, resulting in an areal increase of the shallow-sea environment in which most marine invertebrates live. Such transgression appears to have resulted from progressive displacement of sea water onto the continents by active oceanic ridge systems.

2. As previously mentioned, continental fragmentation leads to morphological divergence because of genetic isolation.

3. Some environments are more stable than others with regard to physical factors such as temperature, rainfall, and salinity. Studies of modern organisms indicate that stable environments lead to intensive partitioning of organisms into well-established niches and to correspondingly high degrees of diversity. Environment instability, decreases in environmental area, and competition of various groups of organisms for the same ecological niche can cause extinctions. Although continental collisions would seem to lead to these types of changes, other factors may also be important.

Paleoclimatic Evidence

Paleoclimatic evidence has been used to both support and negate continental drift. The fact that such evidence can be used to argue either way is related in part to ambiguities in the interpretation of paleoclimatic indicators, and in part to the lack of sensitivity of some climates to plate motions. Besides being latitude-dependent, the distribution of climates today is controlled by a complex interaction of wind currents, ocean currents, and topographic barriers, none of which are well known in the geologic past. Although climatic belts on the Earth today roughly parallel lines of latitude, many microclimatic regions cross-cut such lines. Hence, for ancient climates to be used to reconstruct continental land masses, only widespread paleoclimatic indicators of a given age should be employed (Frakes, 1979). Even then, paleoclimatic data are not sensitive to longitudinal motions of continents, such as those characterizing the opening of the Atlantic Ocean during the last 200 m.y.

Some paleoclimates have been inferred from rock associations that do not unambiguously reflect a particular climate. Some sediments reflect primarily precipitation regimes rather than temperature regimes (Robinson, 1973). Coal, for instance, requires abundant vegetation and a good supply of water but can form at various temperature regimes, excepting arid hot or arid cold extremes. Aeolian sand dunes may form in cold or hot arid (or semi-arid) environments (as exemplified by modern dunes in Mongolia and the Sahara, respectively). Evaporites may form in both hot and cold arid environments, although they are far more extensive in the former. Red beds may form in a variety of temperature and precipitation regimes. Laterites and bauxites seem to unambiguously reflect hot, humid climates. There have been false reports of continental glaciation based on erroneous identification of glacial deposits or on glacial deposits that reflect only localized glacial activity (Fairbridge, 1973). It is now recognized that subaqueous (and some subareal) slump, mudflow, and landslide deposits can be mistaken for glacial moraines. No single criterion should be accepted in the identification of continental glacial deposits. Only a convergence of evidence from widespread locations such as tillities, glacial pavement, faceted and striated boulders, and glacial dropstones should be considered satisfactory. The distribution of various organisms is also sensitive to climate and can be used to reconstruct ancient climatic belts. Modern hermatypic hexacorals, for instance, are limited to warm surface waters (18–25°C) between latitude 38°N and 30°S (fig. 8.7). Late Jurassic hexacorals, plotted on the same reconstruction display a different distribution displaced northwards about 35 degrees. Because paleomagnetic and other data indicate the continents during the late Jurassic were at about the same latitude as at present, the coral distributions indicate a warmer ocean (and climate) with sub-

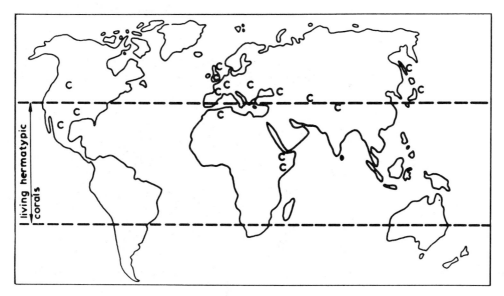

8.7. Distribution of modern and Jurassic hermatypic corals (Frakes, 1979). C = Jurassic assemblages.

tropical conditions extending to high latitudes. It is also possible to estimate paleoclimatic temperatures with oxygen isotopes in carbonates and cherts, provided the rocks have not undergone isotopic exchange since their deposition and provided several other requirements are met (Frakes, 1979).

It has been suggested by Meyerhoff (1970) that major evaporite deposits ranging in age from late Precambrian through Miocene occur in belts that today receive less than 100 cm of annual rainfall. From this observation, Meyerhoff concludes that planetary wind-circulation patterns have not changed since late Precambrian time nor have continents moved. Others, however, have interpreted the same data (together with other paleoclimatic data) as indicating that the paleoequators during most of the Phanerozoic passed through North America and Europe (Ahmad, 1973). Meyerhoff has further pointed out that if one or two supercontinents existed in the Permian, the central parts of these continents would be vast deserts due to their great distances from the sea, and the lack of evidence for such deserts would, therefore, seem not to favor the existence of such supercontinents. However, Meyerhoff overlooks the fact that these supercontinents were partially flooded with shallow seas and that an increase in the ratio of the area of oceans to continents moderates and stabilizes climate, due to the large heat capacity of water. Many other paleoclimatic arguments have been advanced by Meyerhoff and his colleagues to favor continental stability. Most or all of these, however,

can also be adequately interpreted in terms of plate tectonics. The vast Permo-Carboniferous glacial deposits in the Southern Hemisphere, for instance, give strong support to the existence of a southern supercontinent centered over the South Pole at this time (fig. 8.6). Analysis of modern wind circulation patterns in terms of climatic belts applied to Permian supercontinents also predicts remarkably well the climatic belts inferred from the distribution of aeolian sands, evaporites, coals, and glacial deposits of this age (Robinson, 1973).

An example of inferred paleoclimatic distribution during the Cretaceous is given in fig. 8.8. Paleobiogeography, oxygen isotope results, and climatic rock indicators all show that the Cretaceous was a time of worldwide warm climates. Tropical and subtropical conditions extended at least to 45°N and possibly to 70°S latitude. A zone of high aridity existed between 45°N and 55°S (except in western Europe) and frigid polar climates were probably nonexistent. The widespread occurrence of coal in the northern and southern continental extremities and the dry climates near the equator probably reflected the warm-moist ocean currents which flowed around the supercontinent into polar regions.

Paleomagnetic Evidence

As was briefly discussed in Chapter 6, if certain assumptions are met, it is possible to determine the relative locations of the magnetic poles of the Earth

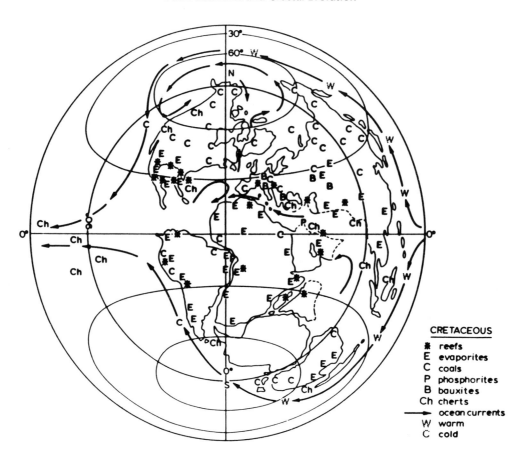

8.8. Distribution of paleoclimatic, fossil, and rock indicators on a paleomagnetic reconstruction of the continents during the Cretaceous (after Frakes, 1979).

in the geologic past by measuring remanent magnetization in rocks (McElhinny, 1973a; Tarling, 1971). Under ideal conditions, it is possible to determine both the inclination and declination of the Earth's magnetic poles at the time a rock becomes magnetized. From this information, pole positions can be plotted on equal-area or stereographic nets. The position of the magnetic poles varies on a time scale of a few thousand years and is known as *secular variation*. The fact that the magnetic pole has remained within a few degrees of the rotational pole for the last few thousand years suggests that both poles may be approximately coincident for longer periods of time. Secular variation is corrected for in paleomagnetic studies by averaging pole positions determined in several-to-many samples collected from the same rock unit. In general, good agreement exists on estimates of paleolatitudes made by paleomagnetic and geologic methods, supporting the idea that the magnetic and rotational poles have been approximately coincident in the geologic past. These data also support a dipole field for the Earth, which is a necessary as-

sumption for the paleomagnetic method. Another problem that is not always insignificant is that of removing the soft magnetization in a rock without affecting the hard magnetization. In dealing with rocks that are deformed, which includes most Precambrian rocks, it is also difficult to orient the rock structurally as it was at the time it was magnetized. The most reliable pole positions come from undeformed, nearly flat-lying rock units. Metamorphism can partially or completely reset either the magnetization direction or the radiometric age of a rock, or both, making it very difficult to define precisely the age of the pole positions deduced from the magnetization measurements. Again, this problem is especially important in Precambrian terranes.

The paleomagnetic method involves the determination of magnetic pole positions for each continent (or portion thereof) as a function of geologic time. Because of the symmetry of the dipole field, however, paleolongitudes cannot be determined. Also, it is not always clear whether a given pole position represents a normal or reversed pole. Confidence in

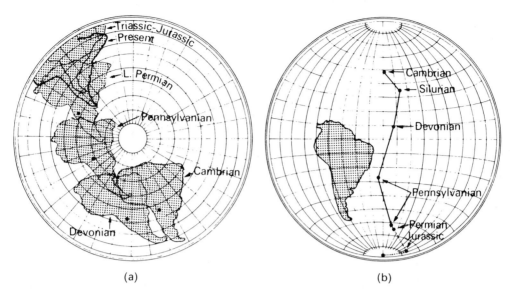

(a)   (b)

8.9. Phanerozoic paleomagnetic results for South America (from Creer, 1965): (a) paleolatitudes and orientations assuming a fixed South Pole; (b) polar wandering curve assuming a fixed continent.

the method, at least for Phanerozoic rocks, comes from the consistency of pole positions obtained from rocks of the same age within a given continent (Bott, 1971). A *polar wandering path* is defined by the loci of poles relative to a fixed continent over some period of geologic time. Pole positions from a given continent as a function of time may be interpreted in terms of continental drift or polar wandering. An example of the two methods of interpretation for South America is given in fig. 8.9. In (a) it is assumed that the Earth's pole of rotation has remained fixed and that South America has drifted as indicated, while in (b) it is assumed that the continent remains fixed and the pole has migrated as shown by the polar wandering path. Estimates of true polar wander seem to indicate that it is insignificant compared to the magnitude of plate motions (McElhinny, 1973b).

Results from different continents are usually compared by plotting *apparent polar wandering* (APW) paths for each continent relative to the present positions of the continents (fig. 8.10). If all of the continents showed the same APW path it would be clear that they did not move relative to each other. The fact that the polar wandering paths diverge from each other going back in time, however, indicates that the continents have drifted separately. The pole may or may not have wandered over the same period of time. It is noteworthy that Europe and Siberia drifted independently of each other until Triassic time, when their APW paths converged. The fact that the North American and European paths are roughly superim-

posed when the continents are reassembled according to the Bullard geometric fit (fig. 8.1) supports this geometric reconstruction.

Perhaps the most valuable application of polar wandering curves is to reconstruct pre-Permian plate

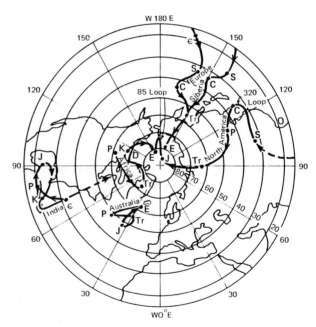

8.10. Apparent polar wandering curves for North America, Europe (with offshoot for Siberia), Africa, Australia, and India; compiled from many sources. E = Eocene; K = Cretaceous; J = Jurassic; Tr = Triassic; P = Permian; C = Carboniferous; S = Silurian; and Є = Cambrian.

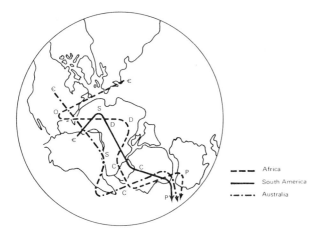

8.11. Reconstructed polar wandering paths for Australia, South America, and Africa (after Tarling, 1971). Continental fit around Antarctica (after Smith and Hallam, 1970). D = Devonian; O = Ordovician; other symbols given in Fig. 8.10.

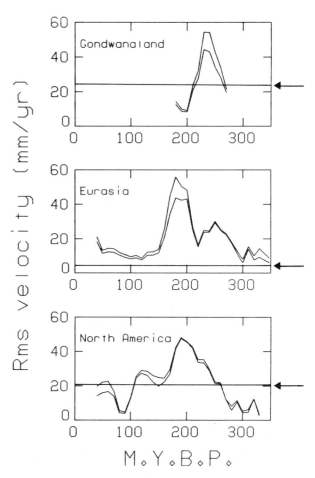

8.12. Minimum continental drift rates deduced from APW paths (after Gordon *et al.*, 1979). Horizontal lines indicated by arrows are present rms velocities. (rms = root mean square)

motions. If pre-Permian pole paths are roughly coincident when plotted on a Permian reconstruction of the continents, it may be concluded that pre-Permian plates (at least those carrying continents) did not move independently of each other. If, on the other hand, they are not coincident, independent plate motions must have occurred. As an example, pre-Permian polar wandering paths of Africa, South America, and Australia are shown in fig. 8.11 on the Permian continental reconstructions based on computerized geometric fits. It is clear that the polar wandering paths are not coincident, nor can they be superimposed by any continent configuration. It is of interest, however, that all three paths are very roughly coincident back to the Carboniferous and the African and South American paths, back to the Devonian. Such a pattern suggests that all three continents drifted independently of each other from the Cambrian to the Devonian when Africa and South America became part of the same land mass. Sometime during the Carboniferous or early Permian, Australia appears to have joined this accumulating supercontinent in the Southern Hemisphere.

Provided true polar wander has been small compared to plate motions, it is possible to calculate minimum continental drift rates from paleomagnetic data (Gordon et al., 1979). Present velocities of most plates carrying continents as deduced from magnetic anomaly distributions on the seafloor relative to fixed hotspot (or plume) references are 5–30 mm/y (rms velocity) while oceanic plates are much larger (60–70

mm/y) (Minster and Jordan, 1978). Minimum continental drift velocities deduced from paleomagnetic results for the last 300 m.y. show a peak in velocity at least twice as large as the present velocity of the corresponding continental plate (fig. 8.12). It is noteworthy that the average velocities of most oceanic plates today are only slightly higher than the minimum velocities of ancient continents. These results indicate that continents have sometimes moved much more rapidly than present continents and that rapid continental motion lasts for short periods of time (≤ 50 m.y.). Data from modern plates indicate that both continental and oceanic plates move fast when a large proportion of their boundary is subduction zones. Maxima in continental drift rates of North America tend to correspond to times of widespread orogeny, at least since the early Proterozoic.

Sea-Floor Spreading Reconstructions

From calculated sea-floor spreading directions and rates and poles of rotation for lithospheric plates, it is possible to reconstruct plate positions in the last 200 m.y. and to enhance our knowledge of the rates of plate separation. One way of illustrating such reconstructions is by the use of flow or drift lines, as shown in fig. 8.13 for the opening of the North Atlantic. The arrows indicate the relative directions of movement. Earlier positions of Africa and Europe relative to North America are also shown, with the corresponding ages in millions of years. In general, the reconstruction agrees well with proposed geometric fits across the North Atlantic (fig. 8.1). It is clear from the lines of motion in the figure that poles of rotation are different for Europe–North America and Africa–North America, thus defining two subplates.

Changes in spreading rates (which range from 2 to 5 cm/yr), however, occur on both subplates at the same time. As shown by the flow lines, changes in spreading direction occur in both subplates at about 60 and 80 m.y. ago. The results indicate that the separation of Africa and North America occurred primarily between 80 and 180 m.y., whereas separation of Eurasia from North America occurred chiefly in the last 80 m.y.

Petrotectonic Assemblages

Because the geometries of continental margins prior to 200 m.y. ago are not preserved and only small fault slices of possible oceanic crust > 200 m.y. in age exist (as ophiolites), it is not possible to employ geometric matching or sea-floor spreading methods to reconstruct early plate histories. If fossil plate

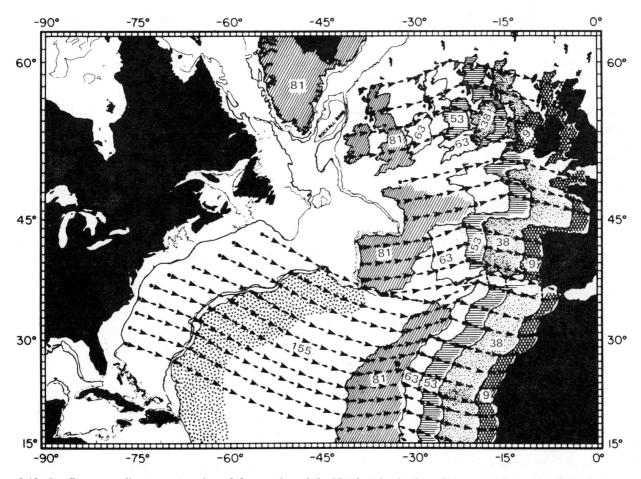

8.13. Sea-floor spreading reconstruction of the opening of the North Atlantic (from Pitman and Talwani, 1972). Black continents represent present positions; dates for earlier positions are indicated in millions of years. Arrows represent flow lines.

boundaries could be identified, much information would become available related to older plate tectonic histories. Rock assemblages that characterize plate boundaries or specific plate interior settings are known as *petrotectonic assemblages* (Dickinson, 1971c). Sandstone compositions have been shown to be particularly sensitive to provenance types governed by plate tectonics (Dickinson and Suczek, 1979). With existing data from modern plate tectonic settings, it is possible to define five petrotectonic assemblages, two characteristic of plate boundaries and three characteristic of plate interiors in continental areas:

1. Oceanic assemblages
2. Subduction-related assemblages
3. Cratonic rift assemblages
4. Cratonic assemblages
5. Collision-related assemblages

Oceanic Assemblages. Geological and geophysical evidence suggest that *ophiolite complexes* represent fragments of older oceanic crust that have been tectonically emplaced in continental orogenic belts (Coleman, 1977). If so, ophiolites represent samples of oceanic crust and upper mantle produced at ancient oceanic ridges. In a completely developed ophiolite complex such as the Troodos complex in Cyprus (Moores and Vine, 1971), the following rock types occur, in order of ascending stratigraphic position:

1. Ultramafic rocks, usually exhibiting a metamorphic-tectonic fabric;
2. Low-K tholeiitic gabbros and diabases, commonly exhibiting cumulus textures;
3. Low-K tholeiitic sheeted diabase dike complex; and
4. Low-K tholeiitic flows, commonly pillowed.

Some ophiolite complexes also have one or more of the following: an overlying section of deep-sea sediments or graywackes; associated podiform chromite bodies; and intrusive trondhjemite (high-Na granites and felsites). Ophiolites are generally in fault contact with their surroundings, and many are incomplete in the sense that they do not show the complete sequence of rock types outlined above.

Several ophiolite complexes are shown in fig. 8.14 together with a typical seismic section of oceanic crust. The gabbro-ultramafic contact in ophiolite complexes is usually interpreted as a "fossil Moho." Measurements of seismic wave velocities in samples

from several ophiolite complexes indicate that the distribution of travel-times of body waves in these complexes is very similar to that characteristic of the oceanic crust and upper mantle (Peterson et al., 1974; Poster, 1973). The abundance and age relationships of diabase dikes characteristic of ophiolites reflect a tensional environment similar to that existing along the axial zone of oceanic ridges. Compositionally, the mafic igneous rocks found in ophiolites are enriched in K, Rb, Cs, Sr, and Ba, and sometimes have higher $^{87}Sr/^{86}Sr$ ratios than fresh ridge tholeiites. They are, however, similar in composition to altered ridge tholeiites, which they are thought to represent. Although most evidence is consistent with an oceanic crustal model for the origin of ophiolites, some may also be consistent with an immature arc model (Miyashiro, 1975). It is important to note that ophiolites may be older than the orogenic belt in which they occur.

Dewey and Kidd (1977) have proposed a model for the development of ophiolites at mid-ocean ridges. The model involves a region of partially melted ultramafic rock beneath the axial zone of the rift, which from the results of experimental studies (Green, Hibberson, and Jaques, 1979) must be 60–70 km deep. Basaltic magmas rise from this region into a wedge-shaped magma chamber with a flat bottom, leaving residual ultramafic material behind (layer one of an idealized ophiolite complex). Fractional crystallization in this magma produces layered cumulates (layer 2) above the residual ultramafic rocks. Diabase dikes are injected upwards from the chamber and feed pillowed basalt flows at the surface thus producing layers 3 and 4, respectively, of ophiolites. Experimental data indicate the parent basaltic magma is not picrite and that it must be high in SiO_2 and low in TiO_2 (Green et al., 1979).

Pelagic limestones and cherts are the most common sediments overlying ophiolites (Siever, 1978). These accumulate in ocean basins as the crust spreads away from the evolving ridge system and sinks in response to cooling. Eventually it sinks beneath the carbonate compensation depth, and pelagic carbonate ceases to accumulate. After the sediments leave the central rift zone, where they may be altered by hydrothermal activity, authigenic components (such as zeolites, smectites, and manganese nodules) begin to form. As the crust nears a subduction zone, arc-derived turbidites may accumulate on top of the pelagic sediments, or alternately, if the original oceanic ridge was close to a continental margin, graywacke turbidites may accumulate directly on the pillow basalts.

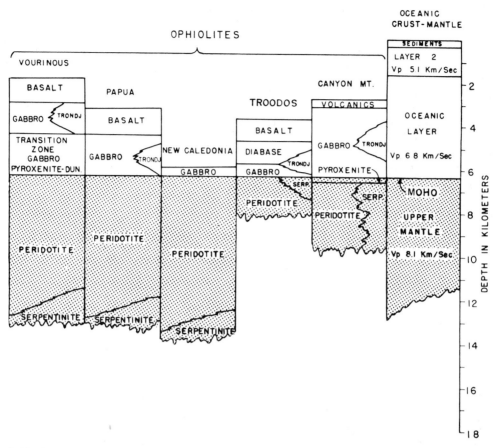

8.14. Comparison of ophiolite complexes with average oceanic crust (after Coleman, 1977).
Serp. = Serpentinite; Trondj. = trondhjemite; Dun. = dunite.

Ophiolites may be emplaced by overthrusting of oceanic crust onto continental or arc areas at convergent plate boundaries by a process known as *obduction* (Coleman, 1971, 1977) or, alternately, by being faulted into place at shallow levels within subduction zones. Several obduction mechanisms for the emplacement of ophiolites are illustrated in fig. 8.15. Fracturing of the oceanic crust as two plates carrying continental (or arc) crust close on each other (A) may cause oceanic slices to be thrust upwards, eventually reaching the margin of the continent on the converging plate. Many of the ophiolites in the eastern Mediterranean area are thought to have been emplaced by this mechanism. Ophiolites may be emplaced by obduction during a continent-continent or arc-continent collision, where they represent the remnants of a closing ocean basin (B). Examples of such ophiolites occur north of the Himalayas and in the Urals. If a plate carrying a continent or an arc is being consumed beneath an oceanic plate, resistance will occur when the continent (or arc) reaches the

subduction zone because of the buoyancy of the continent. Subduction will cease and compressive stresses may build up, resulting in the obduction of oceanic crust over the continent or arc and perhaps the initiation of a new convergent boundary dipping in the opposite direction (C). Tertiary ophiolites in Papua and New Caledonia are thought to have been emplaced by this mechanism. Initiation of a subduction zone in the oceanic crust may result in the uplift and exposure of the crust as islands and eventual obduction onto an arc or continent as the ocean basin on the edge of the leading plate diminishes in size (D). The Macquarie Island ophiolites south of New Zealand may represent an early stage in this process.

Subduction-Related Assemblages. A generalized cross section of a convergent plate boundary is shown in fig. 8.16, with various regions and petrotectonic assemblages labeled. An arc-trench system can generally be divided into four areas which are, as a func-

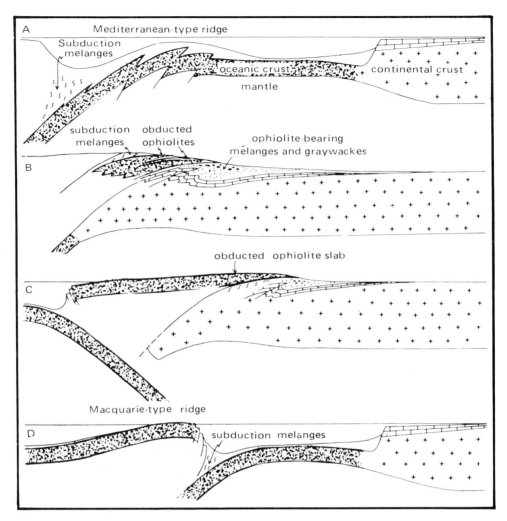

8.15. Possible obduction mechanisms for emplacement of ophiolites (after Dewey and Bird, 1971).

tion of distance from the open ocean, the trench, an arc-trench gap, the volcanic arc proper, and an arc-rear area. Each of these environments exhibits characteristic petrotectonic assemblages. Trenches usually contain small amounts of mixed deep-sea sediment and arc-derived sediment which, in part, may be dragged down the subduction zone. Much of this sediment, however, appears to be deformed and plastered on the arc-side trench wall to form the *subduction complex*. Parts of older exhumed arcs are composed of large bodies of broken and sheared rocks of different origins set in a highly sheared matrix. These terranes, known as *melanges*, are thought to represent exposures of subduction complexes and to have formed at shallow levels in subduction zones where tectonic movements and gravity sliding result in the mixing of rocks of different origins (Silver and Beutner, 1980). Chaotically

arranged blocks of rock in melanges commonly range from meters to tens of kilometers across. Some blocks are composed of ophiolites formed at oceanic ridges; others are composed of graded graywackes, which were originally deposited in a trench or in a *forearc basin* in the arc-trench gap; still others are composed of cherts and carbonates that were deposited on an abyssal plain of an ocean basin. The Franciscan Formation in California is an example of a mixed melange and forearc basin assemblage.

Rock assemblages associated with arc systems are deposited in four environments (Dickinson, 1974a; Dickinson and Suczek, 1979) (fig. 8.16): the trench proper, forearc basins within the arc-trench gap, small intra-arc basins within the arc proper, and on the arc side of back-arc basins. Successions in these basins are dominated by calc-alkaline volcaniclastic sediments, thus forming a eugeoclinal assemblage

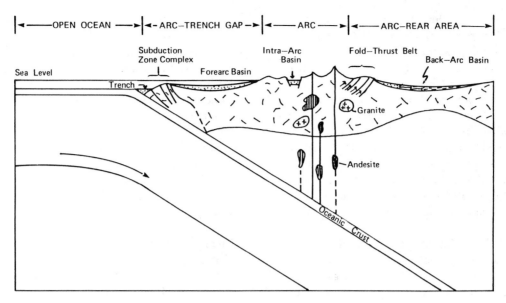

|◄──────OPEN OCEAN──────►|◄─ARC–TRENCH GAP─►|◄───── ARC ─────►|◄──── ARC–REAR AREA ────►|

8.16. Diagrammatic cross-section of an arc system showing major components and sediment basins.

(Chapter 5). The arc and intra-arc basins also are characterized by an abundance of calc-alkaline volcanic rocks. Sediment provenances may be classified into three broad categories: undissected arc, dissected arc, and recycled arc. Sediments derived from undissected arcs are comprised chiefly of volcanic debris, and those from dissected arcs of mixed volcanic, granitic, and metamorphic debris. Recycled arc material derived chiefly from uplifted and eroded portions of subduction complexes is characterized by sediments rich in chert grains and, in some cases, mafic rock fragments.

Back-arc assemblages can vary depending on whether an ocean basin or continent exists behind the arc. If a marginal sea occurs behind the arc, eugeoclinal sediments dominate on the arc side merging with either deep-sea sediments (if it is a large marginal sea) or miogeoclinal sediments, which are derived from an adjacent continent and at least in part rest on sialic basement. The Philippine Sea behind the Izu-Bonin-Mariana arc is an example of the former, and the Japan Sea is an example of the latter. The miogeoclinal association may also vary somewhat, depending upon relief in the adjacent continental source area. It can range from mixtures of marine-quartz-rich sediments, shales, and carbonates if relief is low and the basin is inundated with sea water, to terrestrial conglomerates and arkoses if the relief and basin elevation are relatively high. Thick back-arc assemblages of Tertiary age occur behind the East Indies arc and behind the Andes in South America.

Cratonic Rift Assemblages. Cratonic rifts, as exemplified by the East African rift system and by the Basin and Range Province, are characterized by immature terrestrial clastic sediments and bimodal volcanics. The sediments are chiefly arkoses, feldspathic quartzites, and conglomerates derived from rapidly uplifted fault blocks comprised of granitic and quartzo-feldspathic metamorphic rocks. Evaporites also characterize the early stages of rift formation. If the rift becomes inundated with sea water, as exemplified for instance by the Rhine graben, quartzites, shales, and carbonates may also be important. Bimodal volcanics are generally subaerial tholeiite flows and rhyolites, the latter emplaced either as ash-flow tuffs or domes. In some rifts, alkali basalt and phonolite end members are characteristic and calc-alkaline volcanics are important in only a few rifts. Cratonic rift assemblages may also form in rift basins associated with arc systems, collisional boundaries, and faulted continental margins if the sources are dissected deep enough such that provenance is chiefly granitic rocks. Hence, the cratonic rift assemblage must be used in conjunction with other assemblages and other geologic data to accurately identify a cratonic rift setting in the geologic record.

Cratonic Assemblages. Cratonic assemblages are equivalent to miogeoclinal assemblages (Chapter 5) comprised chiefly of mature clastic sediments (quartz-ites and shales) and carbonates. Such assemblages occur in three tectonic settings: (1) rifted continental margins, (2) platform basins and shelves, and (3) cratonic margins of back-arc basins. Rifted continental margins represent an advanced stage of cratonic rifting, and hence the rift assemblage described above generally underlies rifted continental margin successions. Cratonic assemblages merge with eugeoclinal successions in back-arc basins (fig. 8.16) and with deep-sea sediments along rifted continental margins. Cratonic sandstones are relatively pure quartzose sands reflecting intense weathering, low relief in the source areas, and prolonged transport across subdued continental surfaces.

Collision-Related Assemblages. Collision-related assemblages develop along collisional boundaries where continental crust resists subduction (figs. 8.17 and 8.34). Rock assemblages change both with time and space as the collision progresses and along strike (fig. 8.17). Sediments accumulate in peripheral *foreland basins* which develop in response to uplift and erosion of the collisional zone. Collision-derived sediments are shed chiefly longitudinally from the uplifted area and enter remnant ocean basins as turbidites or foreland basins in marine or subaerial environments (Dickinson and Suczek, 1979). Sandstone provenances are variable, ranging from

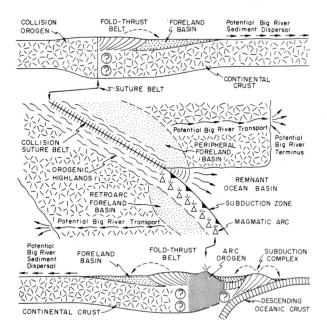

8.17. Cross-section and plan view of an evolving continental collisional boundary (after Dickinson and Suczec, 1979). Retroarc foreland basin is equivalent to a back-arc basin. Dashed arrows indicate sediment transport directions.

volcanic-plutonic to recycled sediments. Foreland basins are commonly shielded from the arc proper by the fold-thrust belts and hence do not usually receive volcaniclastic sediments derived directly from the arc. Various combinations of arkose, feldspathic sandstone, and graywacke characterize these basins. Sediments and volcanics may also accumulate in fault-bounded basins in the thickened overriding continental block (fig. 8.34). These are generally similar to cratonic rift assemblages with abundant felsic ash-flow tuffs.

Hotspot Traces

Chains of seamounts and volcanic islands characterize parts of the ocean floor and are especially common in the Pacific Basin. Major Pacific chains include the Hawaiian-Emperor, Marquesas, Society, and Austral islands, all of which are subparallel and approximately perpendicular to the axis of the East Pacific Rise (Plate I; fig. 8.18). Closely spaced volcanoes form *aseismic ridges* such as the Ninety-East ridge in the Indian Ocean and the Walvis and Rio Grande ridges in the South Atlantic (Plate I). Radiometric dates demonstrate that the focus of volcanism in the Hawaiian chain has migrated to the southeast at a linear rate of about 10 cm/yr for the last 30 m.y. (McDougall and Duncan, 1980). A major change in spreading direction of the Pacific Plate is recorded by the bend in the Hawaiian-Emperor chain at about 45 m.y. and this change is also demonstrated by magnetic anomaly distribution in the northern Pacific Plate. Similar linear decreases in the age of volcanism occur toward the southeast in the Marquesas, Society, and Austral islands in the South Pacific, with rates of migration of the order of 11 cm/yr and in the Pratt-Welker seamount chain in the Gulf of Alaska at a rate of about 4 cm/yr. These volcanic chains have been interpreted in terms of *hotspots* in the upper mantle which are approximately fixed relative to plate motions (Jarrard and Clague, 1977) or in terms of propagating cracks in the lithosphere (Hofmann et al., 1978). In the *hotspot model*, mantle plumes undergo adiabatic melting as they rise to the base of the lithosphere and linear volcanic chains are produced from plume-derived magmas as the lithosphere moves over a given plume. In the *propagating crack model*, lithospheric cracks develop in response to intraplate stresses, cooling and thickening of the lithosphere, or topographic irregularities on the asthenosphere. Magmas from the asthenosphere are erupted as the cracks propagate. The major difficulties with the plume model are (1) maintaining the heat source

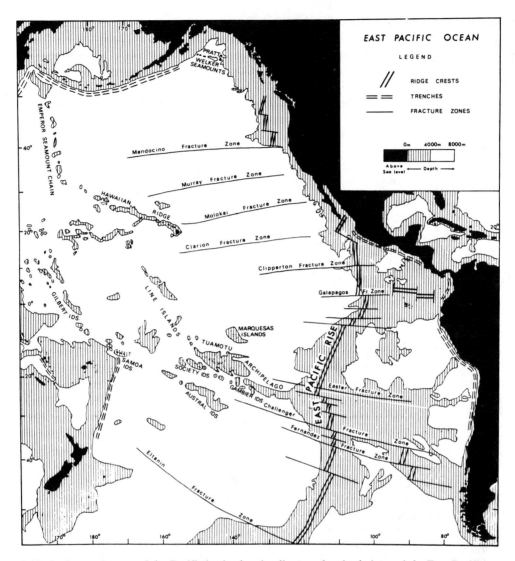

8.18. Bathymetric map of the Pacific basin showing linear volcanic chains and the East Pacific Rise (after McDougall and Duncan, 1980).

in a given plume for long periods of time (> 10 m.y.) and (2) the necessity of having large numbers of short-lived plumes to explain the large numbers of seamounts on the ocean floor not included in chains. The propagating crack model is also faced with problems (Hofmann et al., 1978). The model does not explain: (1) the origin of volcanic islands on or near oceanic ridges; (2) how to maintain an undepleted asthenosphere, which is necessary for the production of oceanic-island basalts (Chapter 7); (3) how the depleted material from the deep mantle that upwells beneath oceanic ridges returns to the lower mantle without mixing with the asthenosphere; and (4) why cracks in the Pacific Plate trend in the same direction and develop at the same rate.

The almost perfect fit of volcanic chains in the Pacific Plate with a pole of rotation at 70°N, 101°W and a rate of rotation about this pole of 1 deg/m.y. for the last 10 m.y. provides strong support for the hotspot model. This also indicates that Pacific hotspots have remained fixed relative to each other over this period of time. Also suggestive that the Hawaiian hotspot has remained fixed is the fact that the sedimentation rate determined in cores of deep-sea sediments just north of the equator has its highest rate when the model predicts that they were at the equator, where maximum sedimentation rates are observed today. If all hotspots have remained fixed with respect to each other, it should be possible to superimpose the same hotspots in their present positions on their positions at other times in the last 150–200 m.y. (Burke et al., 1973; Molnar and Atwater, 1973). Except for hotspots in the near proximity of each other, it is generally not possible to do this, suggesting that hotspots do move in the upper mantle. Rates of hotspot motion are more than an order of magnitude less than sea-floor spreading rates. Data suggest that hotspots in the Atlantic and Indian Ocean basins, for instance, move relative to each other at the rate of 1–2 cm/yr. Absolute plate motions calculated relative to the Hawaiian hotspot, and assuming a rate of volcanic migration along the Hawaiian chain of 10.2 cm/yr, indicate that the North American plate is moving SW at 3–4 cm/yr, the South American plate WNW at 3–3.5 cm/yr, and the African plate N to NNE at 1–2 cm/yr (McDougall and Duncan, 1980). Several lines of evidence suggest that the African plate has been relatively stationary for the last 25 m.y. (Thiessen et al., 1979).

Hotspot traces also have manifestations in the continents, although less well-defined than in ocean basins. For example, North America moved northwest over the Great Meteor hotspot in the Atlantic Basin between 100 and 150 m.y. ago. The calculated trajectory of the hotspot is shown in fig. 8.19. It is noteworthy that the New England seamount chain and Cretaceous kimberlites and alkalic complexes lie on the hotspot path; dated samples fall near the calculated position of the hotspot at the time they formed (Crough, Morgan, and Hargraves, 1980). Geologic data and paleotemperatures indicate that southeastern Canada and New England were elevated at least 4 km as they passed over the hotspot. When dated post-Triassic kimberlites are rotated to their position of origin relative to present Atlantic hotspots, results indicate that the majority formed within 5 degrees of a mantle hotspot (Crough et al., 1980). As an example, the calculated trajectory of the Trin-

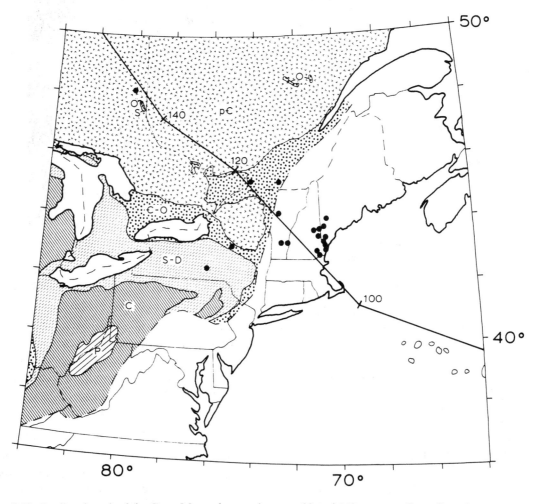

8.19. Predicted track of the Great Meteor hotspot between 80 and 160 m.y. ago (from Crough, 1981). Solid dots are dated intrusions that lie near the calculated position of the hotspot at the time of their formation.

dade hotspot (east of Brazil—see Plate I) matches well the locations of three dated kimberlites from Brazil and also roughly coincides with the distribution of alluvial diamond deposits (which are derived from kimberlites) (fig. 8.20). High heat flow, low seismic velocities and densities at shallow depth, and high electrical conductivity at shallow depth beneath Yellowstone National Park in Wyoming have been interpreted to reflect a mantle hotspot at this locality (Smith and Christensen, 1980). The movement of the North American Plate over this hotspot during the past 15 m.y. has been accompanied by the development of the Snake River volcanic plain (Plate I), with the oldest volcanics occurring at the southwest extremity of the plain in southwest Idaho. As the plate moved southwest, volcanism migrated northeast. Other examples of hotspot traces, at least in part in continental areas, are the Mesozoic granites in Nigeria; the Thulian volcanic chain extending from Iceland to Ireland and recording the opening of the North Atlantic; and the central European volcanic province extending from western Germany to Poland.

Hotspots may also be important in the breakup of continents (Dewey and Burke, 1974). Mantle plumes, for instance, may be responsible for the initial rupture of continents producing RRR triple junctions (fig. 8.21A). Hence, major irregularities along rifted continental margins may be inherited from the original plume distribution (B). Failed arms

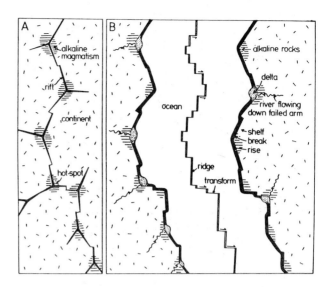

8.21. Early stages in the opening of an ocean basin showing the possible influence of mantle plumes (hotspots). From Dewey and Burke (1974).

of the plume triple junctions may localize major drainage systems, thus resulting in delta progradation at their mouths. Examples are the major river systems entering the Atlantic Ocean from South America and Africa.

ARC SYSTEMS

Seismic reflection profiling combined with geological studies of uplifted and eroded arc systems have led to a greater understanding of arc evolution and of the development of subduction zone complexes and fore-arc basins. Widths of modern arc-trench gaps (75–250 km) are proportional to the ages of the oldest igneous rocks exposed in their adjacent arcs (Dickinson, 1973). As examples, the arc-trench gap width in the Solomon Islands is about 50 km, with the oldest igneous rocks about 25 m.y. in age, and the arc-trench gap width in northern Japan (Honshu) is about 225 km, with the oldest igneous rocks about 125 m.y. in age. The correlation suggests progressive growth in the width of arc-trench gaps with time. Such growth appears to reflect some combination of outward migration of the subduction zone by accretionary processes and inward migration of the zone of maximum magmatic activity. *Subduction zone accretion* involves the addition of masses of sediment to the margin of an arc or continent. Seismic profiling suggests that many subduction zone complexes are comprised of sediment wedges separated by high-angle thrust faults produced by offscraping of oceanic sediments (Karig

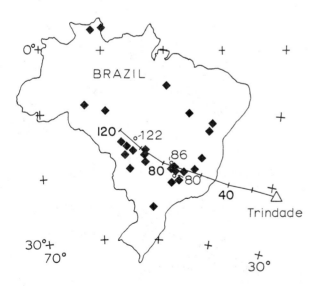

8.20. Locations of dated kimberlites (o) and major alluvial diamond deposits (♦) in Brazil compared to calculated path of the Trindade hotspot (from Crough *et al.*, 1980).

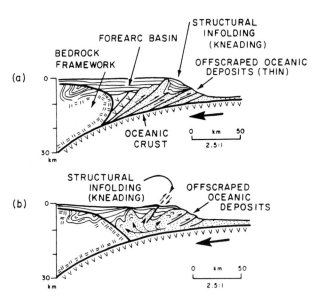

8.22. Models of offscraping accretion in arc-systems (after Scholl *et al.*, 1980).

that sediments have been subducted beneath some arc systems (such as the Aleutian arc). *Sediment erosion*, which involves mechanical plucking and abrasion along the top of the descending lithosphere, causes a trench's landward slope to retreat shoreward (fig. 8.23b,c). Subduction erosion may occur either along the top of the descending slab (b) or at the leading edge of the overriding plate (c). Subduction accretion, kneading, and erosion and sediment subduction are all important processes at underthrust margins and any one of them may predominate at a given place and evolutionary stage.

It is also of interest to consider what happens when an oceanic ridge approaches a subduction zone. Two models have been suggested. One, which is discussed in Chapter 9, involves the annihilation of the ridge and consequent development of a transform fault system; the other involves ridge subduction. If a ridge is subducted, the arc should move "uphill" and become emergent as the ridge crest approaches, and

and Sharman, 1975) (fig. 8.22a). *Offscraping accretion* involves the tectonic skimming of oceanic sediments and of mafic and ultramafic oceanic lithosphere and their addition to the continental (or arc) bedrock framework (Scholl et al., 1980). Offscraping results in outward growth of the subduction zone complex and also controls the location and evolutionary patterns of overlying forearc basins. Some reflection profiles suggest much more complex deformational patterns than simple thrust wedges. Deformation may include large-scale structural mixing and infolding of forearc basin sediments, a process referred to as *subduction kneading* (fig. 8.22b). Geological evidence for subduction kneading comes from uplifted and exposed underthrust margins, exemplified by parts of the Franciscan Complex in California.

Some oceanic and continental sediments may be subducted and, as we mentioned in Chapter 7, geochemical and isotopic data from some arc systems necessitate sediment subduction. Sediments are presumably dragged down by the descending slab (fig. 8.23a). Also favoring sediment subduction is the sparsity of continental and arc sediments along the edge of continents that are now subduction zones or which were subduction zones in the recent geologic past. For instance, the volume of sediment along the Atlantic coast of the United States is at least six times as great as that off the Pacific Coast (Gilluly, 1969), which was a convergent plate boundary until about 30 m.y. ago (Chapter 9). Also, seismic reflection profiling and deep-sea drilling seem to suggest

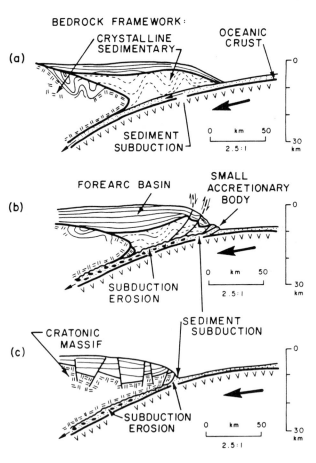

8.23. Models of sediment subduction and sediment erosion at convergent plate boundaries (from Scholl *et al.*, 1980).

it should move "downhill," becoming submerged, as the ridge passes down the subduction zone (DeLong and Fox, 1977). Corresponding changes in sedimentation should accompany this emergence-submergence sequence of the arc. Ridge subduction may also lead to cessation of subduction-related magmatism as the hot ridge is subducted. This could be caused by reduced frictional heating in the subduction zone or progressive loss of volatiles from the descending slab as the ridge approaches. Also, the outer arc should undergo regional metamorphism as the extinct or dying ridge crest is subducted. It is noteworthy that all three of these phenomena are recorded in the Aleutian arc and hence support the subducted ridge model in this area. It has also been suggested that ridge subduction will result in a change in stress regime in the overriding plate from dominantly compressional to extensional, and may result in opening of a back-arc basin (Uyeda and Miyashiro, 1974). It is noteworthy also that subduction of active ridges may lead to formation of new ridges in the descending plate at great distances from the convergent margin. For instance, the rifting of Antarctica from Australia which began almost 50 m.y. ago corresponded to the subduction of a ridge system along the northern edge of the Australian-Antarctic Plate (Chapter 9; fig. 6.25).

When buoyant crust, such as oceanic islands, seamounts, and aseismic ridges, arrives at subduction zones, drastic changes in subduction geometry occur. Formerly straight subduction zones become indented where intersected by aseismic ridges or seamount chains, exemplified by the Emperor and Caroline chains in the Pacific (Plate I). Notable cusps in the subduction zones, and in some instances polarity changes in subduction zone dip, occur at these intersections. Generally, continents or arcs are thought to be too buoyant to be subducted, and hence when a fragment of continental crust arrives at a subduction zone, a *continental collision* takes place. Such a collision involves the loss of an intervening ocean basin and the welding of the two continental fragments together along a suture zone. Buoyancy and temperature constraints seem to allow subduction of only minor amounts of continental crust (tens to hundreds of kilometers) (Molnar and Gray, 1979).

It is of interest in interpreting the history of fossil subduction zones to define properly the polarity (dip) of the subduction zone. Three approaches may be used in this regard. The first involves determining the compositional polarity of the volcanic and/or plutonic rocks in the arc and relating it to the dip of the subduction zone (Chapter 7). Two practical

problems are encountered in applying this method. First, it is necessary to have rocks *of the same age* exposed for sampling over a significant cross-sectional area of the arc. Second, complex thrust faulting and other deformation, which generally characterize older arc terranes, must be unraveled to reconstruct the original relative positions of the igneous terranes.

The second method involves the asymmetry of nappes and thrust faults in the fold-thrust belt behind arc systems. These structures commonly dip in a direction opposite to the dip of the inactive subduction zone along continent-continent, continent-arc, and arc-arc collision boundaries (see figs. 8.16 and 8.17). Orogenic belts that are associated with subduction zones not involving collisions, however, are often characterized by divergent (symmetric) thrusting.

Paired metamorphic belts and zonation within metamorphic belts may also be useful in identifying fossil arc polarities. Juxtaposed belts of blueschist-greenschist facies (high P–low T) and amphibolite facies (intermediate P-T) terranes are known as *paired metamorphic belts* (Miyashiro, 1973) and are well represented in most island-arc systems in the circum-Pacific area. The blueschist-greenschist facies belts almost always occur on the seaward side of arc systems. Melange and eugeoclinal arc-trench gap assemblages are usually associated with low-grade metamorphism of the zeolite, greenschist, and blueschist facies (fig. 4.24). Rocks of the zeolite facies are formed at very shallow depths at low temperatures, and characterize the trench side of arcs. At greater depths in and adjacent to the upper face of the descending slab, blueschist- and/or greenschist-facies assemblages form. In the crust beneath the arc-trench gap and arc, middle- to high-grade metamorphic assemblages are formed (amphibolite facies) at temperatures of 400–600°C and pressures of 5–10 kbar. Excellent examples of this metamorphic-zone polarity have been described by Ernst (1971) in the Sanbagawa terrane in Japan, in the Alps, and in the Franciscan terrane in California (fig. 8.24). On these maps, the implied dip-directions of the subduction zones are shown by large arrows. In each case metamorphic zones show an increase in grade (depth of burial) from the zeolite facies (laumontite-bearing) to a mixed greenschist-blueschist assemblage (lawsonite-bearing) in progressing landward from the inferred position of the trench. In addition, the Sanbagawa and Alpine terranes exhibit an amphibolite-facies zone (albite-amphibolite on the maps) still further inland. The mechanisms of juxtapositon of these belts in the crust are poorly understood but may involve buoyant re-

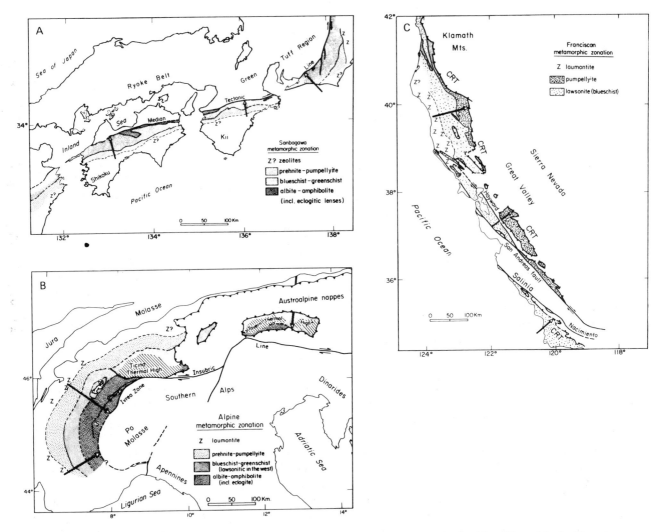

8.24. Zones of increasing metamorphic grade in the Sanbagawa terrane of Japan (A), in the Alps (B), and in the Franciscan terrane of California (C) (after Ernst, 1971).

bound of blocks from the upper part of the descending slab after subduction ceases.

RIFTS

Major features of continental rift systems, some of which have been discussed in Chapters 4 and 7, are as follows (after Neumann and Ramberg, 1978):

1. Rift systems range from those with abundant volcanics to those with little volcanic component. Volcanism may or may not be closely related to extensional faulting.

2. Rift magmas are generally bimodal (alkali basalt-phonolite or tholeiite-rhyolite), although calc-alkaline volcanics occur in some rift systems.

3. Cratonic rifts (such as the East African rift system) are commonly associated with domal uplifts. A characteristic sequence of development for cratonic rifts may be doming and rifting of domes, followed by triple junction development.

4. Rift systems may be narrow and long, such as the East African system, narrow and short, such as the Baikal and Rhine rifts, or very broad, such as the Basin and Range Province.

5. Rift systems commonly show evidence of several periods of rifting and volcanism.

6. Crust and lithosphere beneath rifts are thin, and rifts are underlain by hot, upper-mantle material at shallow depths.

7. Many rift systems are characterized by decreasing alkalinity of volcanism with time (such as

the Kenya rift) and increasing alkalinity with distance from rift axis (*viz.*, the East African rift system).

8. Volcanic rocks from many continental rift systems commonly reflect undepleted or enriched mantle sources.

Towards a Classification of Rifts

It is clear from the data summarized above that rifts represent a diverse group of tectonic features. Studies of the geologic history of rifts indicate complex evolutionary patterns with no simple sequence of events characterizing all rifts. Clearly rifts are of different origins and occur in different regional tectonic settings. Although the immediate stress environment of rifts is extensional, the regional stress environment may be compressional, extensional, or nearly neutral.

As a starting point for understanding rift development and significance, rifts can be classified into two categories, depending on the mechanisms of rifting (Condie, 1981b; Sengör and Burke, 1978). One group of rifts, herein referred to as *mantle-activated rifts*, are produced by doming and cracking of the lithosphere, where the doming is produced by upwelling asthenosphere or rising mantle plumes (fig. 8.25). Another group of rifts, *lithosphere-activated rifts*, are produced by stresses in moving lithospheric plates. Lithosphere-activated rifts may evolve into mantle-activated rifts if asthenosphere upwells and spreads beneath the rifts. Mantle-activated rifts are characterized by relatively large volumes of volcanic rock, while in lithosphere-activated rifts, immature clastic sediments generally exceed volcanics in abundance. In a general way, mantle and lithosphere-activated rifts each can be subdivided into three categories (table 8.1).

Mantle-activated rifts are represented by oceanic ridges, cratonic rifts and aulacogens, and some back-arc basins. Both of the latter two types of rifts may evolve into the former. *Cratonic rifts* develop in stable continental cratons, the most notable modern example of which is the East African rift system (Plate I). *Aulacogens* are failed arms of evolving cratonic-rift triple junctions, perhaps developing over mantle plumes (Burke and Dewey, 1973). Examples of young aulacogens are the Ethiopian and Benue rifts in Africa (fig. 8.26). Although the Ethiopian rift (ii) is not a failed arm of the Afar triple junction, it is much less active than the other two arms, which have opened into oceanic crust. Spreading began in the Red Sea and Gulf of Aden 15–25 m.y. ago and may

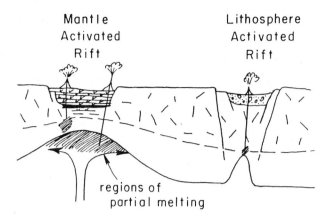

8.25. Diagrammatic cross-section of mantle and lithosphere-activated rifts.

be beginning today in the Ethiopian rift. The Niger triple junction (i) developed in the Cretaceous, and the Benue trough became a failed arm as the other two rift segments opened with the Atlantic basin.

A possible sequence in the breakup of a continent by continued spreading of a cratonic rift is shown in fig. 8.27. A mantle plume or upwelling convection current causes the lithosphere and crust to thin and fracture into a series of grabens, which collect clastic sediments from intervening horsts (A, B), and basaltic magma is injected into the axial portion of the graben system. Eventually new oceanic crust (and lithosphere) is produced as the continent further separates and an oceanic ridge system comes into existence (C). Remnants of basaltic flows and sills and clastic sediments reflecting early stages of the rupturing may be preserved in grabens on retreating continental margins (D). An example of a youthful ridge system and associated ocean basin is the Red Sea (E), which is widening as Africa and Arabia separate.

Lithosphere-activated rifts develop along faulted continental margins, in zones of continental collision, and perhaps in arc systems. Examples of rifts associated with faulted continental margins are those associated with the San Andreas and related faults in California, the Basin and Range Province, and the multiple-rift system in western Turkey. Both of the latter examples appear to have developed in response to imperfect strike-slip faulting (see Chapter 9). Rifts can be generated along continental collision boundaries by irregularities in continental margins and by strike-slip and normal faulting caused by nonperpendicular collision. The Rhine graben in Germany has been suggested as an example of a rift that

Table 8.1. Rift Classification

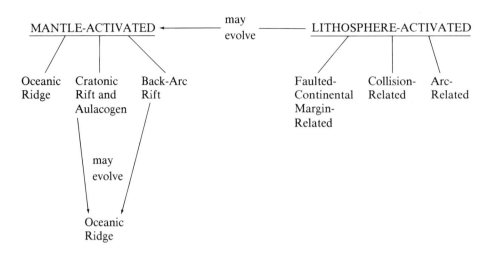

developed at a steep angle to a collisional boundary, and the term *impactogen* has been applied to this type of rift (Sengör et al., 1978). It is proposed that rifting occurred at an irregularity in the European continent as Africa collided with this irregularity (fig. 8.28). An aborted attempt to subduct the irregularity results in depression and rifting. Molnar and Tapponnier (1975) have suggested that stresses associated with the Tibet-India collision (Chapter 9) were transmitted as strike-slip and normal faulting into Eurasia, forming such rifts as the Baikal Rift in Siberia and the Shanshi graben system in China.

It is clear that various petrotectonic assemblages characterize rifts. Oceanic rifts are represented by ophiolites. Small arc-related rifts contain calc-alkaline volcanics and graywackes, and moderate-to-large back-arc basins are characterized by mixed

assemblages including some combination of ophiolites and deep-sea sediments, arc-derived graywackes, and cratonic sediments. Continental-margin back-arc basins, cratonic rifts, and aulacogens contain arkoses, feldspathic sandstones, conglomerates, and bimodal volcanics. Lithosphere-activated rifts may contain a variety of volcanics and sediments, although bimodal volcanics and immature terrestrial sediments gener-

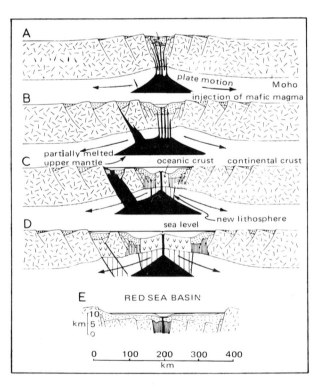

8.27. Stages in the rupture and separation of a continent and development of an accreting plate margin (from Dewey and Bird, 1970b).

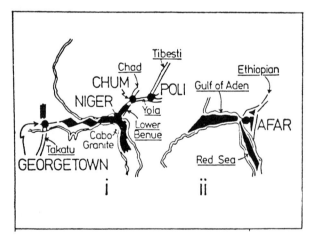

8.26. The Niger and Afar triple junctions (after Burke and Dewey, 1973).

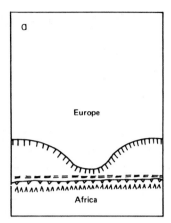
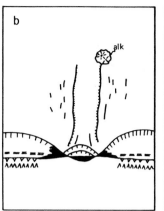
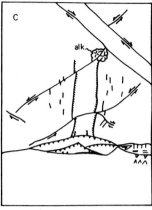

8.28. Idealized sequence of events leading to development of the Rhine graben (after Sengör, 1976). a) Convergence of Africa and Europe; b) Collision of irregularity on Europe with Africa and beginning of rift development; c) Suturing of continents complete, deformation characterized by strike-slip faulting. Key: ᴠᴠᴠ subduction zone dipping beneath Africa; ═══ trench; ᴛᴛᴛᴛ continental margin of Europe; ᴧᴧᴧᴧ rifts; ——— mafic dikes; alk, alkaline volcanism; black area in b) is ophiolite nappe.

ally dominate. Relative proportions as well as kinds of rocks change as rifts evolve—as, for instance, a cratonic rift evolves into a stable continental margin.

CRATONIC BASINS

It is of interest to examine the mechanisms of subsidence of cratonic basins in terms of plate tectonics. Included in cratonic basins are both platform basins and basins at stable continental margins. One of the most significant observations of cratonic basins in this regard is that they exhibit the same exponential subsidence as ocean basins (Sleep, 1971). Their subsidence can generally be considered in terms of two stages. In the first stage, subsidence rate varies greatly, whereas the second stage is characterized by widespread and general subsidence. After about 50 m.y. the depth of subsidence decreases exponentially to a constant value. The similarity to subsidence in ocean basins strongly suggests that the temperature distribution at depth is similar and that the lithospheric thickness beneath oceans and continents is about the same (~ 100 km) (Sclater, Jaupart, and Galson, 1980).

Several models have been suggested to explain cratonic subsidence (Bott, 1979; Sleep et al., 1980). Sediment loading, lithosphere stretching, and thermal doming followed by contraction are the most widely cited mechanisms. Although the accumulation of sediments in a depression loads the lithosphere and causes further subsidence, calculations indicate that the contribution of sediment loading to subsidence must be minor compared to other effects. Subsidence at stable continental margins may result from thinning of the continental crust by progressive creep of a relatively ductile lower crust toward the suboceanic upper mantle. As the crust thins, sediments accumulate in the overlying basins. Alternately, the lithosphere may be domed by upwelling asthenosphere, or a mantle plume and uplifted crust is eroded. Thermal contraction following doming results in a platform basin, or a series of marginal basins around an opening ocean, which fill with sediments. Models involving one or both doming-contraction and stretching may account for the observed thermal subsidence of cratonic basins.

EUSTACY

Eustacy, or changes in sea level with time, can be evaluated from the study of stratigraphic successions. Continental sediments record *transgression*, or advances of the sea onto the continents, and *regression*, retreats of the sea into the ocean basins. A major long-term transgression occurred in the late Cretaceous, with smaller maxima in the late Devonian and Ordovician (fig. 8.29). Major regression is recorded in the late Paleozoic and early Mesozoic. Superimposed on the long-term sea-level variation curves are short-term variations which probably reflect chiefly the extent of glaciation (Chapter 11). Long-term eustatic changes in sea-level have been attributed to many

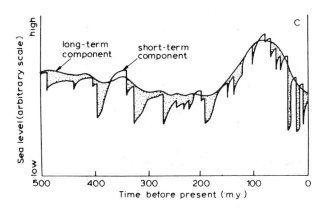

8.29. Sea level variations during the Phanerozoic (after Frakes, 1979).

factors (Frakes, 1979; Pittman, 1978). The most likely cause of large, long-term transgressions is changes in volume of mid-ocean ridges. The cross-sectional area of an oceanic ridge is dependent upon spreading rate, since its depth is a function of its age (fig. 8.30). For instance, a ridge that spreads 6 cm/yr for 70 m.y. has three times the cross-sectional area of a ridge that has spread at 2 cm/yr for the same amount of time. The total ridge volume at any time is dependent upon spreading rate (which determines cross-sectional area) and total ridge length. Times of maximum sea-floor spreading (fig. 9.19) correspond well with the major transgression in the late Cretaceous, thus supporting the ridge volume model for eustacy (Pittman, 1978). Calculations indicate that the total ridge volume in the late Cretaceous would displace enough sea water to raise sea level by about 350 m. This figure is in good agreement with paleogeographic data for the height of sea level during the late Cretaceous.

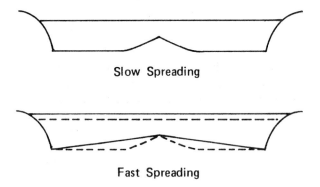

8.30. Change in sea level caused by a greater volume of a fast-spreading ridge (solid line) compared to a slow-spreading ridge (dashed line) (after Turcotte and Burke, 1978).

PLATE TECTONICS AND OROGENY

The classical geotectonic cycle encompasses a progressive sequence of events related to the geosynclinal theory of mountain building. In this model, elongate, fixed belts of subsidence and sedimentation known as *geosynclines* are necessary precursors of orogeny (Coney, 1970). A typical orogenic cycle involves accumulation of geosynclinal sediments ($\pm$ volcanic rocks), intense deformation and plutonism, followed by differential uplift and volcanism. Such a simple sequence of events is rarely, if ever, documented in an orogenic belt.

Orogeny occurs along convergent plate boundaries in plate-tectonic models. Unlike the geosynclinal theory, plate tectonics does not require a regular sequence of events to occur in the same order in space or time in all orogenic belts. The sequence of events can vary significantly, depending upon the various possible interactions that may occur at convergent plate boundaries. In general, orogeny can begin in two ways, known as the activation and collision models (Dickinson, 1971a). The *activation model* involves the initiation of plate convergence at a previously stable continental margin of miogeoclinal character. The *collision model* involves an arc-continent, continent-continent, or arc-arc collision, and is characterized by cessation of subduction or relocation of a subduction zone along a nearby oceanic interface.

In order to indicate the complexity of events that may affect orogeny at convergent plate boundaries, schematic cross sections of existing plate boundaries are shown in fig. 8.31. The first section (A) extends from the Nasca Plate through South America, the Atlantic Ocean, and Africa and into the Indian Ocean. This section shows an expanding ocean basin, the Atlantic, without convergent plate boundaries. The second section (B) is across the Pacific Basin, which, as was previously discussed, is contracting at the marginal subduction zones as the Atlantic grows. The Japan arc and the Andes are active orogenic belts. A northeasterly trending section across the Pacific from the Philippine arc to California (C) illustrates a double subduction system in the western Pacific, in which the Philippine Plate is being consumed in the Philippine-Taiwan subduction zone. In a small region between Luzon and Taiwan, the Philippine slab dips towards the Mariana slab (D), and Asia is approaching and may eventually collide with the Philippines in this area. Another example of a plate carrying a continent that is being consumed beneath an arc is the Australian Plate in

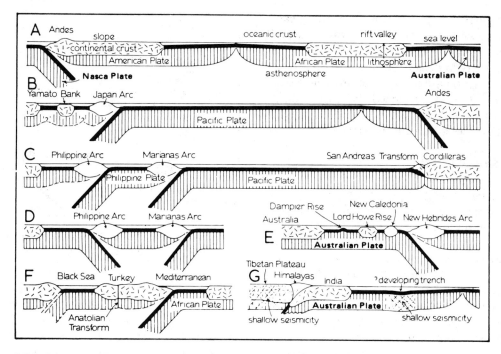

8.31. Schematic cross sections of existing plate boundaries (after Dewey and Bird, 1970a).

the New Hebrides subduction zone (E). Although the continents are passive passengers on lithospheric plates, they play a significant role in plate interactions because their buoyancy prevents any significant consumption at subduction zones, resulting in collisions of continental crust. Two sections of areas in which continent-continent collision is taking (or has taken) place are in southern Europe and in south-central Asia (F and G).

The Activation Model

Examples of activation plate-tectonic orogenic models are illustrated in fig. 8.32. Regions subjected to activation orogeny are characterized by pronounced polarity in the composition of igneous rocks and in the distribution of metamorphic zones. As orogeny proceeds, deformation spreads toward the continent (A). The oldest sedimentary rocks involved in the orogeny may be a miogeoclinal assemblage that recorded a retreating continental margin prior to the onset of convergence. Whether a back-arc basin develops or not during an activation orogeny seems to depend, in part, on the age of the descending slab. If the underthrusting plate is less than 50 m.y. in age, it tends to be buoyed up and the subduction zone migrates in the direction of underthrusting, inducing compression in the overlying plate (A) (Molnar and Atwater, 1978). If the descending plate is older than

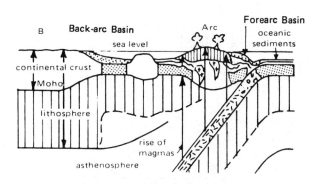

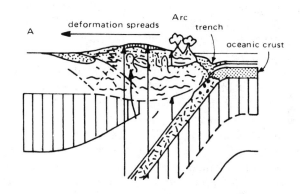

8.32. Activation plate-tectonic orogeny models (after Dewey and Horsfield, 1970). A, continental margin and B, back-arc basin.

50 m.y., the subduction zone seems to migrate ocean-ward, inducing extension in the overlying plate, re-sulting in the opening of a back-arc basin (B). The activation orogeny in the second case is characterized chiefly by normal faulting and volcanism, whereas in the first case it is characterized by large nappes or/and thrusts (the thrust-fold belt) directed toward the continent together with arc magmatism.

The Collision Model

The opening and closing of an ocean basin is known as a *Wilson cycle* (Burke, Dewey, and Kidd, 1977). It begins with the rupture of a stable continent along a cratonic rift system, followed by the opening of an ocean basin. The ocean basin may be small, such as the Red Sea, or large, such as the Atlantic Basin. At some point, subduction begins on one or both margins of the basin and closure commences. Because sialic crust is buoyant and tends to resist significant sub-duction, the net result of the closure is a continent-continent collision with associated orogeny. Similar orogenies, although less widespread, can be produced by arc-continent or arc-arc collisions. Wilson cycles in the geologic past can be recognized by distinctive APW paths. Fig. 8.33 shows the expected APW paths for continents A and B now facing each other across an ocean basin. Going backwards in time (from 14 to 1), there is a common APW path after suturing of the colliding continents at 12. Prior to collision each continent has its own path, with complimentary di-verging and converging phases. Prior to rifting at 4, both continents have similar paths, although they are not identical, because the geometric fit across the initial rift and final suture do not, in general, coin-cide. The evolution of petrotectonic assemblages from cratonic rift to stable continental margin successions followed by collisional assemblages also records the Wilson cycle.

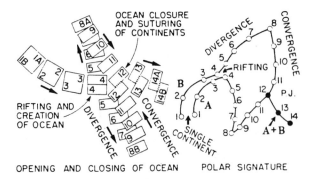

8.33. Idealized APW path characteristic of a Wilson cycle (from Irving *et al.*, 1974).

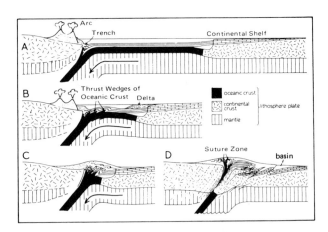

8.34. Schematic sequence of events in a continent-continent collision (after Dewey and Bird, 1970a).

A schematic sequence of events in an orogeny caused by a continent-continent collision is il-lustrated in fig. 8.34. As convergence continues, the intervening ocean basin closes (A) and slices of oceanic crust may be thrust toward the converging continent (B). Upon collision, the buoyancy of the converging continent prevents its consumption by the mantle, and large nappes and thrusts are produced as it resists subduction (C and D). Remnants of the intervening ocean basin (ophiolites) are emplaced along a suture zone together with melange and deep-sea sediments. The *suture zone* marks the boundary where the two plates are welded together, and is interpreted as evidence for a closed ocean basin. Such suture zones characterize the northern flank of the Himalayas (the Indus suture) and the central Urals. A continent-continent collision results in the termina-tion of a convergent plate boundary, and worldwide spreading rates must adjust to compensate for the loss of a subduction zone or a new subduction zone must develop. There is some evidence today that a new convergent plate boundary, defined by a shallow seismicity zone extending southeast from India, may be in the early stages of development to compensate for the India-Eurasia collision (fig. 8.31G).

An idealized model of a continental collisional boundary showing major rock types and structures is given in fig. 8.35. Major thrusts and nappes are directed toward the converging plate (Dewey and Burke, 1973). The suture zone becomes less distinc-tive at greater crustal depths (i.e., a cryptic suture), where it is represented by a shear zone with similar metamorphic rocks on both sides. Continued conver-gence leads to thickening of the crust and lithosphere accomplished by ductile creep or thrusting at shallow levels. Seismicity along some continent-continent col-

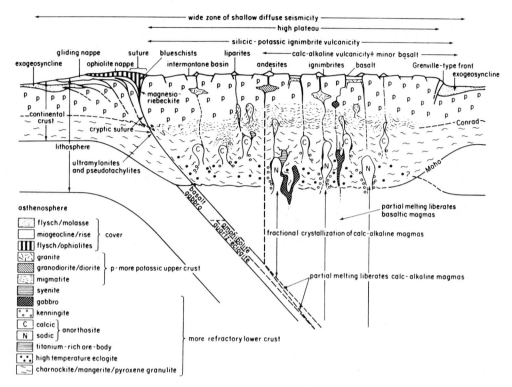

8.35. Hypothetical cross-section of a continent-continent collisional boundary (from Dewey and Burke, 1973).

lision boundaries (such as the Swiss Alps) suggests partial subduction of continental crust. Models in the Alps involve the underthrusting of successive crustal wedges and overthrusting of decollement structures (fig. 8.36). Thickening of the lithosphere above collisional subduction zones depresses it into hotter upper mantle. Partial melting may begin and produce granitic magmas that are erupted as ash-flow tuffs or intruded as granite plutons (fig. 8.35). Fractional crystallization of mantle-derived basalts may produce anorthosites in the lower crust, and losses of granite and volatiles from the thickened crust may leave behind granulite-facies mineral assemblages. Isostatic recovery is marked by the development of continental rifting with associated deposition of arkose-conglomerate and bimodal volcanism. Calc-alkaline volcanism may also continue for a short time in the overriding plate.

It is possible when two continents collide that large sheet-like masses, commonly referred to as *flakes*, are sheared from the top of the converging plate and thrust over the adjacent continent. The study of such interactions is commonly referred to as *flake tectonics* (Oxburgh, 1972). In the Eastern Alps, for instance, Paleozoic metamorphic rocks (C) have been thrust more than 100 km north over the

Bohemian Massif (A) (fig. 8.36). A zone of highly sheared Mesozoic metasediments lies within the thrust zone (B). The overriding crust or flake is less than 12 km thick and appears to represent part of the Carnics Plate (fig. 9.29) that was detached and moved northward as convergence continued in the Miocene. Other

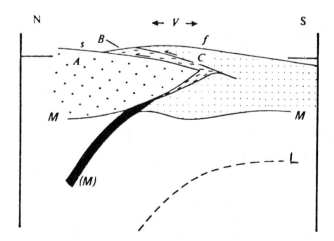

8.36. Schematic cross-section of the Eastern Alps (after Oxburgh, 1972). A, Bohemian massif; B, peripheral Schieferhülle; C, Altkristallin Decke; S, Salzburg; f, Italian frontier; M, Moho; V, intrusive activity; L, base of the lithosphere.

examples of crustal flakes in California, Nevada, and the Himalayas have been described. Flake tectonics appears to be an important deformational style along collisional boundaries.

A possible sequence of events for a continent-arc collision is illustrated in fig. 8.37. When the plate carrying the continent reaches the subduction zone (B and C), the subduction zone becomes choked and thrusting and nappes are directed toward the continent. As with the continent-continent collision model, spreading rates may adjust to compensate for the loss of the subduction zone, or the subduction zone may move to the ocean side of the arc, dipping in the opposite direction (D). The latter change, known as a *subduction-zone flip*, results in the accretion of the arc to the continent and a reverse in the compositional polarity of subsequent magmas associated with subduction.

Several features distinguish activation from collision orogenies and may be helpful in unraveling complex plate-tectonic histories:

1. Symmetry of thrusting. Thrusting and nappe motion are either divergent or directed toward the overriding plate in activation-type orogenies, whereas they are directed toward the plate being consumed in collision-type orogenies (fig. 8.34D).

2. Suture zones characterize collision-type orogenies, although they may not always be exposed or readily identified as such.

3. Activation-type orogenic belts commonly exhibit compositional polarity, paried metamorphic belts, and paired arc-trench gap and arc-rear assemblages, whereas collision types do not.

Combinations of collision and activation mechanisms characterize most orogenic belts, thus giving rise to very complex plate-tectonic histories. Unlike the geosynclinal model of mountain building, the plate-tectonic model can accommodate variation in the order of events; indeed, each orogenic belt, or in some cases segments thereof, may have unique plate-tectonic histories.

PLATE TECTONICS AND ENERGY AND MINERAL DEPOSITS

Plate tectonics provides a basis for understanding the distribution and origin of mineral and energy resources in space and time (Rona, 1977). The occurrence of energy and mineral deposits can be related to plate tectonics in three ways: (1) geological processes driven by energy liberated at plate boundaries control the formation of energy and mineral deposits; (2) such deposits form in specific tectonic settings, which are in turn controlled by plate tectonics; and (3) reconstruction of fragmented continents can be used in exploration for new mineral and energy deposits.

Several requirements must be met in any tectonic setting for the production and accumulation of hydrocarbons such as oil or natural gas. The preservation of organic matter requires restricted seawater circulation to inhibit oxidation and decomposition. High geothermal gradients are needed to convert organic matter into oil and gas. Finally, tectonic conditions must be such as to create traps for the hydrocarbons to accumulate in. Several tectonic settings meet these requirements.

Deposits at Oceanic Ridges

Discoveries over the last decade have shown that oceanic ridges are important sites of metallic mineral deposition (Corliss et al., 1979; Rona, 1978). Hydrothermal activity at ridges is responsible for the formation of sulfide deposits and metalliferous sediments on the flanks of ridges (fig. 8.38d). Important metals are Fe, Zn, Cu, Pb, Au, and Ag. Sampling of the upper 10 m of metalliferous sediment in the Red Sea indicates the presence of about 8×10^6 tons of ore comprised chiefly of Fe, Zn, and Cu. Mn oxide deposits are also important at some ridge locations, such as at the TAG hydrothermal field on the Mid-Atlantic Ridge. Deposits formed at oceanic ridges are accessible in ophiolite complexes in Phanerozoic orogenic belts. Ultramafic rocks in ophiolites contain

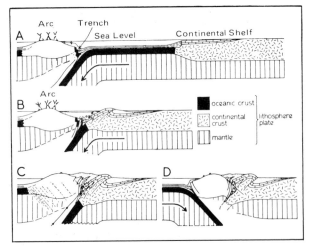

8.37. Schematic sequence of events in a continent-arc collision (after Dewey and Bird, 1970a).

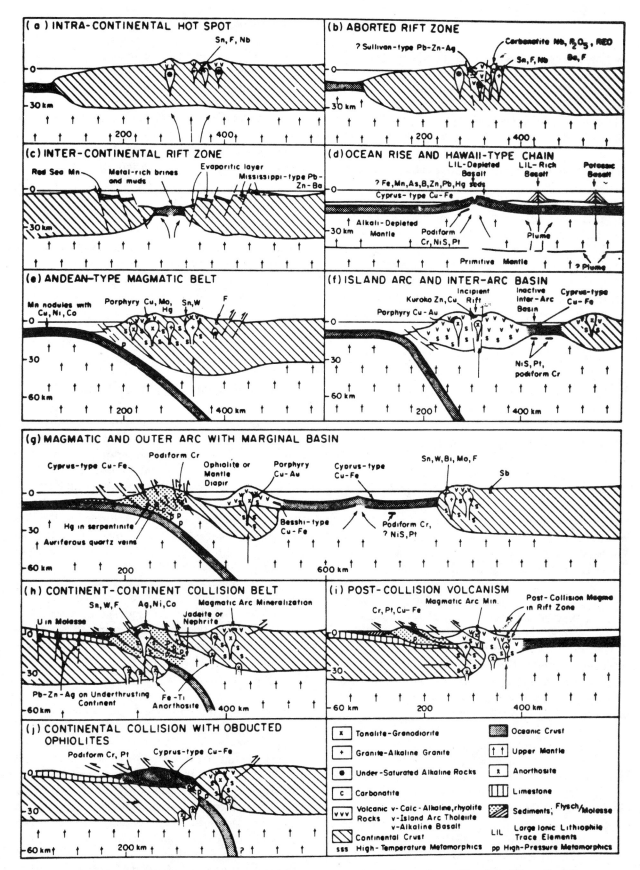

8.38. Schematic cross-sections showing relationship between tectonic setting and metallic mineral deposits (after Mitchell and Garson, 1976).

asbestos, chromite, and Ni ores. Podiform chromite deposits are commonly associated with serpentinized ultramafics. Cyprus-type massive sulfides (Cu-Fe rich) also occur associated with ophiolites and appear to represent hydrothermal deposits formed at ocean ridges.

Deposits at Convergent Plate Boundaries

Metalliferous mineral deposits are important at both continental and arc convergent plate boundaries (fig. 8.38e,f,g,). A major part of known base-metal (Zn, Cu, Mo, Pb), precious-metal (Ag, Au, Pt), and other metal deposits (Sn, W, Sb, Hg) are found associated with present or former convergent margins. In the circum-Pacific area, for example, major metal deposits occur along convergent margins in western North and South America, Japan, the Philippines, New Zealand and Indonesia. More than half of the world's Cu production comes from copper porphyries (large-volume, low-grade deposits of disseminated Cu sulfide) in this region, the most important occurrences of which are in the southwestern United States and in the Andes. Red-bed uranium deposits are also found associated with convergent margins, for example in the southwestern United States. Zonation of metallic mineral deposits has been reported in late Tertiary rocks from the Andes (Sillitoe, 1976). Going east from the coast toward greater depths to subduction zone, the major zones encountered are: contact metasomatic Fe deposits, Cu-Au and Ag veins, porphyry Cu-Mo deposits, Pb-Zn-Ag vein and contact metasomatic deposits, and Sn deposits. Belts are believed to be caused by progressive liberation of metals from the descending slab, with Sn coming from an extreme depth of about 300 km. Geochemical and isotopic data support the general concept that metal deposits associated with subduction are derived from some combination of the descending slab and the overlying mantle wedge. Metals move upward in magmas or in fluids and are concentrated in late hydrothermal or magmatic phases.

Petroleum occurs in back-arc basin environments, and extensive exploration is underway for petroleum accumulations in back-arc basins of the Southwest Pacific. Back-arc basins trap organic material and restrict water circulation so organic matter is not oxidized. Geothermal heat beneath back-arc basins facilitates conversion of organic matter into oil and gas, and later deformation accompanying activation orogeny forms traps for the accumulation of the hydrocarbons. Potential geothermal energy deposits (such as geysers and volcanoes) also occur along convergent plate margins. The heat at these locations is generally associated with recently active volcanism.

Deposits at Collisional Boundaries

Metalliferous deposits are also abundant at collisional boundaries where a variety of local tectonic settings exist, depending on distance from the boundary zone and stage of evolution (fig. 8.38h,i,j). Mineral deposits associated with oceanic ridges (ophiolites), convergent plate margins, and cratonic assemblages and continental rifts may all occur in collisional zones. Hydrocarbons may accumulate in foreland basins associated with collision. Oil and gas rise into sediments of these basins from cratonic-margin sediments that are deeply buried near the suture zone (Dickinson, 1974b). The immense accumulations in the Persian Gulf southwest of the Zagros suture in Iran may have this origin (Plate I).

Deposits in Cratonic Rift Systems

As was previously mentioned, crustal doming precedes rifting and appears to be related to hotspot activity in the upper mantle. Granites intruded at this stage have associated Sn and fluorite deposits (fig. 8.38a). At more advanced stages of cratonic rifting, evaporites accumulate in the rifts, and eventually oceanic metalliferous deposits are formed as the rifts open into a small ocean basins (fig. 8.38c). Aborted rifts (fig. 8.38b) are characterized by fluorite and barite deposits, carbonatites (with associated Nb, P, REE, U, Th, etc.), and Sn-bearing granites. Pb-Zn-Ag deposits occur in limestones during the early and middle-stages of cratonic rifting. The high heat flow from upwelling asthenosphere or mantle plumes provides a source for geothermal heat in cratonic rifts.

Deposits in Cratonic Basins

Cratonic basins, both marginal and intracontinental, provide an ideal setting for the accumulation of oil and gas. The opening of a cratonic rift system into an ocean basin involves a stage when sea water moves into the rift valley and evaporation exceeds inflow; evaporites are deposited at this stage (Rona, 1977). This environment is also characterized by restricted water circulation in which organic matter is preserved. As the rift continues to open, water circulation becomes unrestricted and accumulation of organic matter and evaporite deposition cease. High geothermal gradients beneath the opening rift and

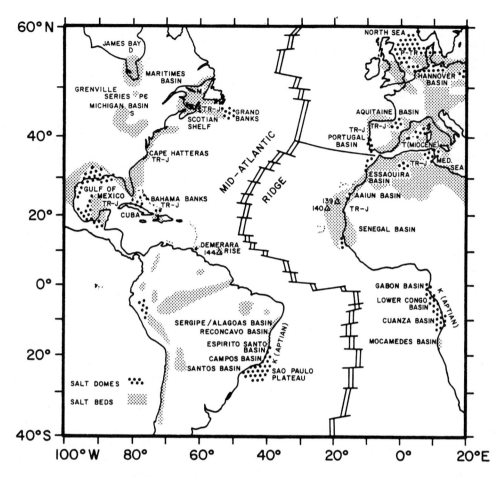

8.39. Distribution of salt deposits around the Atlantic basin (from Rona, 1977). Areas of high hydrocarbon potential correspond to those of salt deposition.

increasing pressure due to burial of sediments facilitates the conversion of organic matter into oil and gas. Finally, at a later stage of opening, salt beds in the evaporite succession, because of their gravitational instability, may begin to rise as salt domes and trap oil and gas. Oil and gas may also be trapped in structural and stratigraphic traps as they move upward in response to increasing pressures and temperatures at depth. Supporting this model are data from wells in the Red Sea, which represents an early stage of ocean development. Wells in this area have encountered hydrocarbons associated with high geotherms and rock salt up to 5 km thick. Also, around the Atlantic Basin there is a close geographic and stratigraphic relationship between hydrocarbon and salt accumulations (fig. 8.39).

Hydrocarbon production in intracratonic basins may also be related to plate tectonic processes (Rona, 1977). As previously discussed, increases in sea-floor spreading rates and oceanic ridge lengths may cause

marine transgression, which results in marine deposition in intracratonic basins. Decreasing spreading rates cause regression, which results in basins with limited circulation, and hence organic matter and evaporites accumulate. Unconformites also develop during this stage. Burial and heating of organic matter in intracratonic basins facilitates hydrocarbon production and salt domes, and unconformities provide major traps for accumulation.

SUMMARY STATEMENTS

1. A large amount of data indicate that the continents have drifted apart from one or two parental supercontinents in the last 200 m.y. The evidence for continental drift has also been interpreted in terms of an expanding Earth. Most data, however, do not support an expanding Earth model.

2. The following methods can be used to reconstruct lithospheric plate positions in the geologic past: (a) matching of continental borders, (b) matching of stratigraphic sections and crustal provinces on opposite continental margins, (c) matching of fossil assemblages and paleoclimatic data, (d) apparent polar wandering paths, (e) sea-floor spreading models, (f) studies of petrotectonic assemblages, and (g) traces of mantle hotspots defined by chains of related igneous rocks.

3. Continental fragmentation leads to diversification and specialization of living organisms, while continental collision leads to competition and extinction.

4. The distribution of paleoclimates on the Earth can be deduced from paleobiological indicators (such as corals and coal), climate-sensitive sediments (such as tillites, evaporites and bauxite), and oxygen isotope results. Paleoclimate distributions constrain the positioning of continents in the geologic past.

5. Plate reconstructions based on paleomagnetic pole positions have four important limitations: (a) structural orientations of the sample are often difficult to estimate; (b) soft magnetization may be difficult to remove without affecting hard magnetization; (c) it is not always possible to accurately date remanent magnetization; and (d) it is not possible to determine paleolongitudes. The fact that polar wandering paths of the continents diverge from each other going back in time indicates that the continents have drifted independently since the Permian. Continents have moved more rapidly for short periods of time (≤ 50 m.y.) in the geologic past than present continental drift rates.

6. Geological, geochemical, and geophysical evidence suggest that ophiolite complexes represent tectonically emplaced fragments of oceanic and/or back-arc basin crust. A typical ophiolite is composed of the following in ascending stratigraphic order: ultramafic rocks, gabbros, sheeted diabase dikes, and pillowed tholeiites. Ophiolites are emplaced by overthrusting of oceanic crust onto continental or arc crust (obduction) and by vertical faulting at shallow levels in subduction zones.

7. Convergent plate boundaries are characterized by the following subdivisions (progressing from trench toward overriding plate) and petrotectonic assemblages: (a) trench (graywacke turbidites); (b) subduction-zone complex (melange, large blocks of broken rock of diverse origin set in a sheared matrix); (c) arc-trench gap (forearc basins with eugeoclinal assemblages and local uplifts); (d) the arc proper (calc-alkaline volcanic rocks and intra-arc basins with eugeoclinal assemblages merging with deep-sea sediments or miogeoclinal assemblages); and (e) the back-arc basin (eugeoclinal assemblages).

8. Cratonic rifts are characterized chiefly by immature terrestrial sediments of granitic provenance and bimodal volcanic assemblages. Cratonic (miogeoclinal) assemblages are comprised chiefly of quartzites, shales, and carbonates; collision-related assemblages are characterized by immature clastic sediments of mixed provenances in foreland basins, graywacke turbidites in remnant ocean basins, and cratonic-rift assemblages in fault-produced basins in the thickened overriding continental blocks.

9. Linear chains of seamounts and volcanic islands in ocean basins and linear distributions of some igneous terranes on the continents are most easily interpreted in terms of lithospheric plates moving over mantle plumes (hotspots).

10. The arc-trench gap may grow by subduction accretion and kneading on the landward slope of the trench and by inward migration of the zone of maximum arc magmatism. Subduction erosion and sediment subduction also characterize some convergent margins.

11. Subduction-zone polarity in ancient arc systems, in some instances, may be deduced from compositional polarity of igneous rocks in the arc, asymmetry of nappes and thrusts along collision boundaries (directed toward the converging plate), and zonation of metamorphic belts (increasing P-T assemblages in dip direction of descending slab).

12. Mantle-activated rifts (oceanic ridges, cratonic rifts and aulacogens, and some back-arc basins) are produced by upwelling asthenosphere or rising mantle plumes. Lithosphere-activated rifts

(related to faulted continental margins, collisions, or arcs) result from stresses in moving lithospheric plates. Cratonic rifts develop in stable cratons and aulacogens represent failed arms of evolving cratonic rift triple junctions.

13. Cratonic basins subside at approximately the same exponential rate as ocean basins. Subsidence is probably related to one or both thermal doming and eroding off the top of the lithosphere followed by contraction or thinning and stretching of the lithosphere.

14. Eustacy is probably controlled chiefly by sea-floor spreading rates (which determine the cross-sectional area of oceanic ridges) and by the total length of oceanic ridges. Rapid spreading and long ridge length result in transgression, whereas slow spreading and short ridge length result in regression.

15. In terms of plate tectonics, orogeny occurs at convergent plate boundaries and may result from (a) activation, the onset of a convergent plate boundary at or near a previously stable continental margin, or (b) collision, involving two continents, two arcs, or an arc and a continent. Unlike the geosynclinal theory of orogeny, the plate-tectonic theory can accommodate a variation in sequence of events, and each orogenic belt may have a unique history.

16. Activation orogenies are characterized by a pronounced polarity in the composition of igneous rocks, metamorphic zones, and paired arc-trench gap and arc-rear assemblages; thrusting, however, is commonly divergent. Collision orogenies exhibit nappe and thrust asymmetry dipping toward the plate being consumed, and have a suture zone (composed of ophiolites and melange) defining the collision boundary.

17. The origin and occurrence of metallic mineral deposits, oil and gas, and geothermal energy are related to plate tectonics. Metalliferous sediments, massive sulfides, and Ni and Cr deposits

form at oceanic ridges. A variety of metallic mineral deposits and geothermal energy occur in the overriding plate at convergent boundaries, and oil and gas are formed in many back-arc basins; cratonic rift systems contain geothermal energy, Sn, U, Th, REE and other rare element deposits; collisional boundaries are characterized by a variety of metallic mineral deposits and by oil and gas occurrences; and cratonic basins are sites of major hydrocarbon accumulation.

SUGGESTIONS FOR FURTHER READING

Bird, J.M., editor (1980) *Plate Tectonics*, Second ed. Washington, D.C.: Amer. Geophys. Union. 992 pp.

Coleman, R.G. (1977) *Ophiolites*. New York: Springer. 229 pp.

Dickinson, W.R., and Yarborough, H. (1976) Plate tectonics and hydrocarbon accumulation. *Amer. Assoc. Petrol. Geol.*, Continuing Education Course Notes Ser. 1. 109 pp.

DuToit, A. (1937) *Our Wandering Continents*. London: Oliver and Boyd. 366 pp.

Frakes, L.A. (1979) *Climates Throughout Geologic Time*. Amsterdam: Elsevier. 310 pp.

Hughes, N.J., editor (1973) Organisms and Continents Through Time. *Spec. Papers Paleont. Assoc.* No. 12. London: Paleont. Assoc. London. 334 pp.

McElhinny, M.W. (1973) *Paleomagnetism and Plate Tectonics*. Cambridge: Cambridge University Press. 358 pp.

Strong, D.F. (1976) Metallogeny and Plate Tectonics. *Geol. Assoc. Canada Spec. Paper No. 14*. 660 pp.

Talwani, M., and Pitman, III, W.C. (1977) Island arcs, deep-sea trenches, and back-arc basins. *Amer. Geophys. Union, Maurice Ewing Series* 1. 470 pp.

Chapter 9

Phanerozoic Orogenic Systems

The major Phanerozoic orogenic belts, as was discussed in Chapter 5, are the Appalachian-Caledonian in eastern North America and western Europe; the Hercynian in central Europe; the Cordilleran around the Pacific basin; and the Alpine-Tethyan extending from western Europe to southeast Asia (Plate I). These and other Phanerozoic orogenic belts have formed in response to complex plate interactions at convergent plate boundaries involving both activation and collision orogenies. Before outlining the plate histories of portions of these orogenic systems, it is instructive to summarize continental reconstructions during the last 500 m.y.

CONTINENTAL RECONSTRUCTIONS

Paleozoic

Although the positioning of continents during the Paleozoic is based chiefly on paleomagnetic data, paleontologic and paleoclimatic results and the distribution of petrotectonic assemblages provide important constraints on reconstruction of specific tectonic settings (Irving, 1979b; Ziegler et al., 1979). The largest supercontinent during the Phanerozoic is *Pangaea*, which includes most of the existing continents (fig. 9.7). *Gondwana* is comprised principally of South America, Africa, Arabia, Madagascar, India, and Australia. Parts of southwestern Eurasia and central Europe may also have been part of Gondwana during some periods of time. Several reconstructions of Gondwana have been proposed (fig. 9.1). Two alternate positions for Madagascar have been proposed—one off the coast of Kenya (A, B, D) and one

adjacent to Mozambique (C). Paleomagnetic and sea-floor spreading reconstructions favor the Kenya position (Norton and Sclater, 1979), and geometric matching of continental margins and correlation of crustal provinces and rock type allow either fit. *Laurasia* consists of North America, Europe, and most of Asia. *Laurentia* includes most of North America, Scotland and Ireland north of the Caledonian suture, Greenland, Spitzbergen, and the Chukotsk peninsula of eastern Siberia (fig. 9.2). *Baltica* consists of northern Europe (except the British Isles) bounded on the south by the Hercynian suture and on the east by the Uralian suture. *Laurussia* includes Laurentia, Baltica, England, and Nova Scotia and adjacent areas. Asia north of the Himalayas and east of the Urals consists of three or more continents, the best defined of which are Siberia, Kazakhstania, and China. Siberia is bordered on the west by the Uralian suture and on the southwest by the Irtysch crush zone. Paleozoic continental reconstructions based chiefly on paleomagnetic results are given in figs. 9.2 to 9.7. Because it is not possible to determine paleolongitudes from APW curves, longitudinal separations between continents are estimated from faunal distributions in space and time (Ziegler et al., 1979).

By 1000 m.y. ago, three or more supercontinents were in existence (Chapter 10, fig. 10.8). These continents are fragmented beginning in the late Precambrian, and by the Cambrian Laurasia splits into Laurentia, Baltica, Siberia, Kazakhstania, and China (fig. 9.2). A striking feature of both the late Cambrian and Ordovician also supported by paleoclimatic data, is the low latitude of the continents. Convergent plate boundaries during the Cambrian and Ordovician existed along the northeastern coast of Laurentia

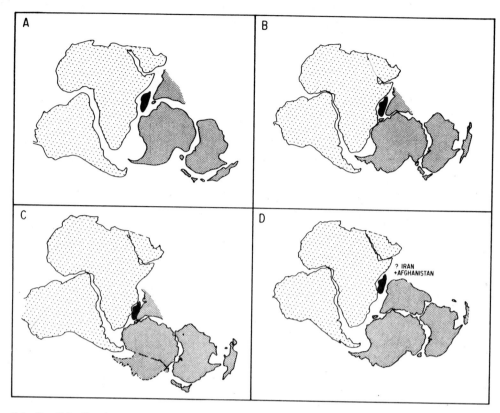

9.1. Possible Gondwana reconstructions (after Powell *et al.*, 1980).

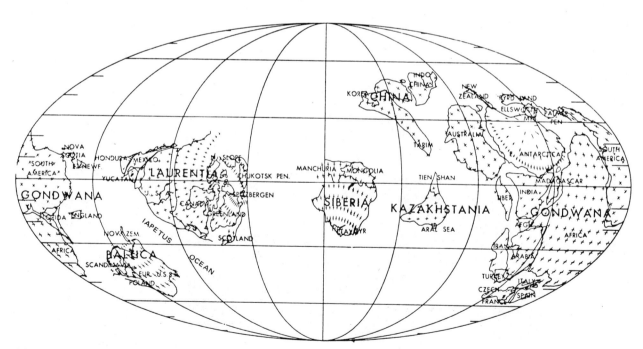

9.2. Late Cambrian continental reconstruction (after Scotese *et al.*, 1979).

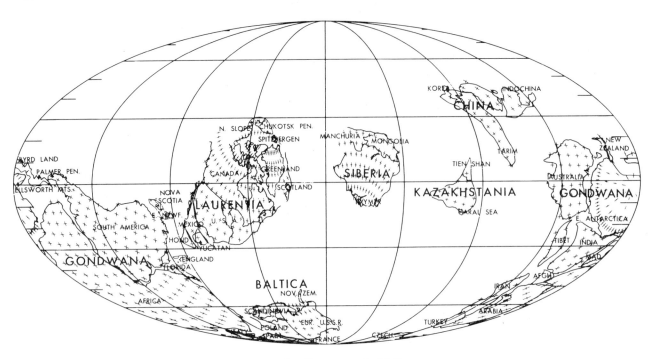

9.3. Middle Ordovician continental reconstruction (after Scotese *et al.*, 1979).

9.4. Middle Silurian continental reconstruction (after Scotese *et al.*, 1979).

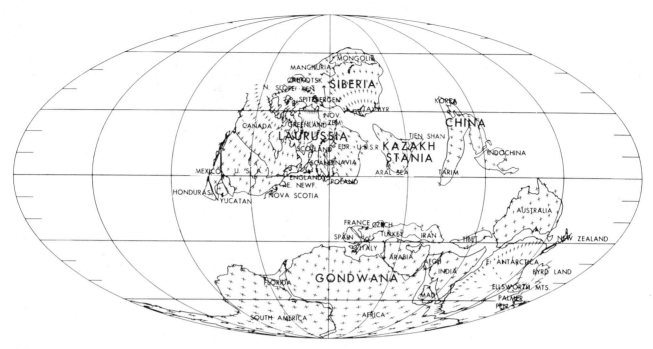

9.5. Early Devonian continental reconstruction (after Scotese *et al.*, 1979).

and the western coast of Baltica, along the south-western margin of Siberia, along the eastern and western margins of Kazakhstania, and along the eastern Australia-New Zealand-Antarctica portion of Gondwana*. By middle Ordovician (fig. 9.3), Gondwana is moving toward the South Pole, which resulted in glaciation in Africa. From upper Cambrian through Ordovician, Laurentia undergoes counter-clockwise rotation but remains at low latitudes.

During the Silurian, Baltica and Siberia move toward Laurentia and Gondwana is approximately centered over the South Pole (fig. 9.4). The Silurian is a time of extensive transgression, presumably reflecting rapid sea-floor spreading and extensive oceanic-ridge systems. Cambro-Ordovician subduction zones continue and the Antarctic-Australian subduction system extends along the western margin of South America. Active subduction systems lengthen along both margins of the Iapetus Ocean between Laurentia and Baltica. By early Devonian the Iapetus Ocean closes, resulting in the Caledonian and Acadian colli-sion orogenies in northern Europe and eastern North America (fig. 9.5). Baltica collides in a more southerly position than its position at the time of the

opening of the Atlantic Ocean in the Mesozoic, and transcurrent faulting moves it northward by early Triassic. Also, convergent plate boundaries are ini-tiated along the northeastern and southeastern coasts of Baltica during the Devonian.

Consuming plate margins characteristic of the early Paleozoic (except those around the closed Iape-tus Ocean) continue into the Carboniferous, and new ones develop on the facing margins of Laurussia and Gondwana, leading to the collision of these conti-nents in the late Carboniferous, producing the Hercynian orogeny (fig. 9.6). Also by late Carbonifer-ous, Kasakhstania collides with Siberia. These two major collisions represent the first stages in the devel-opment of the supercontinent Panagaea. The *Tethys Ocean*, between Gondwana on the South and Laurus-sia, Siberia, and China on the north, also comes into existence. Another major tectonic event in the late Carboniferous and Permian is the rifting of Tibet, Iran, and Turkey away from the northern margin of Gondwana (fig. 9.7). Extensive late-Paleozoic glacia-tion occurs in Gondwana. The rise of mountains along the Hercynian orogenic belt blocks moist air from reaching Laurussia-Baltica, resulting in widespread dry conditions on these continents in the Permian and early Mesozoic. The Permian is also characterized by extensive activation orogeny along the South America-Antarctica-Australia convergent plate boundary (the Samfrau orogeny).

---

*Directions given relative to present continental configura-tions.

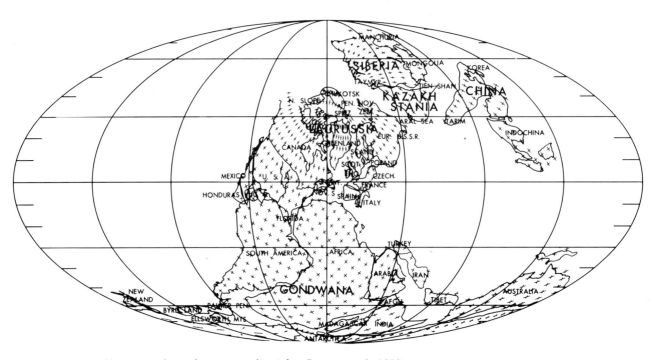

9.6. Late Carboniferous continental reconstruction (after Scotese *et al.*, 1979).

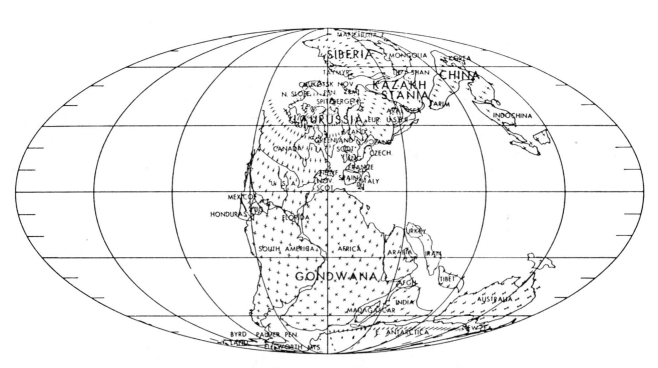

9.7. Late Permian continental reconstruction (after Scotese *et al.*, 1979).

Mesozoic-Cenozoic

In the late Permian and Triassic, Laurentia moves northward relative to Baltica (fig. 9.7). Siberia collides with Baltica and China during the late Permian or Triassic and the Ural Mountains develop along the collisional zone of Baltica-Kazakhstania with Siberia. Pangaea is short-lived and begins to break up almost before it is completed (Irving, 1979b). In the late Triassic China-Malaya collides with Laurussia (fig. 9.8). In the middle and late Jurassic, North America and Africa are rifted apart and the North Atlantic begins to open (fig. 9.9). Major opening of the North Atlantic, however, does not occur until the Cretaceous (fig. 9.10). Also, during the middle to late Jurassic, Africa begins to be rifted away from India-Antarctica. In the middle Cretaceous the South Atlantic begins to open as Africa is rifted away from South America, the western Arctic basin begins to open as Alaska rotates to the west, and the Bay of Biscay opens as Spain rotates counterclockwise. During the late Cretaceous, the Labrador Sea begins to open as Greenland moves away from North America and Africa continues its counterclockwise motion closing the Tethys Ocean (fig. 9.10). North Africa also begins to collide with Eurasia, the South Atlantic becomes a wide ocean, Madagascar reaches its pre-

sent position relative to Africa, and India begins rapid motion northwards toward Asia.

In the early Tertiary, the North Atlantic between Greenland and Norway begins to open and the eastern Arctic basin opens, thus completing the Arctic Ocean, which began to open in the early Cretaceous. India collides with Tibet at about 35 m.y. ago, and the modern Carlsberg Ridge forms in the Indian Ocean. The Ninety-East ridge appears to represent an extinct transform fault associated with the northward migration of India. New Zealand also separates from Australia as the Tasman Sea opens in the early Tertiary. By mid-Tertiary (fig. 9.11) the Tethys Ocean completely closes, leaving the Mediterranean, Black, and Caspian Seas as remnants. Nearly continuous continental crust extends from the tip of South America through Antarctica-Australia to New Guinea during the early Tertiary. In mid-Tertiary time the Scotia arc forms and South America is separated from Antarctica; Australia is also rifted from Antarctica and migrates rapidly northwards. Beginning about 30 m.y. ago, widespread continental rifting results in the initiation of the Basin and Range Province and the East African rift system, including the opening of the Red Sea; also Iceland begins to rise on the Mid-Atlantic Ridge. In the last 10 m.y., spreading directions change in the Pacific basin, the

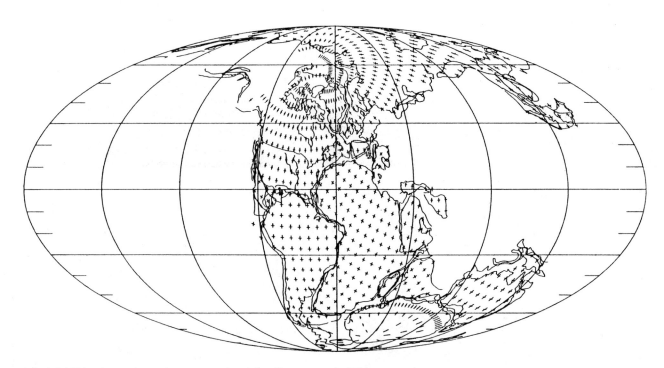

9.8. Mid-Triassic continental reconstruction (after Scotese *et al.*, 1981, personal communication). See Figs. 9.2 to 9.7 for names of land masses.

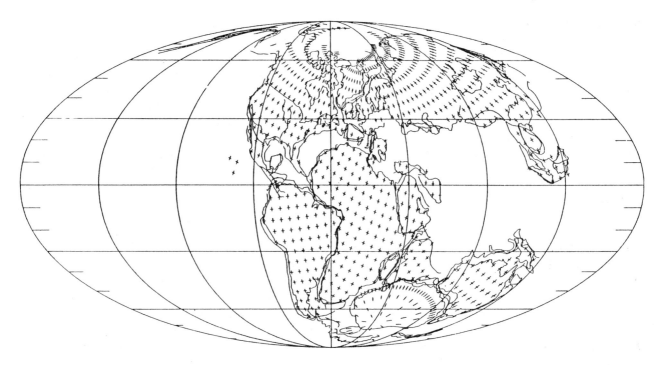

9.9. Late Jurassic continental reconstruction (after Scotese *et al.*, 1981, personal communication). See Figs. 9.2 to 9.7 for names of land masses.

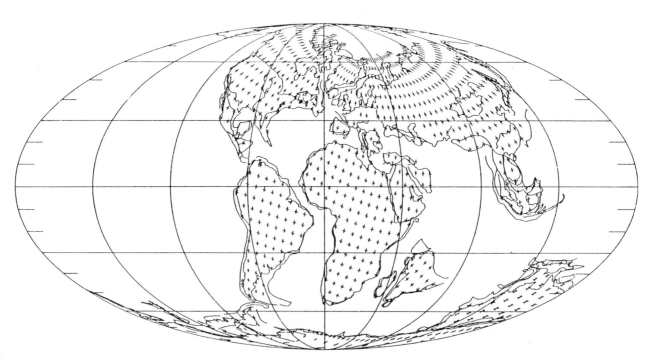

9.10. Late Cretaceous continental reconstruction (after Scotese *et al.*, 1981, personal communication). See Figs. 9.2 to 9.7 for names of land masses.

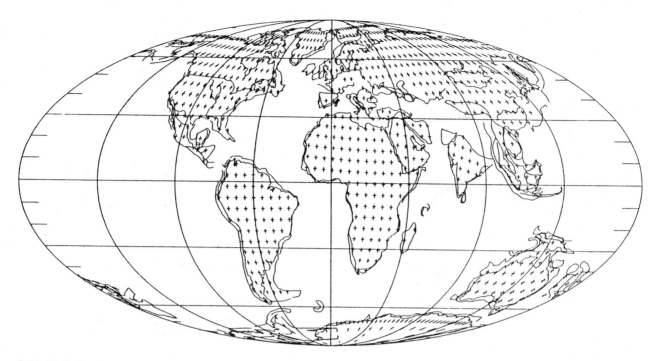

9.11. Early Tertiary continental reconstruction (after Scotese *et al.*, 1981, personal communication). See Figs. 9.2 to 9.7 for names of land masses.

Gulf of Aden completely opens, and the Gulf of California opens as the San Andreas transform system migrates southward.

THE APPALACHIAN-CALEDONIAN SYSTEM

Major Paleozoic orogenies are recorded along both coasts of the North Atlantic. The *Caledonian orogeny*, which occurred in the upper Silurian and lower Devonian, is widespread in the British Isles, Scandinavia, and eastern Greenland. The *Taconic orogeny* in the middle Ordovician, the *Acadian orogeny* in the middle Devonian, and the *Alleghenian orogeny* in the late Carboniferous and Permian are prominent in the Appalachian Mountains in eastern North America. These orogenies can be related to a complex history of the North Atlantic region involving the opening, closing, and reopening of an ocean basin (fig. 9.12). The Iapetus Ocean began to develop in the late Precambrian when North America and Africa were rifted apart (A). At this time, thick sections of cratonic sediments were deposited on both retreating continental margins (Dewey and Bird, 1970a). Before the beginning of the Cambrian a convergent plate margin developed along the African coast and, dur-

ing the Cambrian, along the American coast. The subduction zones may have been composite and, along at least portions of the Atlantic margin, may have dipped oceanward.

Activation and collisional orogeny began in North America, and by early to middle Ordovician major westward-directed thrusts, klippen, and gravity slides formed and plutonic and metamorphic activity occurred, recording the Taconic orogeny (D). At this time eugeocline assemblages formed in the western part of the arc-rear area. During the late Silurian and Devonian, Africa and Europe collide with the northeastern coast of North America. Resulting thrusting, metamorphism, and magmatism record the Acadian and Caledonian orogenies (E). Obducted ophiolite complexes in Newfoundland are interpreted as remnants of the Iapetus Ocean floor. Compressive forces continued into the late Pennsylvanian and Permian in the central and southern Appalachians (and in the Ouachita area), producing large continent-directed thrust sheets and nappes (the Alleghenian orogeny) (F). This deformation may record the final collision of Africa with the southeastern coast of North America. Africa is finally welded to North America along a suture zone which is now buried, in part, by younger sediments in the Atlantic coastal plain. A complex history of Paleozoic deformation and mag-

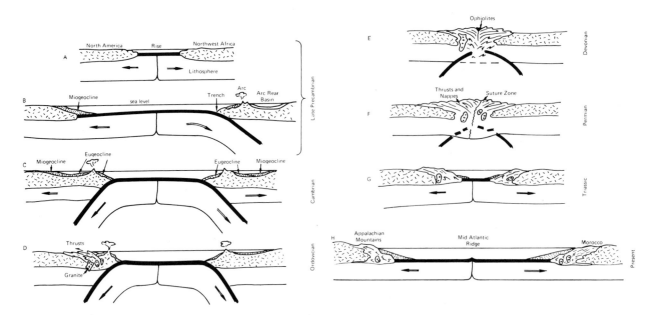

9.12. Schematic cross-sections from North America to Africa illustrating the opening and closing of the Iapetus Ocean and the opening of the Atlantic Ocean (modified after Dietz, 1972).

matism paralleling that recorded in eastern North America is also recorded in northwest Africa, although it is not as well understood (Hurley et al., 1973). The opening of the Atlantic Ocean began in the Triassic (G) and is continuing today (H) with new miogeoclinal assemblages forming on the trailing edges of North America and Africa. Although this simplistic evolutionary scheme describes the overall history of the Appalachian-Caledonian belt, the detailed history at any locality is considerably more complex, as the following two examples illustrate.

The Southern Appalachians

The southern Appalachian Mountains in the southeastern United States can be divided into several tectonic zones, as is shown in figure 9.13. The Valley and Ridge province includes folded and faulted Paleozoic cratonic sediments. The Blue Ridge province is comprised of Precambrian igneous and metamorphic rocks that appear to have been transported westward with the Inner Piedmont and Charlotte and Carolina belts in a decollement thrust. A COCORP seismic-reflection profile in approximately the same location as fig. 9.13 indicates that the decollement is 6 to 15 km thick (Cook et al., 1979). Geologic data indicate that it probably overlies Paleozoic cratonic sediments. The Brevard zone is interpreted as a subsidiary thrust emanating from the main thrust. The

Inner Piedmont province consists of Precambrian metasediments and metavolcanics intruded with a variety of plutonic rocks. The King Mountain belt is a synclinal zone and the Charlotte belt is a region of high-grade gneisses of Precambrian and Paleozoic age. The Carolina slate belt is a synclinorium of somewhat lower-grade rocks. Precambrian basement is unconformably covered with Mesozoic rift and cratonic sediments in the Coastal Plain.

A model for the evolution of the southern Appalachians consistent with available geological and geophysical data is illustrated in fig. 9.14. During the late Precambrian, rifting of Laurasia produces at least three small ocean basins with two continental fragments, the eastern Blue Ridge-Piedmont (EBR-IP) and the Charlotte-Carolina slate belts (CB-CSB) in between (1 and 2). The metavolcanics and metasediments of the western Blue Ridge province are deposited in the ocean basin between North America and the EBR-IP province block. Subduction and related volcanism begins beneath the CB-CSB province by 650 m.y. ago (3) and along the western margin of the EBR-IP province by 500 m.y. ago (4). The small ocean basin between the EBR-IP block and North America closes and cratonic sediments derived from America begin to accumulate by 500 m.y. ago. This closure between 450 and 500 m.y. produces the Taconic orogeny (5). At this time the major thrusting of the Blue Ridge and Piedmont over

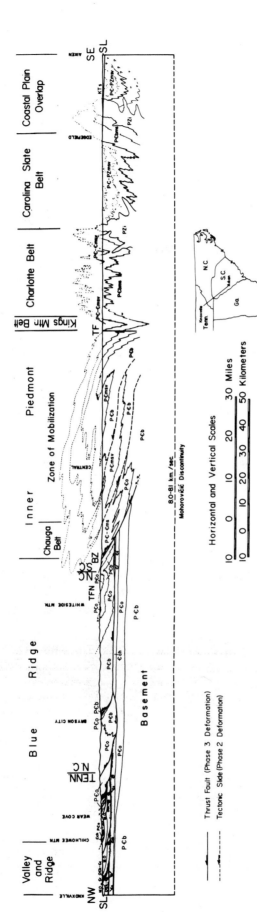

9.13. Diagrammatic cross-section of the southern Appalachians (after Hatcher, 1972). TFN Tallulah Falls Nappe; BZ Brevard Zone; TF-Towaliga Fault; PЄb—earlier Precambrian basement rocks; PЄbms—earlier to late Precambrian basement and metasedimentary and metavolcanic rocks; PЄo—Ocoee Series; PЄmsv—late Precambrian metasedimentary and metavolcanic rocks; PЄ-Єms and PЄ and Єmsv—late Precambrian and Cambrian metasedimentary and metavolcanic rocks; PZi—Paleozoic intrusive rocks; Єch—Chilhowee Group; Єs—Shady Dolomite; Єr—Rome Formation; OЄk-Єc-Knox and Conasauga Groups; M.O.—middle Ordovician Chickamauga Group rocks; KT—Cretaceous and Tertiary Sediments.

197

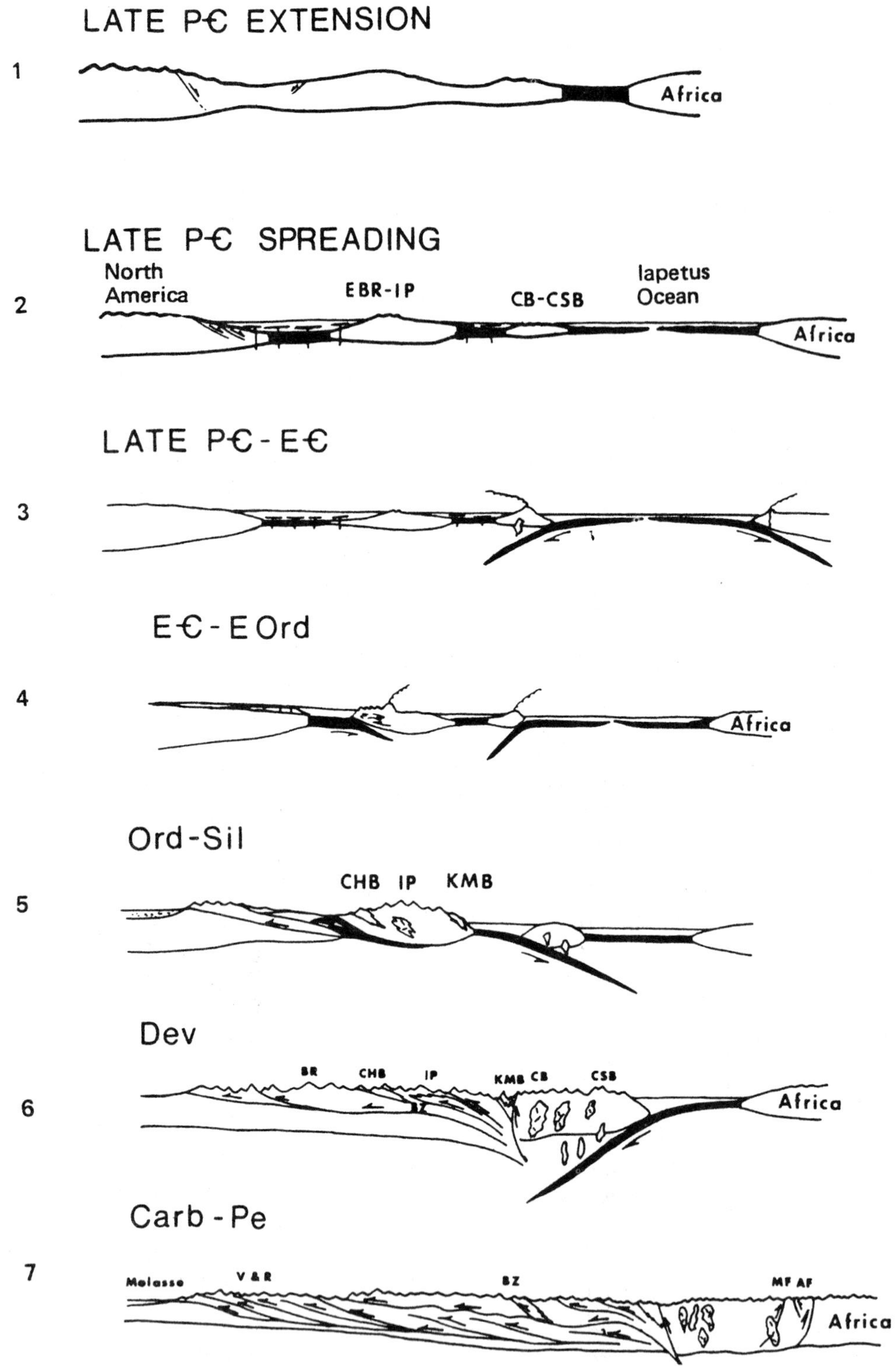

LATE PꞒ EXTENSION

1

LATE PꞒ SPREADING

North
America   EBR-IP   Iapetus
                                                         CB-CSB   Ocean

2

Africa

LATE PꞒ - EꞒ

3

EꞒ - E Ord

4   Africa

Ord - Sil

              CHB IP KMB

5

Dev

          BR   CHB   IP   KMB CB   CSB

6   Africa

Carb - Pe

7   Molasse   V & R   BZ   MF AF

Africa

9.14. Plate-tectonic history of the southern Appalachians (after Hatcher and Odom, 1980). EBR, eastern Blue Ridge; IP, Inner Piedmont; CB, Charlotte belt; CSB, Carolina slate belt; CHB, Chauga belt; KMB, Kings Mountain belt; BR, Blue Ridge; V & R, Valley and Ridge; BZ, Brevard zone; MF, Modoc fault; AF, Augusta fault.

the North America Plate occurs. Such thrusting must involve decoupling of the lower crust and mantle lithosphere from the upper crust, which alone is involved in the thrust. After this closure, sediments are derived primarily from the east. The Acadian orogeny, between 350 and 400 m.y. ago, is characterized by widespread deformation, metamorphism, and plutonism and appears to have been triggered by the closing of the ocean basin between the EBR-IP and Carolina slate belt (6). By this time the Blue-Ridge and Inner Piedmont are accreted to the North American Plate. The King Mountain belt may be a surface remnant of the collisional zone between the Inner Piedmont and Carolina slate belt provinces. After the Acadian orogeny, a new convergent boundary develops beneath the Carolina slate belt dipping to the west (6). The Alleghenian orogeny (250 to 300 m.y. ago) results from the closing of the Iapetus Ocean and the collision of Africa with America (7). This collision produces extensive plutonism and large-scale overthrusting directed toward both the American and the African plates. The African Plate is again fragmented from the North American Plate in the Triassic as the Atlantic basin begins to open. The line of rifting is east of the Carolina slate belt, buried today by Atlantic-Coast cratonic sediments.

Newfoundland

Fragments of both the eastern and western margins of the Iapetus Ocean are preserved in Newfoundland. The succession of rocks from the edge of the Avalon platform northwest into the Central Paleozoic mobile belt is interpreted as the southeast margin of the Iapetus Ocean (Colman-Sadd, 1980). An evolutionary model for this area is summarized in fig. 9.15. In the late Precambrian, a subduction zone dips eastwards beneath the Avalon platform, and the Iapetus Ocean lies to the west (A). The protoliths of the Little Passage gneisses appear to have formed as a subduction complex at this time. By the middle Ordovician, a new subduction zone develops with an extensive arc and back-arc basin system (B). Sediments of the North Steady Pond, St. Joseph's Cove, Riches Island and Isle Galet formations are deposited in these back-arc basins, and include mixed eugeoclinal and ocean-floor sediments. The eastern arc must also be active, in that arc sediments are found in the Isle Galet formation. The closing of the Iapetus Ocean began in the middle Ordovician with the obduction of ophiolites over the Western platform (now in northwestern Newfoundland) (C). Ophiolite obduc-

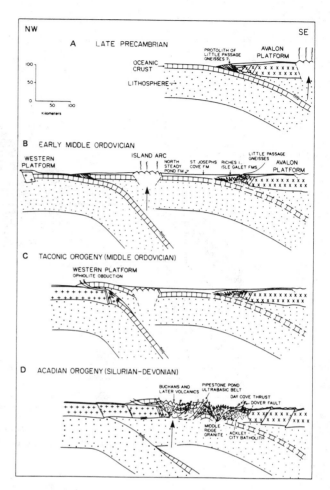

9.15. Evolutionary model for the eastern margin of the Iapetus Ocean as recorded in Newfoundland (after Colman-Sadd, 1980).

tion was separated by major Taconic deformation and metamorphism by about 50 m.y. Actual collision of the Western and Avalon platforms did not occur until late Silurian–early Devonian time and is recorded by extensive thrusting, folding, and volcanism associated with the Acadian-Caledonian orogeny (D). During this time the crust was thickened and the lower portion partially melted, giving rise to granitic plutons in the Central mobile belt.

THE HERCYNIAN SYSTEM

The Hercynian orogenic belt developed in the late Carboniferous in response to the collision of Africa–South America with Europe and North America. The orogenic belt extends from the Ouachita region north of the Gulf of Mexico across northwestern Africa

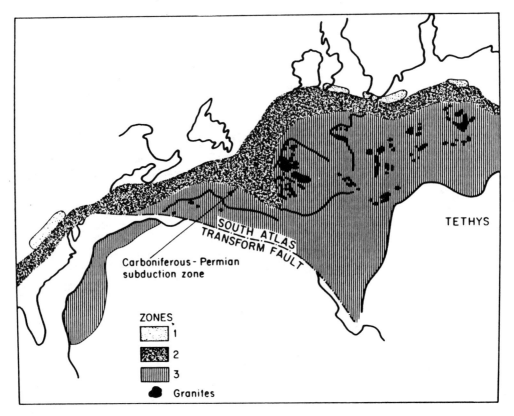

9.16. Major zones of the Hercynian orogenic belt shown on a Pangaea continental reconstruction (after Dewey and Burke, 1973). See text for explanation of zones.

into Europe and perhaps farther east. The belt is comprised of three main zones (fig. 9.16):

Zone 1. This zone is a foreland region that was not affected by Hercynian deformation. It is the site of continental sandstone deposition during the Devonian and of marine cratonic sediment deposition during the early Carboniferous, and is characterized by discontinuous coal basins during the late Carboniferous.

Zone 2. This is a region of mixed marine and nonmarine deposition during the early Devonian and of mafic volcanism and marine sedimentation during the mid-Devonian and early Carboniferous. Mid-Carboniferous sediments are dominantly immature clastic sediments derived from the south.

Zone 3. This zone is comprised of reworked Precambrian basement, examples of which are the Bohemian massif and the Massif Central.

Major periods of deformation in the Hercynian belt occur in the Devonian (345 m.y.) and in the Carboniferous at 325 m.y. and at 300 m.y. The last phase is the major deformation characterized by large decollement thrusts and nappes that developed in response to the continental collision of Europe and Africa.

The many plate-tectonic models that have been suggested to explain the Hercynian orogenic belt (Windley, 1977) fall into activation or collision orogenies. Models for activation orogeny generally involve northward dipping subduction zones in which the European continent is overriding oceanic crust. Most evidence, both geological and paleomagnetic, however, seems to support a collisional model, at least for the late Carboniferous deformational phases. Paleomagnetic data favor a suture zone extending from southern Britain across northern France into east-central Europe (Jones et al., 1979). Although a distinct suture has not been described, ophiolites occur in this zone (such as the Lizard body in southern England), and the sedimentary and tectonic histories are markedly different on both sites of the probable collisional zone.

THE CORDILLERAN SYSTEM

West-Central North America

Late Precambrian and Paleozoic Evolution. Late Precambrian (< 1 b.y.) and Cambrian rocks in the Cordilleran belt of the western United States represent a miogeoclinal assemblage of quartzites, argillites, siltstones, subgraywackes, and small amounts of carbonate and conglomerate which extend from Alaska to northern Mexico (Stewart, 1972). The section thickens in some areas to > 7 km and overlies older sialic basement with an angular unconformity (fig. 9.17A). Minor occurrences of tholeiite flows and sills are found in parts of the succession. In terms of plate tectonics, this miogeoclinal assemblage has been interpreted by some as a trailing continental-edge assemblage that developed during the opening of a proto-Pacific basin. The continental fragment that was rifted away may now comprise part of Siberia. This stable continental margin appears to have persisted until late Ordovician or early Devonian time.

By Ordovician-Silurian time, one or more arcs had formed along and/or nearby the western edge of the continent. Three possible arrangements of subduction zones and arcs are shown in figure 9.17B, C and D. The first case (B) is characterized by a marginal-sea basin and the latter two cases by one or two oceanic basins of unknown size. During the middle Devonian and Mississippian, the *Antler orogeny* occurred in the western United States and the western eugeoclinal assemblage was thrust over the miogeoclinal assemblage along the Roberts Mountain thrust in Nevada (Nilsen and Stewart, 1980). The upper plate consists of Cambrian to Devonian chert, argillite, greenstone, sandstone, and quartzite thrust over late Precambrian to Devonian carbonate, shale, and quartzite. This may have been accomplished in any of the three models illustrated in E, F, and G. In case E, the Antler orogeny is an activation orogeny, whereas in cases F and G it is an arc-continent collision orogeny. The outer arc in G is equated with the Klamath Mountains arc system, which may have been welded to the continent during the early Mesozoic. Minor quantities of calc-alkaline volcanic rocks that erupted during the final stages of the Antler orogeny seem to necessitate a subduction zone dipping beneath the continent (as in E). If either model F or G are accepted, a new subduction zone dipping beneath the continent must have developed after the arc collision, or the subduction zones shown in C–F and D–G must have dipped in the opposite direction. The Pennsylvanian and Permain in the western United States are characterized by deposition in linear basins west of the Antler orogenic belt. This deposition occurs in arc-trench gaps, marginal-sea basins, or both.

Mesozoic Evolution. During the early Mesozoic, the subduction zone beneath the southwestern portion of the United States (south of 38°N) appears to have changed orientation from a northeasterly to a northwesterly strike, as evidenced by the cross-cutting of Paleozoic by Mesozoic structural trends (Burchfiel

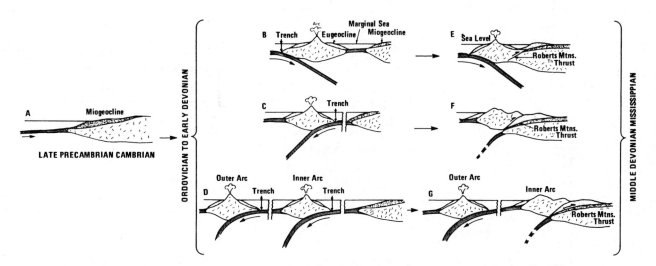

9.17. Schematic representation of the late Precambrian to Mississippian plate-tectonic history of the Cordilleran belt in western North America (modified after Burchfiel and Davis, 1972; Stewart, 1972.) B → E, C → F, and D → G are alternate models that seem equally consistent with existing data.

and Davis, 1972). Paleomagnetic and paleontologic data support this hypothesis and suggest that the drifting plate fragment(s) was rewelded to the continent in western Canada, choking the subduction zone(s) in this area and causing it to jump oceanward (Monger et al., 1972).

Zones of progressively increasing metamorphic grade in an easterly direction (as exemplified in western California—see fig. 8.24C), compositional polarity of igneous rocks, and the general distribution of melanges, arc-trench gap, and arc-rear assemblages indicate for the most part a steep (45–60 degrees), east-dipping subduction system along the western coast of the United States during the Mesozoic. The Franciscan Group in California (of Jurassic-Cretaceous age) is an example of a melange and arc-trench gap assemblage, and it is coeval with the Great Valley sequence, which is a forearc basin assemblage containing, in part, detritus from the unroofing of the Sierra Nevada and related batholiths to the east. The two formations are in thrust-fault contact (fig. 9.18) in western California. The fault zone is generally interpreted as a surface manifestation of the associated subduction zone, as is illustrated in the figure. Fragments of oceanic upper mantle and crust have been tectonically mixed with graywacke in the Franciscan melange.

The early Mesozoic is also the time of collision of the Klamath and related arc systems (outer arc in fig. 9.17G) with the continent (the subduction zone jumping to the west) or, alternately, the complete elimination of the marginal basin behind the peripheral arc system by eastward thrusting (fig. 9.17E). The late Permian-early Triassic *Sonoma orogeny* in the Great Basin, characterized by the eastward-directed Golconda thrust sheet, is an example of this early Mesozoic plate deformation. Mesozoic thrusting differs from Paleozoic thrusting in that it is dominantly divergent. Large westward-directed thrust sheets in the Klamath Mountains appear to reflect underthrusting of the Klamath arc, and the large eastward-directed, Mesozoic thrust belt extending from Alaska to Mexico in the Rocky Mountain area reflects compressive orogenic forces associated with subduction.

It is also noteworthy that increased plutonism and volcanism in the Cordilleran belt at 85–110 m.y. correlates with increased spreading rates in the Farallon Plate at this time (fig. 9.19). Such a correlation suggests that increased spreading and subduction result in increased melting and magma production at convergent plate boundaries. It is also significant that tectonic and magmatic events in the North American Cordillera correlate well with stages in the history of

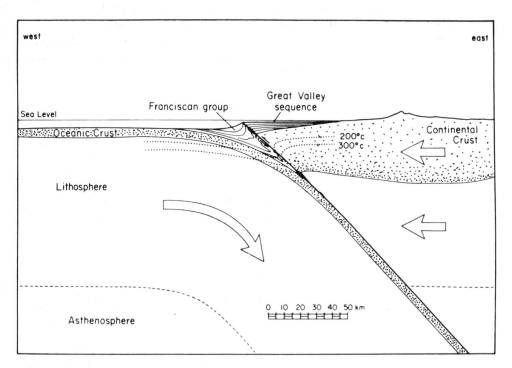

9.18. Diagrammatic cross section of the Mesozoic subduction system in western California (after Ernst, 1970).

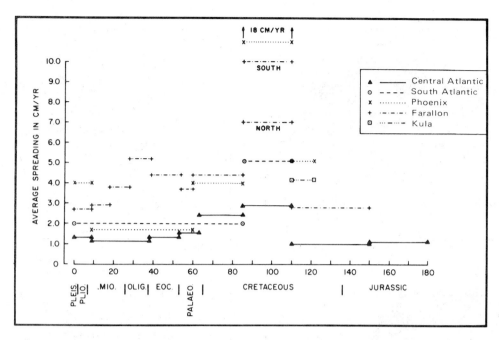

9.19. Average spreading rates in the Atlantic and Pacific Oceans as a function of geologic time (after Larson and Pitman, 1972). For locations of Pacific plates, refer to Fig. 6.16.

the opening of the North Atlantic (Coney, 1973). For example, the changes in spreading directions and rates in the North Atlantic at 135 and 80 m.y. are recorded in the Cordillera by the onset of thrusting and plutonism associated with the Sevier-Columbian and Laramide orogenies, respectively. A reduction in spreading rates at about 50 m.y. also correlates well with a rather abrupt end to the Laramide orogeny between 40 and 50 m.y. These correlations suggest that activation-type orogenies may be controlled more by the motions of an overriding continental plate than by a subducted oceanic plate.

Cenozoic Evolution. Igneous-rock compositional polarity data and age distributions indicate that the Cenozoic subduction zone from about 70 to 30 m.y. ago had a shallow dip of only 20–25 degrees and that it may have extended for 1500 km surface distance beneath the Cordilleran belt (Coney, 1978; Lipman, Prostka, and Christiansen, 1972). A gap in igneous activity from 40 to 60 m.y. between latitudes 34°N and 40°N appears to record temporary cessation of subduction between these latitudes. The transition from a steep to a shallow subduction zone occurs between 70 and 80 m.y. ago and may reflect the overriding of the Farallon Plate (fig. 6.16) by the North American Plate. The eastern segment of the subduction zone may have broken off (partially

or completely) producing imbricate slabs (Lipman, Prostka, and Christiansen, 1972). Polarity data indicate the subduction zone steepened again in the Oligocene along the northwest coast of North America but maintained its shallow dip along the southwest coast until its cessation, starting about 25–30 m.y. ago.

Perhaps the most significant event in Cenozoic plate-tectonic history was the progressive annihilation of the East Pacific Rise in the Cordilleran subduction zone and the concurrent development of the San Andreas transform system (Atwater, 1970). Four stages in this annihilation are illustrated in fig. 9.20. About 30 m.y. ago the rise reaches the trench and the Farallon Plate becomes two Plates, the Juan de Fuca and Cocos Plates (A). As the rise continues to be consumed, the San Andreas transform fault develops, reflecting the relative motions of the Pacific and North American plates (right lateral motion). Each end of the fault is marked by a triple junction that migrates north and south, respectively, as subduction of the rise system continues (fig. 9.20B). About 3–4 m.y. ago, Baja California decouples from the North American Plate and recouples to the Pacific Plate, and the Gulf of California begins to open by a series of offsets along transform faults (fig. 9.20D). Subduction gradually stops along the northwest segment of the trench (north of the northern triple junction) and

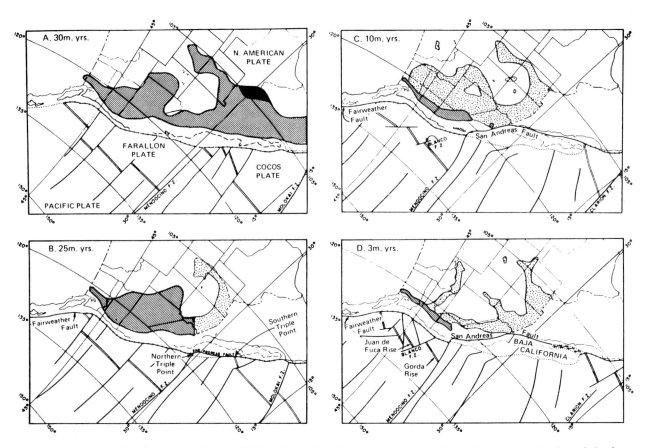

9.20. Progressive annihilation of the East Pacific Rise from 30 to 3 m. y. ago and associated changes in volcanic style in the Cordilleran belt (modified after Atwater, 1970; Christiansen and Lipman, 1972). Key: — = fracture zones: === = oceanic ridge; ⟶ = trench; diagonally lined area = area of predominantly calc-alkaline volcanism; black area = mid-Tertiary, high-K calc-alkaline volcanism; stippled area = basaltic and bimodal volcanism.

the trench fills with sediments. The last major spreading from the Gorda-Juan de Fuca Ridges occurs about 10^5 years ago.

Accompanying progressive annihilation of the ridge system are corresponding changes in tectonic and volcanic style on the continent. As the triple junctions move north and south (and the San Andreas lengthens), volcanic activity on land changes from dominantly calc-alkaline to bimodal and tholeiitic and tectonic style changes from dominantly compressional to dominantly extensional (Christiansen and Lipman, 1972). Some bimodal volcanism and extensional deformation, however, occur simultaneously with subduction of the Farallon Plate (Cross and Pilger, 1978). About 10–20 m.y. ago the extensive Columbia River basalts are erupted behind a marginal arc of Cascade volcanoes (fig. 9.20C) in a continental environment analogous, perhaps, to that found in active marginal basins in the South Pacific today. In fact, the entire Basin and Range Province, which develops opposite the San Andreas transform

system as it lengthens, has many features in common with active marginal basins (i.e., high heat flow, horst-graben topography, thin crust, shallow zone of anomalous upper mantle).

Several models have been suggested for the development of the Basin and Range Province (Stewart, 1978). One model is shown in fig. 9.21. This involves the production of one or several mantle plumes which rise from the stagnating, descending slab to the base of the lithosphere (a). These plumes heat, partially melt, and thin the lithosphere (b), and eventually spread out along its base (at about 30 km depth) (c). As the plumes spread, bimodal-tholeiitic volcanism and Basin-and-Range extensional faulting occur at the surface (c and d). The distribution of radiometric dates in Tertiary volcanic rocks in the Great Basin tends to support the existence of one plume which began beneath east-central Nevada about 50 m.y. ago and spread radially to the present time (McKee, 1971). The most recent volcanic and tectonic activity is concentrated around the edges of

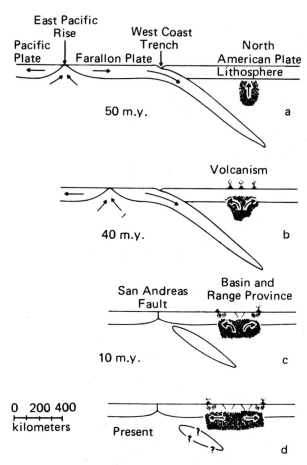

9.21. Schematic cross section of the Cordilleran belt at about 40°N latitude showing possible diapir development of the Basin and Range Province (modified after Scholz *et al.*, 1971).

the Great Basin. The sequence of events depicted in fig. 9.21 attempts to relate the annihilation of the East Pacific Rise and development of the San Andreas transform fault to the spreading of this plume.

Alaska and the Arctic Region

The history of Alaska and the Arctic region has been complex, involving the aggregation of many small plates in southern Alaska (Churkin, Carter, and Trexler, 1980). Although it will be some time before a detailed reconstruction of this area can be made, some of the general features are available. Beginning in the Mesozoic, most plates in the Arctic region were part of Laurasia. As the Atlantic Basin began to open in the Jurassic, the Eurasian Plate moved eastward around the Arctic and the North American Plate moved westward. The overall circum-Arctic movement served to close off the Arctic Basin between

Alaska and the Chukotka peninsula in Siberia. In the Jurassic, the Kula Plate (fig. 6.16) extended north from the Pacific Basin into the Arctic, and a fragment of this plate was later isolated in the Bering Sea and Amerasian basin by initiation of the Aleutian subduction zone.

A suggested reconstruction of the Arctic region is illustrated in fig. 9.22. In the early Jurassic, continental plates begin circumpolar drift that narrows the large polar ocean that separates Chukotka-Alaska from Siberia (a).

Kolyma drifts northward (probably as part of the Kula Plate) as the Pacific Plate grows to the south and the Kula Plate is lost by subduction along the northwestern border (fig. 6.25). By early Cretaceous, the Kolyma subcontinent collides with and is sutured to Siberia along the Verkhoyansk foldbelt (b). During this time, numerous small plates collide with southeastern Alaska and Chukotka and are accreted to the continent. Geologic as well as paleomagnetic results support this interpretation. Also at this time, the Porcupine shear zone develops south of the Brooks Range in Alaska (Churkin, Carter, and Trexler, 1980). By 50 m.y. ago (c), collision and suturing of Alaska-Chukotka with the Eurasian Plate isolate part of the Kula Plate in the Arctic Basin. The present configuration shows three possible remnants of the Kula Plate (d): the Amerasian Basin, the Yukon-Koyukuk Basin, and the Bering Sea floor.

Suspect Terranes

Geological, paleontological, and paleomagnetic data from western North America indicate that much of the Cordilleran belt is a mosaic of foreign terranes. The paleogeographic locations of these terranes, known as *suspect terranes*, are unknown during much of the Phanerozoic (Coney et al., 1980) (fig. 9.23). Most of these terranes appear to have collided with and accreted to North America during the late Mesozoic and Cenozoic. Paleomagnetic results record large post-accretionary horizontal translations of hundreds of kilometers and major rotations about vertical axes. The terranes are characterized by internal homogeneity and similarity in stratigraphy and tectonic style. Boundaries between terranes are faults or shear zones. Most of the terranes display sediments and volcanic rocks of oceanic or arc affinity, and contain rocks chiefly younger than mid-Paleozoic. Ophiolites are common in some terranes such as the Cache Creek (Ch), San Juan (SJ), and Calaveras (C) terranes (fig. 9.23). Permian faunas characteristic of the Cache Creek terrane are distinct from coeval

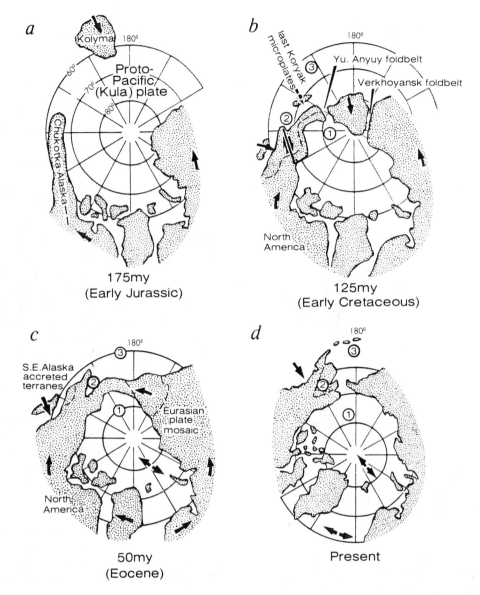

9.22. Possible reconstructions of the Arctic region (after Churkin and Trexler, 1980). (1) Amerasian basin, (2) Yukon-Koyukuk basin, (3) Bering Sea.

faunas in adjacent terranes. Most terranes appear to represent fragments of arcs swept against the western margin of North America. These arcs range in age from late Paleozoic to early Cretaceous. Some terranes such as the Eastern Assemblage in Canada (E), the Yukon-Tanana block (YT), and the Salinia block in California (Sa) contain basement rocks that might represent fragments of other continents.

The original location of most of the suspect terranes is unknown. Some may have been arc systems off the West Coast similar to the arc systems in the Western Pacific today. However, some must have come from great distances to the south as evidenced

by the Tethyan faunas they contain and by paleomagnetic results.

The Caribbean Region

Mexico, Central America, and the Caribbean Basin have undergone an extremely complex plate history whose overall features are just beginning to be understood. Paleomagnetic and paleofaunal distributions indicate that North and South America were connected in the late Paleozoic. The location of central and southern Mexico and Central America at this time is a subject of considerable controversy. Some

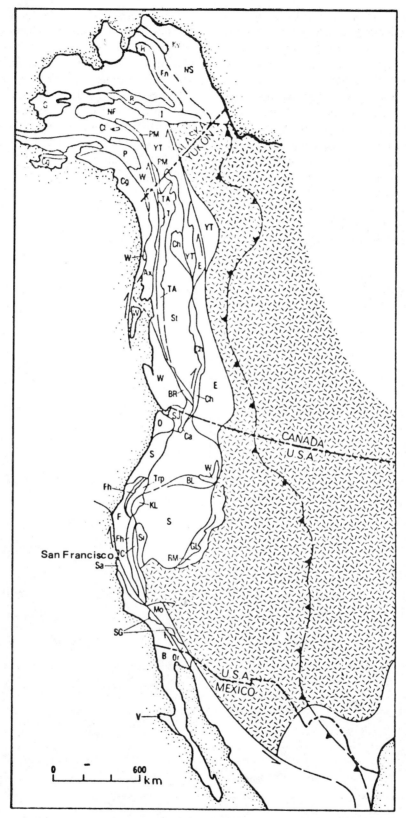

9.23. Suspect terranes in western North America (from Coney *et al.*, 1980
━▼━ eastern limit of Mesozoic-Cenozoic deformation. Symbols refer to
specific terranes, some of which are mentioned in the text.

models involve bending or faulting of this region to the west to allow North and South America to collide in the vicinity of Texas. Others involve fragmentation and rotation of the region and insertion as several "blocks" into the Gulf of Mexico.

Regardless of the location of Mexico and Central America, data suggest that South America–Africa collided with North America along its southern border during the late Carboniferous (Kluth and Coney, 1981). Suturing and deformation progressed from the Ouachita region in southeastern Oklahoma to the Marathon region in southwest Texas. The Ancestral Rockies in Colorado and New Mexico also appear to be related to this collision. This widespread late Paleozoic orogeny in the southwestern United States formed as South America collided with a peninsula including northern and perhaps central Mexico. This configuration persisted into the late Triassic (fig. 9.24a). As shown, Yucatan and Cuba may have been part of the South America plate at this time, while Central America (Costa Rica to Guatemala) was part of southwestern Mexico. Transform faults are shown along both sides of the

northwestern extremity of the South America Plate. By late Juarassic time a branch of the Mid-Atlantic Ridge may have extended between Yucatan and South America as the South American Plate drifted south (b). At this time, Yucatan and Cuba become part of Mexico. A major subduction zone also extends along the western margin of the North American Plate, and is partly responsible for rotating Mexico to the east. By early Cretaceous a new convergent plate boundary dipping south develops north of Cuba with transform faults on each end (c).

At the beginning of the Tertiary, the movement of the Pacific Plate becomes more easterly, causing the westernmost transform fault to reorient itself in a more E-W direction. This results in the shearing off of Central America and eastward migration to its present position along the Motagua Fault. The Cayman Trough also came into existence at this time (Plate I). By mid-Tertiary the Motagua-Cayman transform system had shifted to the east, transferring Cuba to the North American Plate (d). The Puerto Rico Trench also becomes part of this transform as the eastern subduction zone migrates to its present

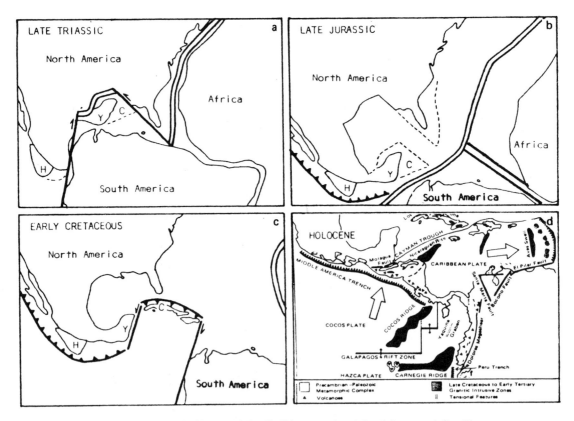

9.24. Diagrammatic plate-tectonic history of the Caribbean region (after Moore and Castillo, 1974). H = Central America (Guatemala to Costa Rica); Y = Yucatan; C = Cuba; ∿∿ = oceanic ridge; — = transform fault; ▼▼▼ = subduction zone.

location in the Lesser Antilles. Right lateral motion begins along the Oca Fault in northern Venezuela at this time, and the Middle America subduction zone propagates southwards with the concurrent growth of the Panama arc which eventually separates the Pacific and Atlantic Oceans. By Oligocene time the Caribbean Plate is well defined. As the ancestral Carnegie Ridge collides with the northern Peru Trench in the middle Miocene, the Galapagos Rift begins to open to the north (d). Blocks of the ancestral ridge are rafted northwards, reaching and sealing the Middle America subduction zone south of Costa Rica as spreading continues adjacent to the Galapagos Ridge (Malfait and Dinkelman, 1972). This subduction-zone deactivation is currently continuing as segments of the Cocos aseismic ridge are being rafted northwards into the Middle America Trench.

The Andes

The Andes have developed in response to the South American Plate overriding the Nasca Plate in the last 100 m.y. One model for the evolution of the central Andes is illustrated in fig. 9.25. The west coast of South America during the late Paleozoic is characterized by a cratonic succession on a stable continental margin (A). To the west of this margin lay a region of continental crust, relics of which are now exposed as metamorphic rocks (~ 2 b.y. in age) along the coast of Peru and northern Chile. This crust probably was part of a continental fragment that collided with the Peru-Chile trench in the late Paleozoic during the development of the Samfrau orogenic belt. Plate convergence began again in the late Triassic, coinciding perhaps with the initiation of the East Pacific Rise. A small continental-margin arc developed by early Jurassic, with a large back-arc basin in which shallow marine sedimentation continued (B). The major phase of orogeny, corresponding with the Laramide orogeny (40–80 m.y.), involves folding and thrusting in the arc-rear area and major granitic plutonism and volcanism in the arc area (C). The enormous Andean batholith is also emplaced at this time.

The late Cenozoic is characterized by cyclical volcanic activity and intense folding and thrusting in the altiplano and eastern Cordillera (D). Back-arc basins received up to 15 km of immature sediments, derived chiefly from the rapidly rising Andes Mountains. Volcanism and orogeny moved inland by several hundred kilometers between the Jurassic and late Cenozoic. Geochemical studies suggest that an appreciable amount of sediment derived from the growing

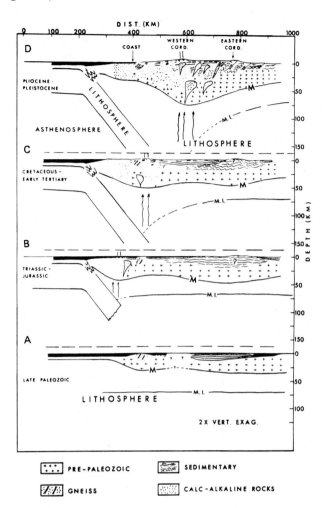

9.25. Schematic cross-sections showing the evolution of the central Andes (after James, 1971).

arc system may have been subducted and contributed to the production of late Cenozoic calc-alkaline magmas. Also, the lithosphere increased greatly in thickness in response to the South American plate overriding the Nasca plate.

THE ALPINE-TETHYAN SYSTEM

The Alpine Orogenic Belt

The Alpine orogenic belt crosses southern Europe and extends into Turkey and Iran, joining with the Hindu Kush–Himalayan belt in Asia. The present distribution of recognized plates in this area is shown in fig. 6.14. Dewey et al. (1973) have proposed plate reconstructions of the western part of the belt for the last 180 m.y. based chiefly on the distribution of convergent and divergent petrotectonic assemblages, paleomagnetic pole positions, and crustal deformation. Because of the lack of unambiguous data related to the motions of small plates, however, the reconstruction must be considered speculative. An outline of their model is given here to illustrate some of the complexities associated with continent-continent collisions. The reconstruction is characterized, in general, by successive fragmentation and collision of minor plates (between Africa and Europe), culminating in the collision of the African and European Plates. The motions of the African and European Plates are determined from sea-floor spreading directions and rates in the North Atlantic (Chapter 8). Throughout its drifting history, Africa is characterized by a counterclockwise motion relative to Europe. Such a motion over the last 180 m.y. caused a progressive, although extremely complex, closure of the Tethys Ocean, which separated Eurasia from Africa about 200 m.y. ago.

At about 180 m.y. ago the North Atlantic begins to open, and complex rifting in southern Europe and northern Africa is initiated (fig. 9.26). Ophiolites and deep-sea sediments suggest that Morocco and Oran were rifted from North Africa at this time. Also, the Apulian and Rhodope-Turkey regions separated from northeast Africa, and the Carnics region was rifted from southern Europe. The Iran Plate, which began to separate from Africa-Arabia in the late Triassic, continued to move northeastwards as an ocean basin opened between Arabia and Iran. Marine transgression in the early Jurassic around the margins of the western Tethys Ocean supports the existence of active oceanic ridge systems in this area. Evidence for subduction zones comes from the distribution of calc-alkaline volcanics and other convergent-plate margin rock assemblages. Major subduction zones are along the northern edge of Tethys Ocean beneath the present-day Major Caucasus Mountains (north-dipping) and along the northeastern edge of Iran (south-dipping). A less certain subduction zone may have existed west of the Moesia Plate. By the middle Jurassic, the accreting plate margin along the southwestern boundary of the Oranaise Plate evolved into a transform fault system, and both the Oranaise and Moroccan Plates closed on North Africa. In the late Jurassic the Mid-Atlantic Ridge shifted eastwards toward Africa. Subduction continued along the

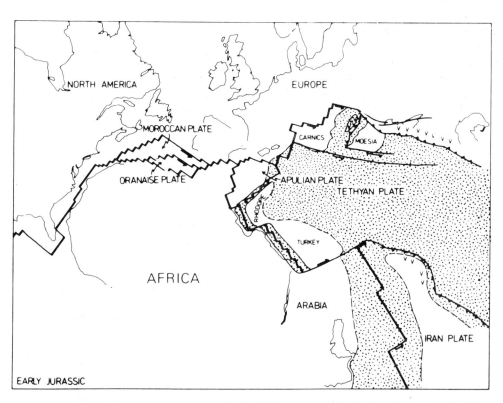

9.26. Early Jurassic plate boundary scheme in the Alpine orogenic belt (after Dewey *et al.*, 1973). Key: shore line —; vvv subduction-zone related volcanic rocks; oceanic ridge ═══ ; transform faults ──⇀; other faults — —; subduction zone (triangles point down dip) ▲▲▲ ; direction of thrusting ⇒.

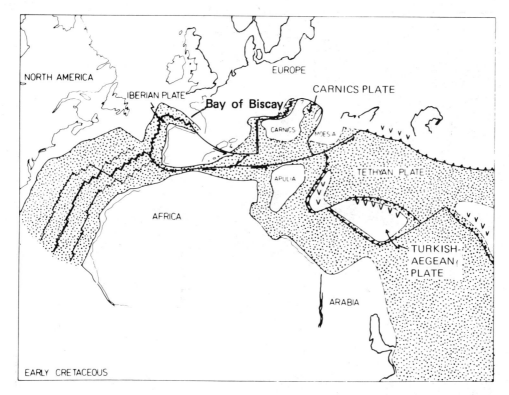

9.27. Early Cretaceous plate boundary scheme in the Alpine orogenic belt (after Dewey *et al.*, 1973). Symbols as in Fig. 9-26.

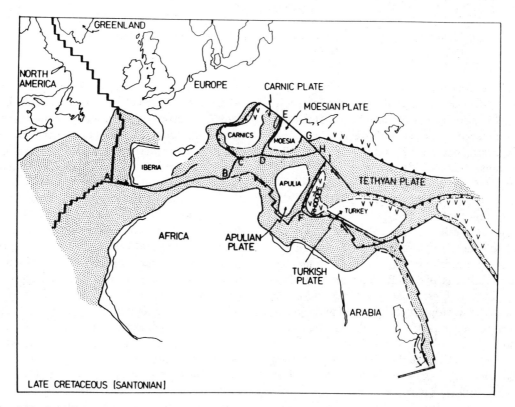

9.28. Late Cretaceous plate boundary scheme in the Alpine orogenic belt (after Dewey *et al.*, 1973). Symbols as in Fig. 9-26.

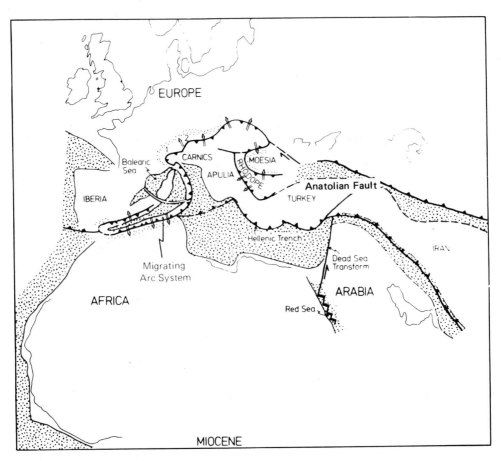

9.29. Middle Teritary plate boundary scheme in the Alpine orogenic belt (after Dewey *et al.*, 1973). Symbols as in Fig. 9.26.

northern and southern margins of the Tethyan Plate and west of Moesia, as inferred form petrotectonic assemblages. It also appears to have begun along the northeastern margin of the Turkish-Aegean Plate and the southeastern margin of the Iran Plate. The Moroccan and Oranaise Plates became sutured to North Africa by the beginning of the Cretaceous.

The Iberian peninsula is rifted away from the American-European Plate chiefly between 100 and 140 m.y. ago as the Bay of Biscay opens (fig. 9.27). New, probably north-dipping subduction zones are recorded by petrotectonic assemblages along the southwestern margins of the Rodope and Turkish-Aegean plates. At approximately 80 m.y. ago, the Mid-Atlantic Ridge propagates northwest and Greenland begins to separate from North America (fig. 9.28). A major change in plate motions at this time results in reorganization of plate boundaries. A new subduction zone develops west of the Carnics Plate and Africa continues its northwestward rotation, further closing the numerous seaways of the western Tethys Ocean. This rotation also changes the motions

of such plates as Apulia, Rhodope, and Turkish-Aegean to a more westward course. During the late Cretaceous, an extremely complex pattern of subduction zones, spreading centers, and transform faults exists in the western Tethyan area, and during the early Tertiary major deformation and volcanism occur throughout this region. Major convergent boundaries exist on both sides of the Iberian Plate and on both sides of the rapidly northward-moving Iran, Turkish-Aegean, and Rhodope Plates. Collision between Iberia and Europe produces major deformation in the Pyrenees and obduction of oceanic crust occurs in Corsica. Collision also begins along the western margin of the Carnics Plate, as is indicated by the early nappes in the Alps. Calc-alkaline volcanism is widespread in Iran during the Eocene, reflecting active subduction zones along both margins of the Iran Plate.

During the Miocene, the Anatolian transform fault develops as the Turkish-Aegean Plate moves west in response to the continued northwestward rotation of Africa (fig. 9.29). Major convergence is

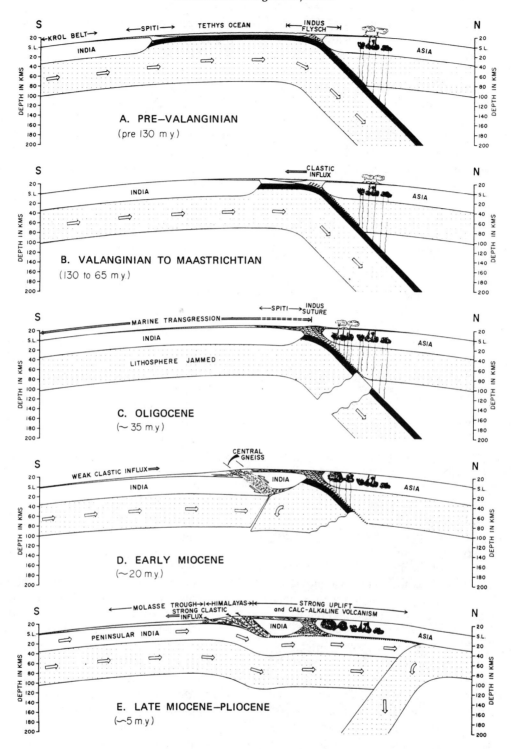

9.30. Schematic cross-sections showing the India-Tibet collision (after Powell *et al.*, 1973).

taken up by "choked" subduction zones along the western and northern margins of the Carnics Plate, by subduction beneath the Apennines (in Italy), and by subduction along the southern margins of the Turkish-Aegean and Iran plates (Dewey and Sengör, 1979). Corresponding deformations include major thrusting and nappe formation in the Alps and Carpathians. In the late Miocene, Italy as a migrating arc system begins to separate from Corsica-Sardinia as the Balearic and Tyrrhenian Seas open. The Calabria subduction zone (Plate I) is a remnant of the once extensive subduction system around the Apennine Peninsula. During the Pliocene, the northern margin of Arabia collides with Iran, and the Zagros crush zone (Plate I) is initiated. This collision leads to intense folding of the miogeoclinal succession along the coast of Arabia and the southward spreading of clastic sediments onto the Arabian Plate. Remnants of the Tethys Ocean are isolated by collision of Iran with Asia, and today are represented by the Black and Caspian Seas. The modern Hellenic subduction zone south of Greece develops at about 6 m.y. ago in response to the choking of subduction zones to the north.

The Himalayas

The rapid northward migration of India during the Tertiary resulted in a major collision between India and Tibet at about 35 m.y. ago. One model for the evolution of this collisional boundary is illustrated in fig. 9.30. Prior to 130 m.y. ago, India is part of Gondwana separated from Asia by the large Tethys Ocean (A). A convergent plate boundary exists along the southern coast of Asia. Between 130 and 35 m.y. ago, India moves north as the Tethys Ocean closes. Evidences of the approach of India at about 50 m.y. are eugeoclinal rocks derived from the Indus arc system which are deposited on the northern margin of India (B). Collision between the two continents is recorded beginning at about 35–40 m.y. by widespread transgression in India, deformation along the collisional boundary, and the development of the Indus suture (C). Continued convergence of the Indian Plate leads to widespread thrusting and folding in the Miocene and Pliocene (D,E), and perhaps to underthrusting of continental lithosphere (E). By mid-Miocene, coarse clastic sediments are shed to the south from the rapidly rising Himalayas. Tibet is elevated by almost 5 km, and calc-alkaline and bimodal volcanism occur on the Tibetan Plateau (fig. 8.35). The Himalayas continue to rise, not reaching

their present elevation until late Pliocene or Pleistocene time. Whether or not continental lithosphere is thrust beneath Tibet (D), however, is a subject of considerable disagreement.

Studies of earthquake first-motions in eastern Asia suggest that stresses resulting from the India-Tibet collision may have been transmitted many thousands of kilometers into northeastern Asia (Molnar and Tapponnier, 1975). Much of the deformation north of the Himalayas is strike-slip and can be related to nonparallel convergence of India with Tibet. It is possible that the Baikal rift zone and Shansi graben system (Plate I) develop as major tension cracks near the end of shear zones produced by the India-Tibet collision.

SUMMARY STATEMENTS

1. APW and geologic-paleontologic data indicate that two or more supercontinents existed by 1000 m.y. ago. During the late Precambrian and Cambrian, Laurasia fragmented into five or more subcontinents. These, in turn, collided, reforming Laurasia, which combined with Gondwana in the late Paleozoic to form Pangaea. Pangaea has fragmented during the Mesozoic and Cenozoic. Major Phanerozoic orogenic systems developed along convergent and collision plate boundaries.

2. The Appalachian-Caledonian orogenic system is characterized by the following simplified history: (a) late Precambrian opening of the Iapetus Ocean between Africa and North America; (b) development of complex convergent plate margins along both continental margins of the Iapetus Ocean during the late Precambrian and early Paleozoic; (c) the Taconic orogeny (Ordovician) caused by subduction and closure of back-arc basins; (d) progressive collision of Africa and North America producing the Caledonian-Acadian orogenies (late Silurian and Devonian); (e) continued convergence, producing thrust and nappes of the Alleghenian orogeny (late Carboniferous-Permian); and (f) opening of the Atlantic basin beginning in the Triassic.

3. The Hercynian orogenic system, extending from the Ouachita region of North America to eastern Europe (and farther east), developed in response to the collision of Africa–South America with Europe and North America.

4. The Cordilleran orogenic belt in western North America records a complex plate-tectonic history, beginning with the development of one or more arc-subduction systems along the western continental border and/or in the proto-Pacific basin by the Ordovician-Silurian. An arc-continent collision orogeny (the Antler orogeny) occurs in the middle Devonian and Mississippian, principally involving eastward-directed thrusting. The Mesozoic is characterized by an east-dipping subduction system. During this time the Klamath arc system either collides with the continent or eliminates a marginal sea by eastward thrusting. Annihilation of the East Pacific Rise in the Cordilleran trench begins about 30 m.y. ago with concurrent development of the San Andreas transform fault. Changes in volcanic style from calc-alkaline to bimodal and basaltic and in tectonic style from dominantly compressional to extensional (Basin and Range faulting) also occur in the Cordilleran belt opposite the growing transform fault system. The Basin and Range Province may have developed in response to mantle plumes rising from a stagnating descending slab.

5. Geological, paleontological, and paleomagnetic data suggest that much of the Cordilleran belt in North America is comprised of a mosaic of foreign terranes. These terranes appear to represent arc systems that were added and sutured to North America during the late Mesozoic and Cenozoic.

6. The Caribbean region underwent a complex plate-tectonic history in which portions of Mexico and Central America may have originally occupied part of the Gulf of Mexico.

7. The plate-tectonic history of the Alpine orogenic belt is characterized by a complex and evolving system of oceanic ridges, subduction zones, and transform faults that give rise to many small plates between Africa and Europe as the two continents close on each other. The major deformation in the form of nappes and thrust faults occurs in the early and middle Tertiary in response to the final continent-continent collision.

8. India collided with Tibet about 35 m.y. ago, resulting in major deformation and suturing of the two continents along the Indus suture. The Himalayas and Tibet were uplifted in the late Tertiary and Quaternary in response to continued convergence.

SUGGESTIONS FOR FURTHER READING

Bird, J.M., and Dewey, J.F. (1970) Lithosphere plate-continental margin tectonics and the evolution of the Appalachian orogen. *Geol. Soc. Amer. Bull.*, **81**, 1031–1061.

Dewey, J.F., Pitman III, W.C., Ryan, W.B.F., and Bonnin, J. (1973) Plate tectonics and the evolution of the Alpine system. *Geol. Soc. Amer. Bull.*, **84**, 3137–3180.

Kanasewich, E.R., Havskov, J., and Evans, M.E. (1978) Plate tectonics in the Phanerozoic. *Can. Jour. Earth Sci.*, **15**, 919–955.

Smith, R.B., and Eaton, G.P., editors (1978) Cenozoic Tectonics and Regional Geophysics of the Western Cordillera. *Geol. Soc. America Memoir* 152, 388 pp.

Van der Voo, R., and Channel, J.E.T. (1980) Paleomagnetism in orogenic belts. *Revs. Geophys. Space Physics*, **18**, 455–481.

Chapter 10

Precambrian Crustal Development

INTRODUCTION

One of the major problems in geology is the role of plate tectonics in the Precambrian. Although many methods of identifying plate boundaries and reconstructing plate motions are available for the last 200 m.y. (Chapter 8), few are applicable to older crustal provinces. Furthermore, in most Precambrian terranes deformation and metamorphism commonly obliterate original rock characteristics and partially or completely reset isotopic dating systems and magnetization directions. The most useful methods for evaluating tectonic settings in the Precambrian are petrotectonic supracrustal assemblages, broad-scale deformation and metamorphic patterns, and paleomagmagmagnetic studies.

There are two extreme schools of thought regarding the role of plate tectonics during the Precambrian. The strictly uniformitarian school suggests that plate-tectonic processes have always operated on the Earth and that the Precambrian can be interpreted in terms of Phanerozoic-type plate tectonics (Burke et al., 1976). The other, a nonuniformitarian view, advocates that the early lithosphere was too hot to be subducted and that thin, rapidly moving plates were jostled around on the Earth's surface like blocks of ice in the Arctic Ocean. These models call upon vertical or horizontal displacements accompanied by buckling and shearing of plates to accomodate new plate growth.

The cooling of the Earth during the last 4.5 b.y. certainly must have been important in governing tectonic processes (Chapter 11). During the Archean, the heat production in the Earth was more than four times greater than at present (fig. 11.5). Convection

would surely have been more vigorous, and the zone of partial melting (i.e., a proto-low-velocity zone) which defines the base of the lithosphere would be at shallower depths. Thus plates also would be thinner and more rapidly moving than today. Whether or not subduction occurred in such a regime is a question of current controversy. Many investigators believe that modern subduction is driven by a negative buoyancy effect caused by the cooling of the lithosphere as it moves away from oceanic ridges and by phase transitions as it descends into the asthenosphere (Chapter 6). The eclogite phase transition (Chapter 3) is particularly important in the buoyancy model. However, estimated Archean geotherms do not fall in the eclogite stability field (fig. 10.1), and hence if eclogite is a major cause of buoyancy-driven subduction, it could not have been important prior to about 1 b.y. (Baer, 1977). Although the oceanic lithosphere may have been thinner in the Archean, as defined by the geotherm-mantle solidus intersection in fig. 10.1 (~ 40 km), the viscous drag of more rapidly moving asthenosphere may have been stronger (Hargraves, 1978). Such forces may have been sufficient to pull lithosphere back into the mantle, thus allowing subduction in the early Precambrian (Chapter 11). Viscous-drag subduction may have evolved into buoyancy-driven subduction with time as the Earth cooled. In addition, cooling and thickening of the mantle lithosphere beneath thick crust could lead to decoupling of the buoyant crust and sinking of the mantle lithosphere into the asthenosphere (fig. 10.22B).

A large amount of data seems to support a nonuniformitarian plate-tectonic model for the Precambrian (Kröner, 1981). Examples are as follows:

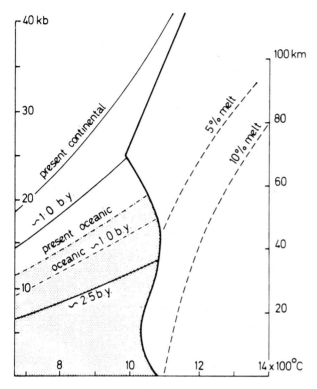

10.1. Changes in geothermal gradient with time in relation to depth of partial melting of the mantle (after Baer, 1977). Heavy solid line is the solidus of wet peridotite and patterned area represents the field where eclogite is *not* stable.

1. Crustal-province boundaries can be matched across Precambrian mobile belts (Chapter 5). This fact cannot be readily explained if mobile belts are the result of continent-continent collisions.

2. Relict-age provinces occur as islands in Precambrian mobile belts and numerous older relict dates occur in younger mobile belts, indicating an ensialic origin for such belts (Chapter 5). Older structural trends can be traced across mobile belts into adjacent older crustal provinces (fig. 5.4). These features cannot readily be interpreted in terms of continental collisions.

3. Ophiolites, unambiguous arc petrotectonic assemblages, and blueschists are either not recognized or not widespread in the geologic record until after 1000 m.y. ago.

4. Even in low-grade Archean and early- to mid-Proterozoic crustal provinces, melanges and sutures have not been widely recognized.

5. Precambrian anorthosite-granite suites define a broad belt across central North America into western Europe when considered on a pre-Permian

continental reconstruction (fig. 10.8). If portions of these continents drifted independently during the Precambrian, it is difficult to explain how they regrouped themselves to preserve these patterns.

6. As reflected by petrotectonic assemblages (discussed later in this chapter), continental rifting did not begin until about 2 b.y. ago and did not become widespread until 1 b.y. ago. Fragmentation of portions of the continents leading to the formation of large ocean basins appears also to have begun about 1 b.y. ago.

7. Paleomagnetic studies, discussed at the end of this chapter, do not allow significant *independent drift* of large portions of the continents prior to late Precambrian time.

Data relevant to tectonic setting from Archean and Proterozoic crustal provinces and Precambrian paleomagnetism are summarized and discussed in the following sections.

ARCHEAN ROCK ASSOCIATIONS

Greenstone Belts

Stratigraphy. Archean greenstone successions are structurally and stratigraphically complex. Estimated thicknesses range from 10 to 20 km and are minimal in that neither the top nor the base of most successions is exposed. An idealized stratigraphic column of a greenstone succession based on studies of the Barberton belt in South Africa is given in fig. 10.2. The lower succession is comprised chiefly of low-K tholeiites, komatiites, and ultramafic flows and sills with minor felsic tuff and layered chert. This succession is commonly referred to as a *bimodal succession*, in that volcanic rocks of intermediate composition (andesites, dacites) are rare or absent. The upper succession, commonly referred to as a *calc-alkaline succession*, contains tholeiites, calc-alkaline volcanics of intermediate composition, a small amount of ultramafic rocks and komatiite, and a dominantly clastic sediment succession. The sedimentary unit, which in some greenstone successions can be divided into lower and upper subunits, generally is conformable with the volcanic unit. The lower sedimentary subunit is characterized by abundant graywackes and tuffs, with minor amounts of chert, and the upper subunit, when present, by quartzites, shales, and conglomerate. Two general trends are observed with increasing stratigraphic height in most Archean greenstone successions: (1) a decrease in the amount of ultramafic rocks, komatiites, and to a lesser extent tholeiites,

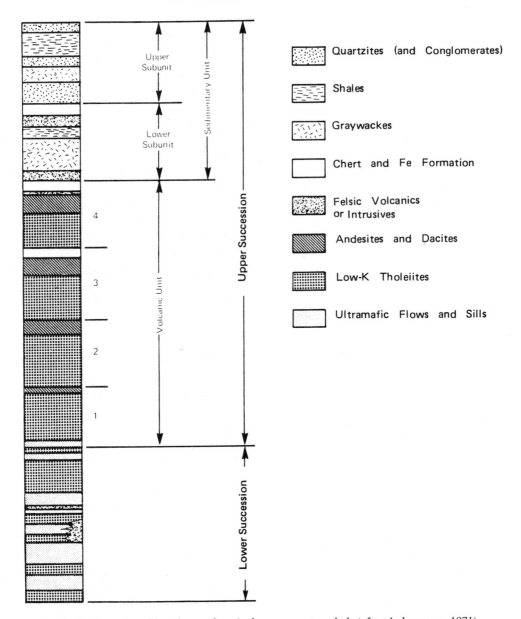

10.2. Idealized stratigraphic column of an Archean greenstone belt (after Anhaeusser, 1971).
Numbers refer to volcanic cycles.

and (2) the ratio of volcaniclastics to flows increases, as does the relative abundance of andesite and other calc-alkaline volcanics. In addition, immature clastic sediments and tuffs may dominate in the upper parts of greenstone successions.

Another feature characteristic of most greenstone successions is volcanic and sedimentary cyclicity (Anhaeusser, 1971). Cycles range in thickness from less than a centimeter in some graywacke turbidites to several thousand meters in volcanic cycles (fig. 10.2). Volcanic cycles commonly begin with ultramafic or mafic flows and end with felsic

tuffs capped with chert. With increasing stratigraphic height, the proportion of ultramafic and mafic rock to felsic and intermediate rock decreases.

Although it is not yet possible to accurately correlate entire sections between adjacent greenstone belts, the existence of trains of greenstone inclusions in intervening tonalitic gneisses which can be traced from one greenstone belt to another strongly indicates that greenstone belts are remnants of larger Archean basin successions. Supporting this interpretation is the existence of distinct chert horizons, such as the Manjeri Formation in Rhodesian greenstones and the Marble Bar chert in the Pilbara Province in Western Australia which can be traced or correlated between greenstone belts over great distances.

Greenstone belts of more than one age may exist in the same geographic area, commonly separated by an unconformity. Examples have been described from Zimbabwe, Western Australia, and India (Condie, 1981a). Where two ages of greenstone exist, the older successions are typically bimodal while the younger successions may be bimodal, calc-alkaline, or mixed. Calc-alkaline successions, however, are not observed to underlie bimodal successions. Greenstone belts also exhibit provinciality, in that bimodal types dominate in some crustal provinces such as southern Africa and western Australia and calc-alkaline types dominate in other provinces such as North American Archean provinces.

Volcanic Rocks. Volcanic and associated hypabyssal rocks dominate in Archean greenstone successions. Although composition ranges from ultramafic to rhyolite, mafic volcanics greatly dominate in most successions. Field relationships and studies of volcanic textures and structures indicate that Archean volcanism was largely subaqueous. Volcanic complexes described from the Superior and Slave provinces in Canada contain many volcanic centers, and range from 100 to 175 km in diameter (Goodwin and Ridler, 1970). Some volcanic complexes emerged above sea level during the late stages of development, erupting large volumes of felsic subaerial and subaqueous ash flows.

In bimodal greenstone successions, komatiites and tholeiites are the dominant volcanics. Although the term komatiite is still used in different ways by different individuals (Condie, 1981a), it is clear that it must be restricted to lavas with clearly defined quench textures, such as spinifex texture. In terms of composition, *komatiites* are generally considered ultramafic or mafic quench-textured volcanic rocks in which the $CaO/Al_2O_3 \gtrsim 0.8$, $MgO > 9\%$, $K_2O < 0.9\%$, and $TiO_2 < 0.9\%$. Although komatiites and ultramafic flows are common in Archean greenstone successions, they are uncommon in the Proterozoic and very rare in the Phanerozoic. Such a secular distribution probably reflects the falling geothermal gradient in the Earth with time.

Ultramafic lavas have distinct characteristics, as fig. 10.3 illustrates. Spinifex texture is commonly preserved in the upper parts of such flows. This texture is characterized by randomly oriented skeletal crystals of olivine or pyroxene, and results from rapid cooling and the near-absence of crystal nuclei. Flows range from those with well-developed spinifex zones that comprise nearly all of the upper half of the flow (A) to flows without spinifex textures but with polygonal cooling fractures (C). Ultramafic and tholeiitic flows range from about 2 m to >200 m thick, are commonly pillowed, and may be associated with either sills or pyroclastic rocks of similar composition, or both. Layered mafic and ultramafic igneous complexes are also found in most greenstone successions. Geochemical studies indicate they are closely related to enclosing volcanic rocks. Archean ultramafic and mafic volcanics are depleted in LIL elements and show variable amounts of light-REE depletion. Tholeiites may be divided into two groups based chiefly on REE patterns (Condie, 1981a): TH1 is characterized by flat REE patterns ($\sim 10 \times$ chondrites) and TH2 by a sloping REE pattern and light-REE enrichment. When considered in terms of geochemical discriminant diagrams, Archean tholeiites are most similar to modern mid-ocean ridge tholeiites or to arc tholeiites.

Andesites and felsic volcanic rocks in greenstone successions occur chiefly as volcaniclastic rocks with minor flows. Tuffs, breccias, and agglomerates are common, and their distribution can be used to define volcanic centers. Many tuffs have been reworked and exhibit cross-bedding, graded bedding, and other sedimentary structures. Felsic porphyry dikes and plugs are of local importance in some successions. Ash-flow tuffs with well-preserved eutaxitic textures (flattened shards and pumice fragments) have been described from felsic volcanics in Australia and Canada. Although in terms of some elements the three main groups of Archean andesites are similar to modern immature-arc, mature-arc, and continental-margin andesites, respectively, differences in many immobile-element concentrations render such correlations improbable. Alkaline volcanic rocks are rare in Archean greenstone successions and when found occur chiefly as volcaniclastics and associated hypabyssal intrusives.

Sedimentary Rocks. Clastic sediments generally comprise between 15 and 30 percent of Archean greenstone successions. Graywacke and associated argillite are the most important sediments, with quartzite, arkose, conglomerate, and shale being of only local significance (Lowe, 1980). Nonclastic sediments, principally chert, are of minor importance but widespread distribution. Graywacke-argillite successions appear to be, in large part, turbidites, as is indicated by the common occurrence of small-scale graded bedding and other textures indicative of

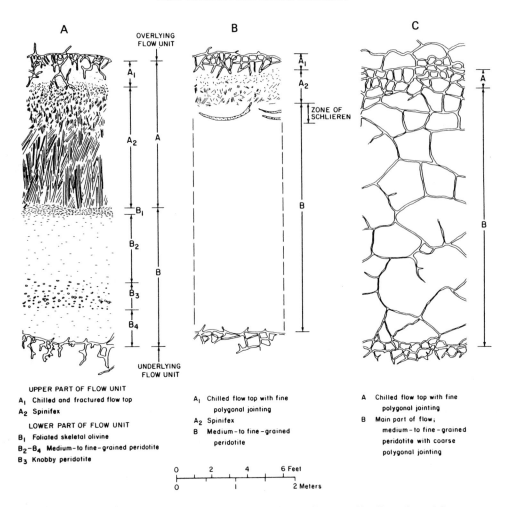

10.3. Diagrammatic sections through three types of ultramafic komatiite flows from Munro Township, Canada (from Arndt *et al.*, 1977).

turbidity-current deposition (Condie, 1981a). Archean graywackes have a remarkably constant composition which is similar to granodiorite.

Archean conglomerates are represented by both contact (pebbles touching) and disrupted framework types. The latter is commonly associated with graywackes and appears to have formed during subaqueous slumping. Contact conglomerates, on the other hand, are associated with arkoses and quartzites and are thought to represent subaqueous alluvial fan deposits formed near rapidly uplifted source areas. Quartzite and arkose are uncommon in most greenstone successions, and when found occur as massive, cross-bedded units.

Provenance studies of Archean clastic sediments indicate that volcanic rocks are important in the source areas. Only locally are granite and gneissic

sources of importance. In particular, andesitic and felsic rock fragments are abundant in most graywackes and conglomerates.

Layered chert and banded iron formation are the most important nonclastic sediments in greenstone successions. The earliest evidence of life on Earth occurs in the form of microstructures in Archean cherts (see Chapter 11). Most chert and iron formations appear to represent chemical or biochemical precipitates in relatively quiet water. Relict detrital textures in some cherts, however, indicate that they are chertified graywackes or tuffs. Carbonates and barite are very minor sediments in some greenstone belts.

Archean sedimentary environments appear to represent, for the most part, tectonically active basins, with slumping and turbidity currents generated by

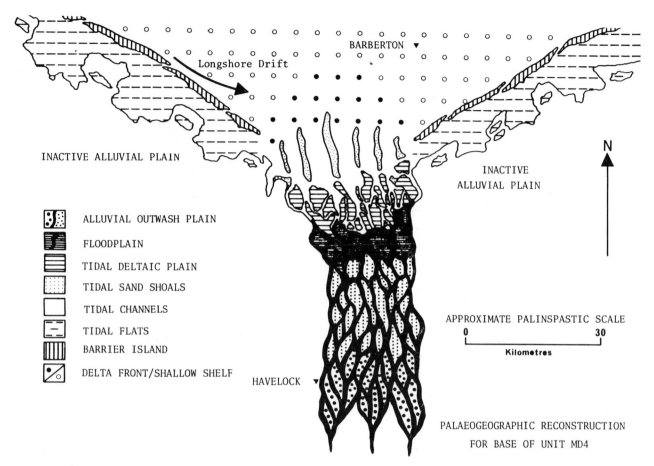

10.4. Paleoenvironment reconstruction of the Moodies tidal sediments (from Eriksson, 1977).

earthquakes and volcanic eruptions. Locally, stable conditions exist when chert, carbonate, and quartzite are deposited. Studies in the Warrawoona Group (~ 3.5 b.y.) in Western Australia indicate widespread shallow-water deposition, in part involving deposition of evaporites. Recent detailed studies of the Fig Tree and Moodies Groups in South Africa have indicated the importance of deltaic and tidal-flat sedimentation (Eriksson, 1977). A typical sandstone-shale tidalite succession is characterized by three facies representing (from the base upwards) flood-tidal delta, lower tidal flat, and upper to mid-tidal-flat environments. A reconstruction of the sedimentary environments during the deposition of the Moodies group shows a southern source area with an alluvial outwash plain followed by a tidal deltaic plain which merges with a submarine delta (fig. 10.4). Studies of graywackes from the Slave Province in Canada indicate deposition by turbidity currents on a large, complex submarine fan (Henderson, 1972).

Cratonic Sediments

Cratonic sediments are not well preserved in Archean crustal provinces. As was already mentioned, a few greenstone belts appear to terminate with cratonic deposition, for instance the Barberton succession in South Africa. Remnants of probable cratonic sediments such as quartzite, pelitic schist and gneiss, and carbonate (marble) occur in Archean high-grade terranes.

The only well-preserved Archean cratonic succession, however, is the Kaapvaal Basin succession in southern Africa (Pretorius, 1974). This succession, which spans at least 1000 m.y. of geologic time, dates back to the deposition of the Pongola Supergroup at about 3.0 b.y. ago. The axis of the Kaapvaal Basin migrated northwestward and the amount of inundation increased over its 1 b.y. lifespan. The Pongola Supergroup is comprised chiefly of quartzite, shale, carbonate, and conglomerate that overlie an unknown amount of mafic and felsic volcanic rocks. Younger rock successions in the Kaapvaal Basin, with their corresponding ages, maximum thickness, and principal rock types, are as follows:

- The Dominion Reef Group: 2.7–2.8 b.y., 3 km, felsic and mafic volcanic rocks;
- The Witwatersrand Supergroup: 2.5–2.7 b.y., 7 km, quartzite, conglomerate, shale;
- The Ventersdorp Supergroup: 2.3–2.5 b.y., 3.5 km, mafic and felsic volcanics, quartzite;
- The Transvaal Supergroup: 2.1–2.3 b.y., 8 km, carbonate, shale, quartzite, iron formation;
- The Waterberg Supergroup: 1.7–1.9, 10 km, quartzite, sandstone, conglomerate, arkose, and shale.

The Kaapvaal Basin development can be considered in terms of a harmonic analysis in which the first harmonic represents a periodicity of formation of successive basins and the third harmonic, broad cyclic variations in rock types to energy level (Pretorius, 1974). Maximum energy levels are accompanied by major volcanism producing the Dominion Reef and Ventersdorp volcanics.

Granitic Rocks

Granitic rocks in Archean provinces fall into one of three categories: gneissic complexes and batholiths, diapiric (syn-tectonic) plutons of variable composition, and discordant granite plutons. Gneissic complexes and batholiths comprise most of the preserved Archean crust. These terranes are complexly deformed, and range in composition from tonalite to granite. They contain large infolded remnants of supracrustal rock as well as numerous inclusions. Inclusions increase in abundance near supracrustal terranes and occur in various stages of digestion and fragmentation, and trains of inclusions commonly interconnect greenstone belts. Gneissic complexes contain rocks that range from banded gneisses and migmatities to faintly foliated homogeneous gneisses. Rocks are chiefly tonalitic to trondhjemitic in composition, although granodiorite and granite compositions are of importance locally. In some high-grade terranes, such as Southwest Greenland, gneissic granodiorites predominate. Contacts between gneissic complexes (and batholiths) and supracrustal terranes are generally intrusive, faulted, or strongly deformed. Unconformable relationships are preserved at the base of some greenstone successions. Field, textural, and geochemical relationships indicate that although

some Archean gneisses have sedimentary precursors, most have igneous precursors. Gneisses with sedimentary parentage have been described in the southwestern Superior Province, where gneissic layering can be traced along strike into graywacke bedding. Intrusive field relationships, trains of inclusions, and geochemical studies strongly indicate, however, an intrusive origin for most Archean gneissic terranes.

Granitic plutons range from foliated to massive and discordant to concordant, and some are porphyritic (fig. 5.9). Geophysical studies indicate that most Archean plutons extend to depths less than 15 km. Some plutons, referred to as *diapirs*, have well-developed concordant foliation around their margins and appear to have been deformed during forceful injection. Others, which are usually granite (*sensu strictu*) in composition, commonly have discordant contacts and massive interiors, and appear to have a post-tectonic origin.

Stratiform Igneous Complexes

Stratiform igneous complexes are found within Archean supracrustal successions and within gneissic complexes. They occur as both post-tectonic, discordant intrusions and as syn- or pre-tectonic deformed intrusions. The largest known Archean body is the Great Dyke in Zimbabwe (2.46 b.y.), which has a strike length of 500 km and an average width of 6 km. Stratigraphic thicknesses in large intrusions range from about 5 to 8 km. Geochemical studies indicate that the original magma composition of most Archean stratiform bodies is that of tholeiite. Rhythmic and cryptic layering are characteristic features and have best documentation in the Stillwater Complex in Montana.

Structure and Metamorphism

Deformational patterns in Archean crustal provinces are complex and reflect polyphase deformation and, commonly, more than one period of metamorphism. Supracrustal terranes are isoclinally folded, in some cases more than once. Deformation is more intense in high-grade than in low-grade areas. Existing data indicate that major folding in greenstone belts reflects primarily vertical forces associated with emplacement of diapiric plutons. Early stages of dominantly compressional tectonics, as reflected by nappes and thrusts, have been described in the Pilbara and Rhodesian Provinces (Coward, 1976). Typical greenstone belts have undergone two or three periods of major deformation in which regional metamorphism to the greenschist or amphibolite facies accompanied one or more of these periods. High-grade terranes, unlike granite-greenstone terranes, are characterized by large nappes and thrust zones and reflect extensive compressional forces and, probably, crustal thickening.

Both regional and contact metamorphism occur in granite-greenstone terranes, whereas regional metamorphism characterizes high-grade terranes. Metamorphic mineral assemblages from Archean crustal provinces reflect the low-pressure type of regional metamorphism (fig. 4.24), with temperature gradients on the order of 30–50°C/km (Condie, 1981a). Only in continental areas of high heat flow (such as the Basin and Range Province) are similar geothermal gradients found today. Assemblages from Archean high-grade terranes record granulite-facies conditions, with temperatures in the range of 800–900°C and pressures of 8–12 kbar.

Relationship of Archean High- and Low-Grade Terranes

The relationship of Archean high-grade terranes to low-grade terranes (granite-greenstone terranes) is a subject of considerable discussion and controversy. Models fall into three categories (Condie, 1981a):

1. High-grade terranes differ in age and tectonic setting from low-grade terranes;

2. High-grade terranes represent the uplifted and eroded root-zones of low-grade terranes; or

3. High- and low-grade terranes record the same ranges of age but reflect different tectonic settings.

Recent radiometric dates from adjacent high- and low-grade terranes (see table 5.1) indicate that such terranes have complex but closely related geologic histories and that neither is necessarily older or younger than the other. The progressive increase in metamorphic grade in going from low-to high-grade terranes, as recorded for instance in going from the Rhodesian Province into the Limpopo belt (fig. 5.11), and the continuity of some greenstone belts from low- into high-grade terranes, also clearly point toward related thermal and geologic histories. Hence, model number one seems unacceptable.

A model dependent upon erosion level relating the two types of terranes is illustrated in fig. 10.5. The diagram shows an idealized greenstone belt which, with increasing depth, passes into high-grade gneisses that are equivalent to shallow-level tonalites. These gneisses contain numerous inclusions of greenstone recrystallized at higher metamorphic grade. The high-grade terrane is exposed at the surface by faulting. Such a relationship has been suggested between the Rhodesian granite-greenstone province and the Limpopo belt in southern Africa, between the high-

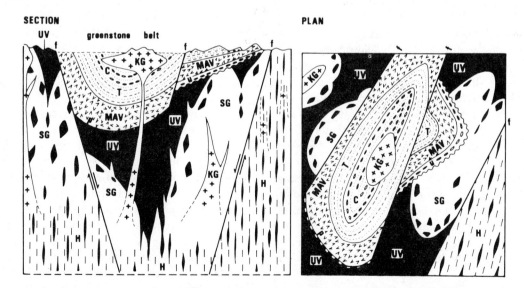

10.5. Diagrammatic representation of possible depth relationships between Archean granite-greenstone and high-grade terranes (from Glikson, 1976). Symbols: UV = ultramafic-mafic rocks; SG = tonalite-trondhjemite; MAV = mafic to felsic volcanics; T = graywacke-argillite; C = conglomerate, arkose, quartzite; KG = high-K granites; H = high-grade gneisses and granulites; u = unconformity; p = paraconformity; f = fault.

and low-grade terranes in the Yilgarn Province in Western Australia, and between the high and low-grade Archean terranes in southern India (Naqvi, Divakara, and Narain, 1978). Despite the appealing attributes of a differential uplift model explaining the relation between high- and low-grade terranes, such a model is faced with three major problems (Condie, 1981a):

1. Many supracrustal rocks preserved in high-grade terranes do not represent high-grade equivalents of greenstone belts, but instead are comprised chiefly of quartzites, pelitic schists, and carbonates.

2. The deformational style in granite-greenstone terranes reflects dominantly vertical forces, whereas that in high-grade terranes reflects subhorizontal forces.

3. Many high-grade terranes exhibit high initial $^{87}Sr/^{86}Sr$ ratios compared to granite-greenstone terranes, indicating either different mantle sources or significant differences in crustal residence time.

Most data from Archean high- and low-grade terranes can best be explained within the context of a model involving different thermal-tectonic settings for each terrane as they evolve side by side.

ORIGIN OF ARCHEAN CRUSTAL PROVINCES

Many models for the origin and tectonic evolution of the Archean crust have been proposed. Some of the earliest models involve the density-driven inversion of greenstone volcanics and underlying tonalitic crust. Tonalites are "reactivated" and diapirically intrude overlying greenstone belts which settle in synclinal keels between diapirs. One of the first models that attempted to explain both high- and low-grade terranes involved mobile belts (high-grade terranes) that develop over primary convective upcurrents in the mantle, while greenstone belts develop over secondary upwellings. Glikson (1976) has proposed a model for greenstone belts based on evolving oceanic crust. The model involves partial melting of the lower part of an early oceanic crust to produce tonalite-trondhjemite magmas which rise and intrude overlying basalts. This is followed by uplift and erosion and partial melting of the lower tonalitic crust to produce granites.

Perhaps the most unique model for the origin of greenstones is the impact model (Green, 1972a). According to this model, greenstone belts are interpreted as large impact scars, similar to lunar maria, which are initially filled with mafic-ultramafic lavas

and are later folded and intruded with tonalites. Two major problems with the impact model, however, have rendered it unpopular. First, most Archean greenstones are younger (chiefly 2.6–2.7 b.y.) than maria impact and volcanism (>3.0 b.y.); second, there is an absence of impact textures, structures, and minerals in Archean greenstones.

Continental rift models for the production of greenstone belts have been popular among many investigators. One model proposed by Condie and Hunter (1976) for the Barberton greenstone belt in South Africa involves an ascending mantle plume (fig. 10.6) beneath early sialic crust and lithosphere. Lateral spreading of the plume opens a rift which is partially filled with plume-derived mafic and ultramafic lavas (2). As the plume subsides, erosion of the flanks of the rift produces detrital sediments (3), and the volcanics and sediments are folded and metamorphosed. Sinking mafic rocks invert to amphibolite and undergo partial melting, producing tonalites which rise as diapirs (4). Continued subsidence leads to partial melting of the lower tonalitic crust and to the production of granite magmas which are intruded at shallow crustal levels (5,6).

In addition to explaining the major sequence of events in granite-greenstone terranes, rift models account for the large volumes of ultramafic-mafic lavas by positioning a "hot" plume source; the increasing ratio of calc-alkaline to mafic volcanics with stratigraphic height in response to increasing importance of an amphibolite source and decreasing geothermal gradient; and the production of granite magmas from lower crust. Rift models fail in that they do not provide an origin for preexisting sialic crust nor for Archean high-grade terranes (assuming they are not depth-equivalents of granite-greenstone terranes).

The overall similarity of greenstone successions to modern arc successions has led to a variety of subduction-related models to explain granite-greenstone terranes (Condie, 1981a; Tarney et al., 1976). The most popular model is the back-arc basin model for greenstone belts (Tarney et al., 1976), with high-grade terranes representing uplifted and eroded portions of adjacent arcs. An idealized sequence of events leading to the production of granite-greenstone and high-grade Archean terranes is illustrated in fig. 10.7. Early stages of development of the back-arc basin are characterized by eruption of ultramafic lavas (1, 2) and later stages by calc-alkaline and mafic volcanics and deposition of sediments derived both from the continent and the arc (3). The back-arc basin succession is deformed and intruded with syn-tectonic

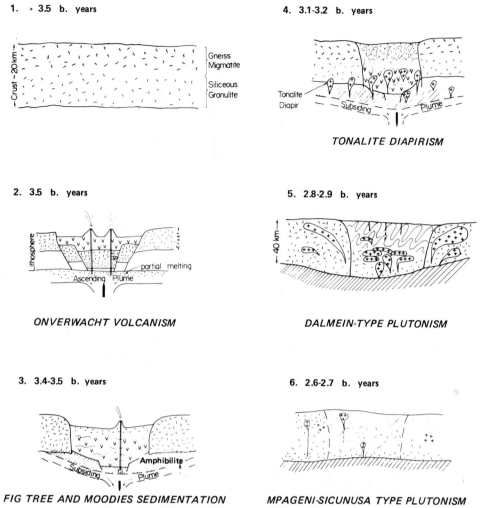

1. > 3.5 b. years

Crust ~20 km

Gneiss
Migmatite

Siliceous
Granulite

4. 3.1-3.2 b. years

Tonalite
Diapir — Subsiding — Plume

TONALITE DIAPIRISM

2. 3.5 b. years

Lithosphere

Crust

partial melting

Ascending Plume

ONVERWACHT VOLCANISM

5. 2.8-2.9 b. years

40 km

DALMEIN-TYPE PLUTONISM

3. 3.4-3.5 b. years

Amphibilite

Subsiding Plume

FIG TREE AND MOODIES SEDIMENTATION

6. 2.6-2.7 b. years

MPAGENI-SICUNUSA TYPE PLUTONISM

10.6. Plume model for the development of the Barberton greenstone belt in South Africa (after Condie and Hunter, 1976).

tonalites during arc activation-type orogeny (4) and later by post-tectonic granites (5). Preferential uplift of the arc is then necessary to expose high-grade terranes.

Convergent-plate boundary models, however, are also faced with significant problems (Condie, 1981a). The age and geographic distribution of greenstone belts within granite-greenstone terranes is not readily explained by such a model. Greenstone belts of the same age (2.6–2.7 b.y.) lie side by side with interconnecting older and contemporary tonalitic gneiss terranes, and often cover an entire crustal province, such as the Superior and Rhodesian provinces. If each greenstone belt represents a back-arc basin, we have the problem of inserting numerous miniature subduction zones without appreciably disturbing interconnecting gneiss terranes. If, on the

other hand, each greenstone belt represents a remnant of a large volcanic basin, as most evidence suggests, then each large basin would represent a back-arc basin. It defies one's imagination to reconstruct descending-slab geometries to account for the distribution of such back-arc basins. Some investigators have called upon a series of arc-arc collisions to bring the back-arc basin successions together (Burke, Dewey, and Kidd, 1976). This model, however, cannot readily account for the continuity of older gneissic terranes and trains of inclusions connecting greenstone belts, the province-wide continuity of chert horizons between greenstone belts, or the rarity or absence of nappes, sutures, melanges, and low-angle thrusts which characterize collisional boundaries (Chapter 8). Paleomagnetic results also do not support this model, as will be discussed later in the

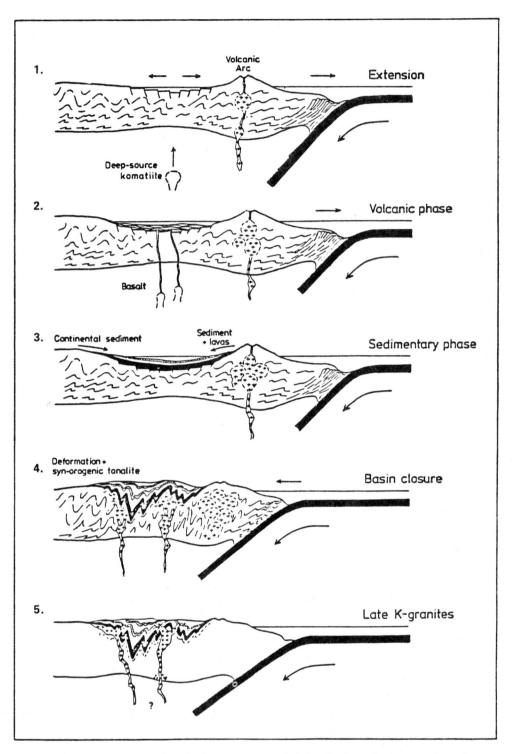

10.7. Evolutionary sequence of an Archean greenstone belt in a back-arc basin tectonic setting (from Tarney *et al.*, 1976).

chapter. Another problem with a convergent-plate boundary model is the apparent absence of the sequential distribution of petrotectonic assemblages from trench, through arc-trench gap and arc, to back-arc basin (Chapter 8).

Archean high-grade terranes may develop over major mantle upwellings which thin the lithosphere and crust and provide heat for extensive high-grade metamorphism. During periods of quiescence, cratonic sediments are deposited; later they are deformed, metamorphosed, and partially melted with the sialic basement. Encapsulation of basalt beneath the ductile sialic crust prevents extensive mantle volcanism in high-grade areas. Another model is that high-grade terranes form over convective down-currents which result in thickening and local stabilization of the crust. Cratonic sediments are deposited, buried, metamorphosed, and deformed within the down-dragged crustal root zone. This model has the advantage of accounting for granulite-facies rocks at the surface today that must have been buried 35–40 km when they underwent metamorphism. Since these areas are underlain by 35–40 km of crust today, the Archean crust must have been on the order of 70–80 km thick if later crustal underplating has not occurred.

PROTEROZOIC ROCK ASSOCIATIONS

Supracrustal Successions

Proterozoic supracrustal rocks are similar in some respects to Archean supracrustals and in other respects to Phanerozoic supracrustals. They vary considerably in degree of deformation and grade of metamorphism. During the Proterozoic, relative proportions of supracrustal rocks change with time (Chapter 11). In a general way, Proterozoic supracrustal rocks can be classified into one or a combination of four lithologic assemblages (Condie, 1981b): (1) quartzite-carbonate-shale; (2) bimodal volcanics-quartzitearkose; (3) calc-alkaline volcanics-graywacke; and (4) ophiolite-deep sea sediments. Assemblage 4 has only been recognized in the late Proterozoic terranes ($\lesssim$ 1000 m.y.). As evidenced by exposed basal unconformities, assemblages 1 through 3 are *ensialic* in that they were deposited, in large part or entirely, on older sialic crust. Most published Proterozoic supracrustal stratigraphic sections lie

within or near the area known to be or probably underlain by Archean continental crust (fig. 10.8).

The Quartzite-Carbonate-Shale Assemblage. Assemblage 1 comprises over 60 percent of known Proterozoic successions. Representative sections, which attain thicknesses in excess of 10 km, are given in fig. 10.9. The overall predominance of quartzite and shale is evident in the first four sections. Quartzites are typically thick-bedded and massive, and commonly show large-scale cross-bedding. Shales may be recrystallized to phyllites. Stromatolitic dolomite is important in the upper portions of the Kibaran Group and Snowy Range Supergroup and local carbonate-rich horizons occur in the Belt Supergroup. Minor sediments include arkose, conglomerate, tillite, and chert. Sandstone provenance studies indicate the importance of granitic-gneiss sources for Assemblage 1. Volcanic rocks are of minor importance in most sections. These are dominantly tholeiitic basalts with minor amounts of felsic or alkaline volcanics. They are bimodal, in that rocks of intermediate composition are rare or absent. In some successions, like the Animikie Group in Minnesota and the Mt. Bruce Supergroup in Western Australia, layered chert and banded iron formation are major rock types. In the Warramunga Group in Australia, subgraywacke is important. Unconformities in Assemblage 1 successions do not appear to reflect major deformation, but only localized uplift and erosion. Lateral facies changes occur in most successions, which aids in defining basin shapes and source directions. Fluvial, tidal, and shallow marine deposits are most commonly reflected by Assemblage 1.

The Bimodal-Volcanic-Quartzite-Arkose Assemblage. Assemblage 2 is the most diverse of the three assemblages and accounts for about 20 percent of described Proterozoic successions. Measured sections range from about 5 to greater than 10 km in thickness. Volcanic rocks, which are typically bimodal basalt-rhyolite, dominate in some successions, such as the Sinclair Group in Namibia, while they are minor in others, such as the East Arm graben succession and Pahrump Group in North America (fig. 10.10). Either volcanic end member may dominate. Basalts are typically tholeiites and occur as both subaqueous and subaerial flows. In some successions, such as the Keweenawan succession in the Great Lakes area, tremendous thicknesses of basalt greatly dominate. Felsic volcanics are rhyolites and related rocks and occur chiefly as volcaniclastics. Ash-flow tuffs are

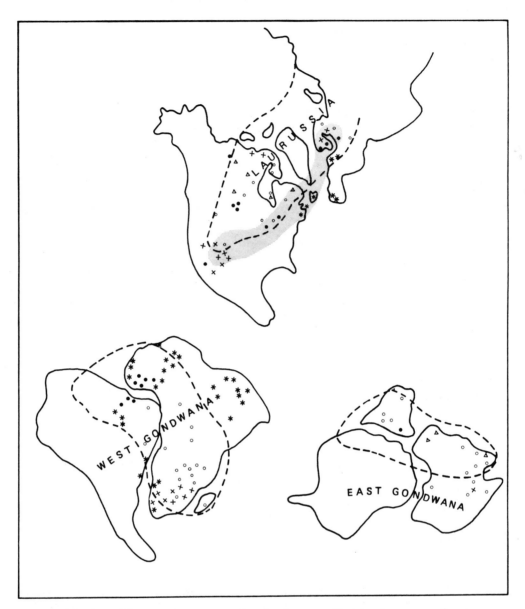

10.8. Distribution of Proterozoic supracrustal rocks shown on Proterozoic supercontinent reconstructions. Dashed line indicates known or probable extent of Archean crustal terranes. Key: o = Assemblage 1, + = Assemblage 2, ● = Assemblage 3, * = ophiolites and probable ophiolites, Δ = combination assemblages; stippled pattern shows Proterozoic belt of massive anorthosites and anorogenic granites. Relative locations of Laurussia, Eastern and Western Gondwana are arbitrary.

particularly common in some successions, such as the Proterozoic successions in the southwestern United States. Sediments in Assemblage 2 are typically immature clastics such as arkose, feldspathic quartzite, and conglomerate which appear to have been derived from rapidly uplifted terranes of dominantly granitic composition. In some successions, shales (phyllites), massive mature quartzites, iron formation, and

carbonate may be of importance. In the Mt. Isa succession in Australia, evaporites have been reported. In general, Assemblage 2 successions are characterized by rapid facies changes over short distances and mixed subaqueous and subaerial volcanics and sediments. Red beds, reflecting a widespread oxidizing atmosphere, first appear in both Assemblage 1 and 2 successions at about 2 b.y. ago.

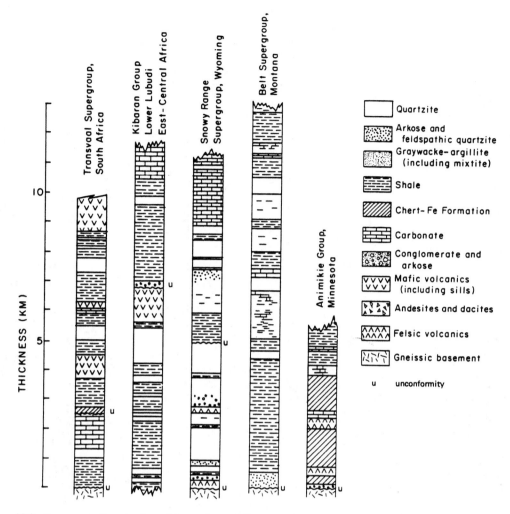

10.9. Representative stratigraphic sections of Proterozoic Assemblage 1.

The Calc-Alkaline Volcanic-Graywacke Assemblage. Assemblage 3 is characterized by a dominance of calc-alkaline volcanics and basalt (fig. 10.11). Most successions are comprised of a continuous suite of volcanics that have evolved along tholeiitic and/or calc-alkaline fractionation trends. Andesites and dacites are common in some successions, illustrated by the Yavapai and Flin Flon–Snow Lake successions in North America (fig. 10.11). Bimodal volcanics are characteristic of some successions, such as the Tewinga Group in northeastern Australia. Basalts are typically pillowed flows and associated sills and calc-alkaline volcanics represent a combination of subaqueous and subaerial volcaniclastic rocks. Sections contain variable amounts of graywacke and associated argillite which exhibit graded bedding and other textures and structures suggestive of a turbidite origin.

Minor sediments include conglomerate (both matrix- and framework-supported), quartzite, chert, iron formation, shale, and, rarely, carbonate.

Some assemblages, such as the Flin Flon–Snow Lake succession in northern Canada and the Birrimian successions in West Africa, resemble and in some respects are indistinguishable from Archean greenstone successions. Four major differences, however, appear to distinguish most Archean greenstones from Proterozoic "greenstones": (1) ultramafic and komatiitic volcanic rocks are uncommon or absent in Proterozoic successions; (2) volcaniclastic rocks, and in particular potassium-rich felsic volcanic rocks, are more abundant in Proterozoic successions; (3) graywacke is proportionally more important in most Proterozoic successions; and (4) chert and iron formation are less common in Proterozoic successions.

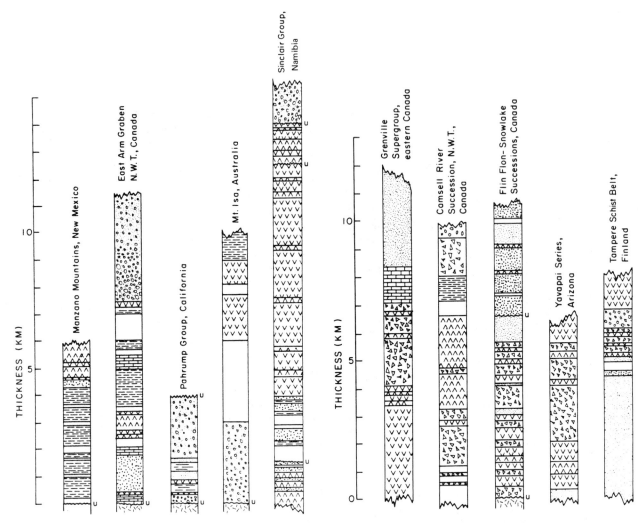

10.10. Representative stratigraphic sections of Assemblage 2. Explanation given in Fig. 10.9.

10.11. Representative stratigraphic sections of Assemblage 3. Explanation given in Fig. 10.9.

Ophiolites. Ophiolite and associated deep-sea sediments (Assemblage 4) are first recognized in terranes 0.5–1.0 b.y. in age. Although some ophiolites may represent immature oceanic arcs, most appear to represent fragments of back-arc oceanic crust. Reported occurrences are from Pan-African belts in Africa, Arabia, and South America and from late Proterozoic terranes in Newfoundland and other North Atlantic locations (fig. 10.8). Estimated thicknesses of late Proterozoic ophiolites range up to 8 km, but most are less than 5 km. They are bounded by thrust faults and appear to have been emplaced by obduction. Although the range in metamorphic grade and degree of deformation is considerable, some ophiolites preserve primary textures and structures. Most, however, represent only partial ophiolite successions.

An example of a complete late Proterozoic ophiolite is the Jabal Ess ophiolite in western Arabia (fig. 10.12). The lower contact of the ophiolite is a melange up to 250 m thick which contains fragments of the ophiolite rock types in a sheared serpentinite matrix. The lower unit in the ophiolite section is a serpentinized peridotite with relict cumulus textures, and this is in fault contact with a layered gabbro sequence. Overlying and gradational with the layered gabbro is a sheeted diabase dike complex with individual dikes ranging from 2 to 30 cm wide. Overlying the dike complex is a section $\gtrsim 300$ m thick of altered pillow basalts, with a single horizon of shale and laminated chert 50 m thick. A small body of plagiogranite (trondhjemite) is intruded into the section. The ophiolite succession is folded and has been

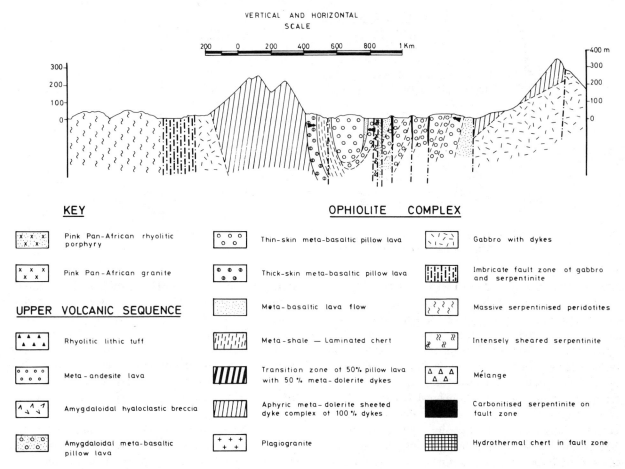

VERTICAL AND HORIZONTAL
SCALE

10.12. Idealized cross-section of the Jabal Ess ophiolite, northwestern Saudi Arabia (from Shanti and Roobol, 1979).

emplaced by thrusting with a basal melange produced along the low-angle thrust.

The Detrital Quartz Problem. One of the most important supracrustal rocks in Proterozoic sections is quartzite. It is of interest to discuss the various possible sources for detrital quartz listed in table 10.1. Mafic and intermediate rocks do not provide adequate sources for quartz—either they do not contain quartz or their quartz is too fine-grained and too minor to serve as a source for quartzites. Derivation of detrital quartz from phenocrysts in felsic volcanics, dikes, and sills requires weathering of vast amounts of volcanic rock, since quartz phenocrysts rarely comprise over 10–20 percent of such rocks; also, quartz phenocrysts are generally smaller than the average grain size of most quartzites. Vein quartz and silicified rocks are very minor in the crust and hence unlikely to serve as major sources for detrital quartz. Most detrital quartz in Proterozoic quartzites occurs as monomineralic grains or as quartzite rock

fragments, and hence cannot represent unmetamorphosed chert. Metachert and recycled quartzite sources also cannot supply the vast amounts of single-crystal quartz found in Proterozoic quartzites. This leaves only a granitic or gneissic terrane as the major source for detrital quartz. Such a source requires intense chemical weathering to provide feldspar-free quartz in one cycle of weathering, erosion, and deposition. Increased decomposition rates for feldspars, however, may be aided in high-energy environments like tidal flats and fluvial systems.

Interpretation of Supracrustal Assemblages

A summary of the average supracrustal abundances in Assemblages 1 through 3 is given in table 10.2. Although some rocks such as quartzite and shale occur in all assemblages, the three rock associations are distinct when the entire package of rocks in each assemblage is considered. As was discussed in Chapter 8, Phanerozoic supracrustal assemblages can be inter-

Table 10.1. Possible Sources of Archean Detrital Quartz

Source Rocks	Comments
1. Basalt, gabbro, nonporphyritic calc-alkaline volcanics	Inadequate source
2. Quartz-phenocryst bearing felsic volcanics	Requires extreme weathering and/or selective concentration of quartz
3. Vein quartz	Local source only
4. Silicified source rocks	No evidence
5. Unmetamorphosed chert	Requires polygonization of fine-grained quartz intergrowths
6. Metachert, recycled quartzite	Requires prolonged weathering
7. Granitic or gneissic terrane	Requires prolonged weathering or high-energy depositional environment

After Donaldson and Ojakangas (1977).

preted in terms of tectonic settings, and it is of interest to consider Proterozoic assemblages in terms of possible Phanerozoic analogues. The closest similarities are with Assemblages 1 and 4. Proterozoic ophiolites are strikingly similar to Phanerozoic ophiolites in both rock succession and structure, and appear to represent obducted fragments of late Proterozoic oceanic crust. Assemblage 1 occurs in three stable tectonic settings: rifted continental margins, cratonic margin of back-arc basins, and intracratonic basins. In rifted continental margin and cratonic margin of back-arc basin settings, sections thicken toward the ocean, as is reflected for instance by the late Proterozoic Ocoee Series in the southeastern United States. Intracratonic basins are represented by the Athabasca and Thelon basins of early Proterozoic age on the Canadian Shield (Plate I).

Assemblages 2 and 3 form in a variety of continental and continental-margin rifts today (Chapter 8). Assemblage 2 is found in modern lithosphere-activated rifts and cratonic rifts. Existing geologic data indicate that most Assemblage 2 successions formed in cratonic rifts or aulacogens. Geochemical and Sr isotope studies indicate that mafic volcanics from Assemblage 2 successions are of mantle origin, and that felsic volcanics and most syn- to post-tectonic granites are of lower crustal origin. Some early Proterozoic cratonic rifts appear to be part of large rift systems, as for example the Dewaras and Waterberg successions in southern Africa (1.7–1.8 b.y.). Other successions, such as the Pahrump and Uinta Mountain Groups in the western United States, may represent aulacogen successions. Proterozoic rift volcanism and sedimentation are characterized by both subaqueous and subaerial contributions. Sediments represent chiefly fluvial and intertidal deposits, although local marine inundations are recorded by thick carbonate and shale sequences in some rift successions. The abundance of mature quartzites in some successions, as for instance in the Jotnian successions in Scandinavia and in the Proterozoic successions of New Mexico, may reflect in-

Table 10.2. Average Abundances of Major Rock Types in Proterozoic Assemblages 1, 2, and 3.

Assemblage	1	2	3
Quartzite	30	28	5
Shale	40	18	2
Bimodal volcanics	5	26	9
Calc-alkaline volcanics	—	5	50
Carbonate	14	4	2
Graywacke-argillite	3	3	28
Arkose-conglomerate	5	14	3
Chert-iron formation	3	2	1

tense winnowing of feldspars in an intertidal environment within the rifts.

Today, Assemblage 3 is found in back-arc, intra-arc, and fore-arc basins. Archean greenstone successions have many overall similarities to Proterozoic Assemblage 3 successions. The same problems described for Archean greenstones, if they are equated with arc systems, also apply to many, or perhaps most, Proterozoic Assemblage 3 successions. If, however, we equate Assemblage 3 with a cratonic rift, as is suggested by the continental rift model for Archean greenstone belts, we are faced with the problem that this assemblage is not recognized as forming in cratonic rifts today. Sedimentary rocks in modern cratonic rifts are typically Assemblage 2. Also, modern cratonic rifts are commonly characterized by alkaline volcanism, or combined bimodal (tholeiite-rhyolite) and alkaline volcanism (Chapters 7 and 8), rather than by tholeiitic and calc-alkaline volcanism. These differences, however, may be accounted for by some combination of steeper geotherms in the Precambrian and absence of marine inundation of most modern cratonic rifts. It is possible that cratonic rifts changed from "hot" unstable basins in the Archean to "cool" rifts with time in response to a falling geothermal gradient in the Earth. Such a trend should be accompanied by a decrease in the degree of melting in mantle plumes responsible for the rifts, and thus mafic, ultramafic, and calc-alkaline magmas should be important in early cratonic rifts while alkaline magmas, requiring only a small amount of mantle melting, should characterize young cratonic rifts. Marine inundation of Archean and Proterozoic rifts provides a suitable environment for graywacke deposition, whereas modern cratonic rifts are chiefly subaerial and thus contain an Assemblage 2 sedimentary package indistinguishable from that found in lithosphere-activated rifts. A greater amount of continental-rift marine inundation in the early Precambrian may be due to the presence of a greater total volume and length of mid-ocean ridges at that time than today. Such increases would result from increased spreading rates in response to greater heat production in the Earth during the early Precambrian. Prior to ~2 b.y., Assemblage 3 may have been produced chiefly in "hot" unstable basins, while after ~1.0 b.y. it formed chiefly in back-arc basins and other basins associated with convergent plate boundaries; between 1.7 and 1.0 b.y. it may have been produced in both environments.

Some Proterozoic assemblages represent combinations of Assemblages 1, 2, and 3. For instance, Assemblage 2 is overlain by Assemblage 1 in the Mt. Bruce and Nabberu Supergroups and in the Kimberley Basin succession in Australia and in the Thelon and Athabasca basin successions in Canada. These successions may represent lithosphere-activated rifts which were aborted very early and evolved into cratonic or stable continental-margin basins. It is possible that many Assemblage 1 successions actually began as Assemblage 2 successions in rift environments, as evidenced by the presence of basal arkoses and volcanics. The Circum-Ungava geosyncline in Canada (fig. 10.16) exhibits a complex lateral facies distribution as well as an evolutionary change with time, as will be discussed later. It is noteworthy that the Kaapvaal basin in South Africa changes from a cratonic succession (the Witwatersrand Supergroup, 2.5–2.7 b.y.) into an Assemblage 2 succession (the Ventersdorp Supergroup, 2.3–2.5 b.y.) and then back into a cratonic succession (the transvaal Supergroup, 2.1–2.3 b.y.). This appears to represent a cratonic basin that was rifted, aborted, and then again stabilized.

The distribution of supracrustal successions with time is summarized in fig. 10.13. By 3.8 b.y. Assemblage 3 is well-established and widespread, forming chiefly in "hot" unstable continental basins analogous to cratonic rifts. At about 3 b.y. the appearance of Assemblage 3 in southern Africa and perhaps elsewhere records the first stabilization of continental crust. Two billion years ago is an important time in geologic history—at this time the lithosphere cooled and became rigid enough to sustain brittle fracture, and continental rifting began. Also at this time we have our first evidence for convergent plate boundaries (see below), suggesting the local onset of buoyancy-driven subduction in response to local cooling of the mantle lithosphere such that eclogite may form (fig. 10.1). Also, the widespread distribution of Assemblage 1 at 2 b.y. reflects widespread stable cratons at this time (fig. 10.8). If Archean greenstone successions and most early Proterozoic Assemblage 3 successions formed in plume-activated basins, 2 b.y. may also have been a time of declining mantle-plume activity. Whereas the Archean and early Proterozoic are characterized by numerous small mantle plumes, the late Proterozoic and Phanerozoic are characterized by fewer but larger plumes, as is suggested by the possible distribution of plumes today (fig. 6.28). As will be discussed later in this chapter, 1 b.y. appears to record the first opening of small ocean basins and localized onset of Wilson-cycle tectonics within continental areas. The first appearance of ophiolites and numerous examples of arc volcanics coincide with this time. Also, 1 b.y. is a

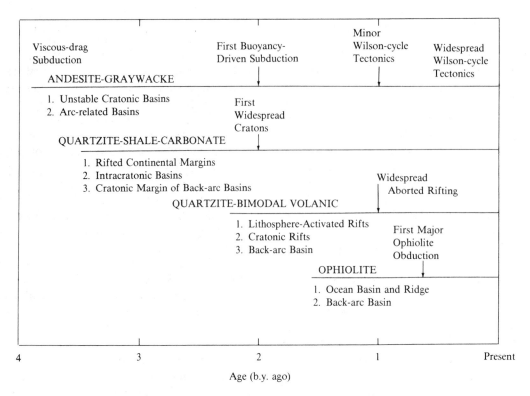

10.13. Summary of the distribution of supracrustal assemblages with geologic time.

time of widespread aborted cratonic rifting (Sawkins, 1976). Widespread Wilson-cycle tectonics involving the opening and closing of large ocean basins, however, did not begin until 500–300 m.y. ago.

Intrusive Rocks

Granitic Rocks. Gneissic complexes and batholiths comprise most of Proterozoic crustal provinces, as Chapter 5 discussed. Tonalitic and granodioritic complexes are widespread and record up to five periods of deformation. Many of these gneissic complexes are reactivated Archean complexes, as is indicated by radiogenic isotope tracer studies. Most granitic rocks which actually crystallized from magmas during the Proterozoic are granite (*sensu strictu*), quartz monzonite, or syenite-monzonite in composition.

The Anorthosite-Granite Suite. Proterozoic massive anorthosite bodies, commonly associated with granite or syenite, lie in a very broad belt extending from the southwestern United States to central Europe on a pre-drift continental reconstruction (fig. 10.8). This belt crosses several crustal provinces, with anorthosites in North America falling chiefly in the age range of 1.4–1.5 b.y. and those in Greenland and Europe chiefly in the range of 1.7–1.9 b.y. (Emslie, 1978).

The greatest number of anorthosite massifs occurs in the Grenville Province (fig. 5.12), where about 20 percent of the province is underlain with anorthosite.

The massive anorthosites, which are composed of more than 90 percent plagioclase, are interlayered with gabbroic anorthosites and exhibit cumulus textures and rhythmic layering. Most massifs range from 10^2 to 10^4 km^2 in surface area. Many bodies are intruded into older granulite-facies terranes and appear to reflect post-tectonic emplacement; these bodies are commonly highly fractured. Others are deformed along with surrounding rocks and appear to represent syn-tectonic plutons. Gravity studies indicate that most bodies are from 2 to 4 km thick and are sheet-like in shape, suggesting they represent portions of stratiform igneous complexes.

Granite, commonly with rapakivi textures, and other high-K$_2$O granitic rocks are closely associated with many massive anorthosite bodies, suggesting a genetic relationship. Geochemical and isotopic studies, however, indicate that the anorthosites and granites are not derived from the same parent magma by fractional crystallization or from the same source by partial melting. Most data are compatible with an origin for the anorthosites as cumulates from the fractional crystallization of high Al$_2$O$_3$ tholeiitic magmas produced in the upper mantle (Emslie, 1978). The granitic magmas appear to be produced by partial

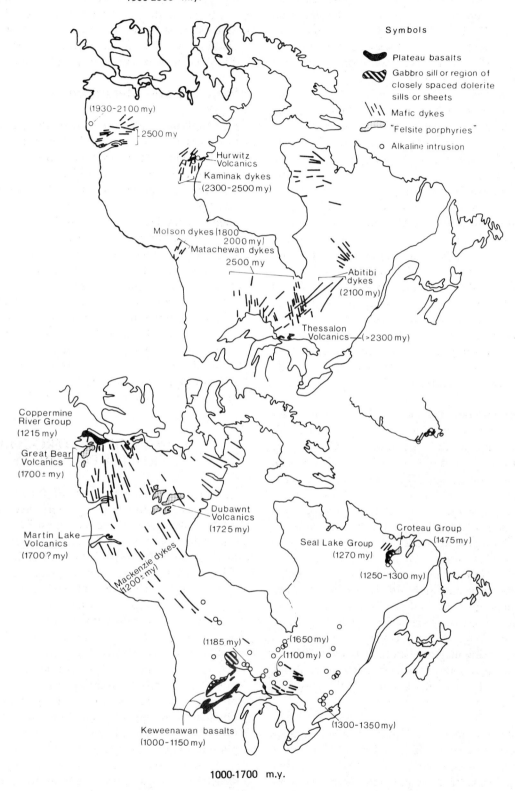

1900-2300 m.y.

Symbols

Plateau basalts

Gabbro sill or region of closely spaced dolerite sills or sheets

Mafic dykes

"Felsite porphyries"

○ Alkaline intrusion

(1930-2100 my)

2500 my

Hurwitz Volcanics

Kaminak dykes (2300-2500 my)

Molson dykes (1800-2000 my)

Matachewan dykes 2500 my

Abitibi dykes (2100 my)

Thessalon Volcanics (>2300 my)

Coppermine River Group (1215 my)

Great Bear Volcanics (1700± my)

Dubawnt Volcanics (1725 my)

Croteau Group (1475 my)

Seal Lake Group (1270 my)

Martin Lake Volcanics (1700? my)

Mackenzie dykes (1200± my)

(1250-1300 my)

(1185 my)

(1650 my)

(1100 my)

(1300-1350 my)

Keweenawan basalts (1000-1150 my)

1000-1700 m.y.

10.14. Precambrian mafic dike swarms and plateau basalts of the Canadian Shield (from Baragar, 1977).

melting of lower crustal rocks of intermediate composition (Condie, 1978). The heat for lower crustal melting may be derived from crystallizing tholeiites in magma chambers in the lower crust. In this respect, the anorthosite-granite association is a product of bimodal magmatism (Chapter 7).

The tectonic setting in which the anorthosite-granite suite forms is a subject of continuing discussion, although most investigators agree upon an extensional environment. Data suggest the following sequence of events which are interpreted within the framework of a continental rift system: (1) intrusion of anorthosite-granite complexes, together, in some areas, with surface rhyolite volcanism (such as in the St. Francois Mountains in Missouri); (2) uplift and erosion followed by or together with rifting; (3) continued rift development accompanied by deposition of immature sediments and bimodal volcanics (Assemblage 2) and/or alkaline volcanics. This model suggests the existence of an extensive mantle heat source (upwelling?) beneath Europe at 1.7–1.9 b.y. and beneath central North America at 1.4–1.5 b.y. Whether such rifting reflects aborted cratonic rifts or rifts associated with continental collisions is currently a subject of controversy.

Stratiform Igneous Complexes. Proterozoic stratiform igneous complexes range in size from small sills ≲ 100 m thick to the vast Bushveld Complex in South Africa, which is over 8 km thick and covers an outcrop area of > 66,000 km^2 (Hunter, 1978). Important complexes which have been studied in some detail are the Muskox and Duluth Complexes in North America, both 1.1–1.2 b.y. in age, and the Bushveld Complex, about 2 b.y. in age. Stratiform complexes are characterized by rhythmic and compositional layering and cumulus textures. Textural and geochemical data indicate these bodies formed by fractional crystallization of tholeiitic parent magmas. Cyclic layering in the bodies is generally interpreted to reflect episodic injections of new magma into fractionating magma chambers. Large stratiform bodies are major sources of many metals, such as Cr, Ni, Cu, and Fe and, in the case of the Bushveld complex, also Pt.

Mafic Dike Systems. As Chapter 5 briefly described, major swarms of mafic dikes, chiefly of Proterozoic age, intrude Precambrian crustal provinces. Some swarms extend only a few hundred kilometers along strike, while others, like the Mackenzie swarm which was intruded at ~ 1250 m.y. is more than 3000 km long and over 500 km wide (fig. 10.14). Such vast

dike swarms indicate widespread extensional environments during the mid-Proterozoic. They also indicate that the crust (and lithosphere) were cool enough to sustain brittle fracture on a wide scale. Many of these dikes may have been feeders for plateau basalt fields.

Structure and Metamorphism

Structure and metamorphism in Proterozoic crustal provinces have been described in Chapter 5. It is worth reemphasizing that polyphase deformational patterns are complex and reflect chiefly compressive forces in most provinces. Also, a large proportion of these provinces is comprised of reactivated Archean gneissic complexes, as is illustrated for example by the Churchill Province in northern Canada.

Metamorphic grade is variable, with amphibolite and granulite-facies terranes most widespread. Although facies series are also variable, medium-pressure series are most important, and high-pressure (blueschist) series make their first appearance at about 1 b.y. in the Gariep belt in southern Africa. This contrasts with Archean provinces, where low-pressure facies dominate. The trend from lower to higher pressure series with time reflects the falling geothermal gradient in the Earth (Chapter 11).

ORIGIN OF PROTEROZOIC CRUSTAL PROVINCES

Although increasing numbers of Proterozoic supracrustal successions are being interpreted in terms of modern plate-tectonic regimes, most Proterozoic mobile belts are not readily accommodated by such models (Kröner, 1981). Most mobile belts between 2.5 and 1.0 b.y. in age appear to have developed on older sialic crust and not to have involved opening and closing of large ocean basins. The criss-crossing pattern of mobile belts without offset; the widespread occurrence of relict-age terranes in mobile belts; the matching of older crustal provinces across mobile belts; the continuity of structural trends of older mobile belts through younger ones; and the absence of ophiolites, arc volcanics and arc sediments, and blueschists all argue against the interpretation of most Proterozoic mobile belts in terms of Wilson-cycle plate tectonics.

As is indicated from supracrustal assemblages, structural geologic studies, and paleomagnetic studies of Proterozoic rocks, convergent plate boundaries are recorded locally at about 2 b.y. ago, but Wilson-cycle tectonics appear not to have begun until between

1000 and 600 m.y. ago. Hence, some modified form of plate tectonics may have operated during the Archean and most of the Proterozoic.

Evidence for Early Proterozoic Convergent Plate Boundaries

Most Assemblage 3 successions, as was previously mentioned, are similar to Archean greenstone successions and probably reflect a similar tectonic setting, whatever that setting may be. Lithologic and structural data from several early Proterozoic Assemblage 3 supracrustal successions and associated gneissic terranes, however, have been interpreted in terms of convergent plate tectonics. Perhaps the most convincing case for an early Proterozoic arc system with a back-arc basin is the Wopmay orogen and Epworth basin in the Bear Province of northern Canada (fig. 5.3). Hoffman (1980) suggests that at about 2.1 b.y. the Archean crust in northwestern Canada was rifted apart over several plumes. Thick bimodal volcanics and arkoses are preserved in the East Arm and Kilohigok aulacogens formed during this rifting. The Epworth basin cratonic succession (Assemblage 1) is deposited along the Wopmay continental margin as a small ocean basin opens. Westward subduction of this oceanic crust (fig. 10.15A) leads to an arc-continent collision at about 1.9 b.y. (B). The Hepburn and Wentzel syn-tectonic batholiths are emplaced during this collision and deformation. The Great Bear plutonic complex (zone 4) is interpreted as part of the overriding plate. In this complex, tonalitic plutons intrude cogenetic calc-alkaline volcanic rocks (LaBine Group). Zone 1 represents relatively undeformed shelf sediments and zones 2 and 3 represent thrust slices or flakes that moved eastward during the collision. Following collision, an east-dipping subduction zone developed, giving rise to calc-alkaline volcanism of the Sloan Group (C). Also, the Wopmay fault, a major transcurrent fault, developed in response to oblique subduction. The final configuration after the deposition of middle Proterozoic sediments over most of the Bear Province is shown in D.

Convergent plate margins have also been proposed at about the same time (1.7–1.9 b.y.) in the western United States and in Scandinavia. In southeastern Wyoming, the major shear zone separating the Archean Wyoming Province from the Proterozoic province in Colorado (fig. 5.7) has been interpreted as a deep-seated manifestation of a suture zone (Hills and Houston, 1979). The volcanic-plutonic terrane south of the shear zone is equated with an arc system that collided with a stable continental margin on the

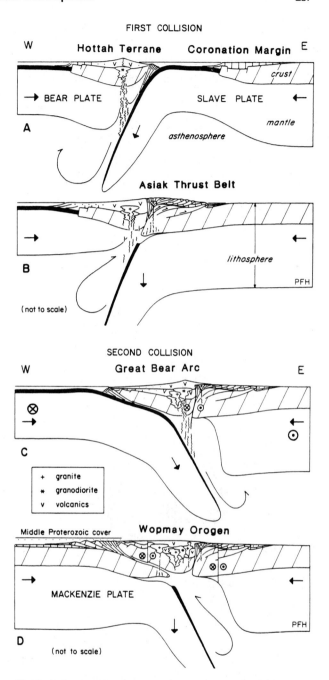

10.15. Diagrammatic cross-sections summarizing a convergent plate margin model for the development of the Wopmay orogen, northwestern Canada (after Hoffman, 1980).

north. The distribution of supracrustal rocks in Scandinavia has been interpreted by some to reflect a northward-dipping subduction zone at about 1.8 b.y. (Hietanen, 1975). A belt of calc-alkaline volcanic rocks and graywackes extending from southern Finland west as far perhaps as southern Norway is interpreted as an arc system. Thick successions of graywacke north of this belt in central Finland and Sweden are interpreted as a back-arc basin, and remnants of cratonic sediments in northern Finland as the stable continental margin of the back-arc basin.

Examples of Proterozoic Rift Systems

As was previously discussed, continental rifting appears to have begun at about 2 b.y., although not becoming widespread until 1.0 b.y. (Sawkins, 1976). One of the earliest documented studies of a cratonic rift system is the circum-Ungava system in Canada (fig. 10.16). Although this system has also been interpreted as a continental collision boundary (Gibb and Walcott, 1971), most data favor the rift model (Dimroth, 1981). The circum-Ungava system is comprised of the Labrador trough, the Cape Smith foldbelt, and the Belcher foldbelt, which surround the Ungava craton on three sides in northern Quebec

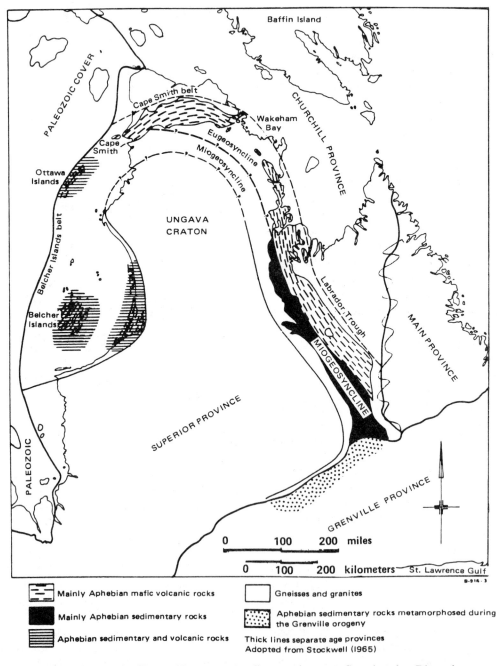

10.16. Map showing the Circum-Ungava geosyncline, northeastern Canada (after Dimroth, *et al.*, 1970).

(fig. 10.16). Additional portions of the system may extend farther west along the southern margin of Hudson Bay. The supracrustal rocks are chiefly an Assemblage 1 succession along the Ungava craton margin, while mafic volcanics and intrusives predominate away from this margin. Cyclic sedimentation and volcanism characterize the succession, which appears to have been deposited or erupted largely or entirely on Archean sialic crust. Rocks in the Labrador trough begin with Assemblage 2, and evolve into Assemblage 1 on the west and into a thick succession of basalt, gabbro, and shale on the east. In terms of the proposed model, this is interpreted as a cratonic mantle-activated rift which opens into a small ocean basin (as represented by the mafic rocks and shale) with a marginal basin on the west. It is clear, however, that the basalt-shale succession is not similar to Phanerozoic ophiolites. The entire succession is deformed and metamorphosed, with major overthrusting and nappes directed toward the Ungava craton.

A currently popular model for the development of the circum-Ungava succession involves thinning of the lithosphere and crust above mantle upwellings or plumes, producing rift-like basins into which Assemblage 2 successions are deposited at 1.9–2.0 b.y. The sediments are derived primarily from the Ungava craton. Extensive heating in the upper mantle produces large volumes of tholeiites which are erupted and intruded in the outer part of the U-shaped basin system. Subsidence of mantle upwellings leads to compression of the basins directed toward the Ungava craton. At this time (~1.8 b.y.) supracrustal rocks are deformed and thrust towards the cratonic interior. Some models suggest that minor subduction may also have occurred during closing of the rift.

Perhaps the best known example of a Precambrian rift system is the Keweenawan rift, which developed between 1.0 and 1.2 b.y. in central North America. The mid-continent gravity high (fig. 4.18) delineates this rift system where it is covered with

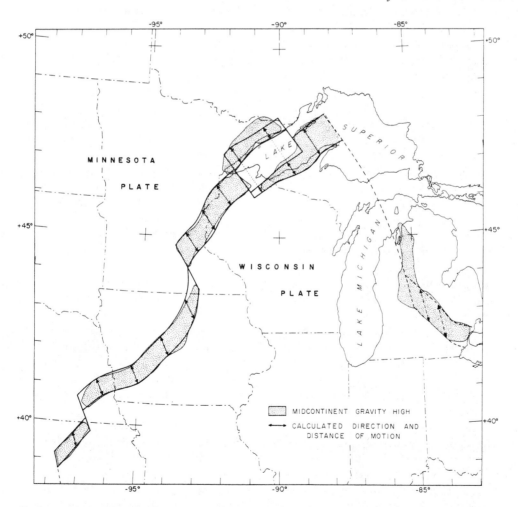

10.17. Gravity model for the opening of the Keweenawan rift system (from Chase and Gilmer, 1973).

Phanerozoic cratonic sediments. Where exposed in the Lake Superior area, the rift contains up to 15 km of basalts and red clastic sediments, filling a trough about 150 km wide and over 1500 km long (Card et al., 1972). The trough is a fault-bounded rift zone which extends as far southwest as central Kansas. Most sections are comprised of more than 70 percent tholeiitic lavas, with the remainder dominantly felsic volcanics, arkose, shale, and conglomerate. Individual basalt eruptive sequences range up to 7 km thick and are of the order of 33,000 km^2 in areal extent. Basalts were fed by fissures, examples of which are extensive dike swarms in the Lake Superior area. Radiometric dates suggest the rift was active over a relatively short period of time between 1120 and 1170 m.y. ago.

Model studies of the mid-continent gravity anomaly associated with the rift suggest that it represents an aborted cratonic rift system and that the width of the gravity high is the width of the rift (Chase and Gilmer, 1973). The results also suggest that the rift basalts do not rest on older sialic crust, but fill a gap between separated edges of the continent. When the rift is opened about the best-fit pole of rotation (located in northern New Mexico), the amount of rifting increases regularly with the angular distance from the pole, and transform faults (which offset the rift) lie on small circles about the pole (fig. 10.17). The division of the gravity anomaly in the western part of Lake Superior is due to the intrusion of the Duluth stratiform complex beyond the northwestern margin of the rift. The magnetic anomaly beneath Michigan probably represents an offset arm of the rift, and a late Proterozoic Assemblage 2 succession in the Michigan basin supports this interpretation.

Late Proterozoic Wilson-Cycle Tectonics

Geologic data indicate that modern-style plate tectonics began about 1000 m.y. ago and become important in the Pan-African system in north Africa and Brazil and in areas around the North Atlantic by 500–700 m.y. ago. Fig. 5.13 shows the areas affected by the Pan-African tectono-thermal event in Africa. As previously discussed (Chapter 5), most of the area represents reactivated Archean crust. However, in the Dahomeyan-Pharusian belt in west Africa and in the Red Sea area in northeast Africa and Arabia, continental fragmentation, subduction, and continental and arc collisions characterize the Pan-African event (Kröner, 1979). A summary of the geologic history of the Hoggar segment of the Pharusian belt was given in Chapter 5 (fig. 5.14). This history can be interpre-

ted in terms of cratonic rifting, with deposition of Assemblage 2 in the rift system followed by extensive extrusion and intrusion of basalts (1000–800 m.y.). A small ocean basin opens (<200 km across) 700–800 m.y. ago and closes again with only minor subduction, reflected by sparse calc-alkaline igneous rocks. Continental collision at 650–700 m.y. is reflected by nappes and thrusts directed toward the west, a suture zone with ophiolites, and emplacement of numerous granites of lower crustal origin. The northeastern margin of the Sahara craton underwent a somewhat different history, involving perhaps several arc systems which became welded to the craton through collisions between 1000 and 600 m.y., thus extending the continental crust in this area. Rocks older than 1000 m.y. are not recognized in Arabia, suggesting that Arabia was added to Africa by arc-continent collisions between 1000 and 600 m.y. ago.

The Damara-Gariep-Ribeira mobile belt system in southern Africa and eastern South America (fig. 5.4) appears to reflect a three-armed rift system, with the Damara arm becoming an aulacogen (Porada, 1979). A small ocean basin appears to have opened between southwest Africa and southern Brazil, closing again with only minor subduction. Strong asymmetry of thrusts and nappes to the east, extensive granite intrusion in the Ribeira (South America) portion of the belt, minor deformation in the Gariep belt, and other features support a collision model in which the descending slab dipped to the west.

Late Proterozoic volcanic rocks in Newfoundland and other areas in the North Atlantic appear to be remnants of late Proterozoic arc systems and obducted ophiolites (Strong, 1979) (fig. 10.18). The abundance of bimodal volcanics and other Assemblage 2 rocks in the Burin peninsula and Avalon platform of eastern Newfoundland reflect cratonic rift systems in this area between 800 and 600 m.y. ago. These rifts eventually opened into the Iapetus Ocean basin between 500 and 700 m.y. (Chapter 9). Remnants of back-arc basins associated with those opening occur as obducted ophiolites (fig. 10.18). The sparsity of calc-alkaline volcanics, however, indicates that subduction was not an important process during the late Proterozoic evolution of the North Atlantic basin.

Ensialic Mobile Belts and the Development of Wilson-Cycle Tectonics

Accepting the likely premise that ensialic mobile belts do not represent convergent plate boundaries or zones

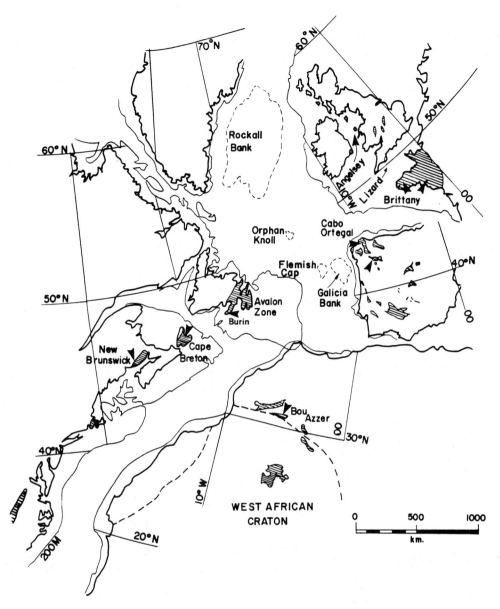

10.18. Late Proterozoic-early Cambrian supracrustal sequences shown on a pre-Mesozoic reconstruction of the North Atlantic (from Strong, 1979). Possible or known ophiolites are indicated with arrows.

of continental collision, how then did they form? One model for their formation, known as the *millipede model* (Wynne-Edwards, 1976), involves the movement of ductile continental lithosphere over convective upcurrents or plumes in the mantle (fig. 10.19). In this model, ductile lithosphere moves away from oceanic spreading centers, retaining cohesiveness and cooling as it moves. It eventually encounters compressive lithospheric forces. Continental lithosphere, hence, may drift over a mantle plume or upcurrent, resulting in reactivation of large amounts of older

sialic rocks. Thinning and subsidence of the lithosphere and crust over the upwelling may result in oceanic inundation of the resulting basin and deposition of sediments (2). Early sediments are chiefly rift sediments (Assemblage 2) and later sediments chiefly cratonic sediments (Assemblage 3). Mafic and felsic magmas are produced in the mantle and lower crust, respectively, and should define trajectories in the crust (from volcanics and plutons) increasing in age as distance from the upwelling increases, in a manner similar to generation of hotspot traces (Chapter 8).

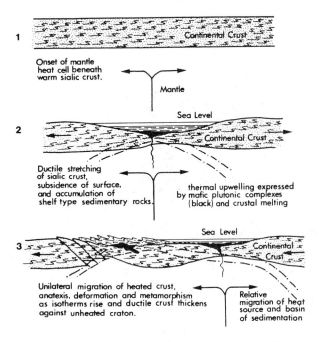

10.19. Diagrammatic sections of the millipede model for the development of Precambrian mobile belts (from Wynne-Edwards, 1976).

The older crust is also updated as the K-Ar (and perhaps other) isotopic system is reset as the plate moves over the upwelling. Some portions of the crust are partially melted to produce migmatite complexes, while others, together with infolded supracrustal rocks, are metamorphosed to the granulite facies. As the spreading lithosphere meets resistance, deformational styles change from extensional to compressional, and major thrusts and nappes develop directed outwards from the center of the mobilized region (3). The width of the mobile belt is controlled by the duration of the underlying spreading system and the rates of plate motion. The contact between the craton and the mobile belt is a tectonic or metamorphic break, such as is typically found between Proterozoic provinces (Chapter 5).

The millipede model of plate deformation may represent one stage in the changing behavior of continental crust (and lithosphere) as the temperature of the Earth cools with time (fig. 10.20). Temperatures are high enough in the Archean such that significant melting occurs in the lower sialic crust, and the crust as a whole is pictured as thin and very ductile. Greenstone belts develop in proto-rift systems. Decreasing temperatures in the Proterozoic lead to millipede tectonics as the lithosphere and crust become more rigid, yet still deform as ductile, subsolidus solids. By late Proterozoic time the lithosphere becomes rigid enough to behave as a brittle solid, and continental fragmentation and collision begin as modern plate tectonics becomes established.

Another modified plate-tectonic model for early to mid-Proterozoic ensialic mobile belts was recently

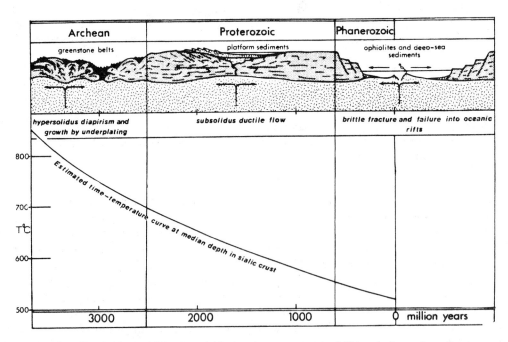

10.20. Changing behavior of the crust with time in response to a falling geothermal gradient (from Wynne-Edwards, 1976).

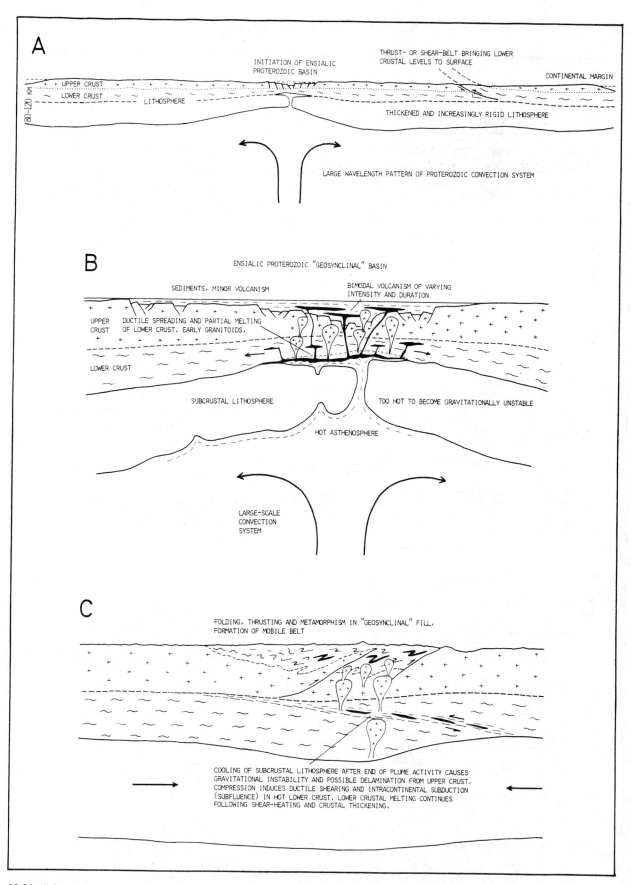

10.21. Schematic sections showing the development of early Proterozoic ensialic mobile belts (from Kröner, 1981).

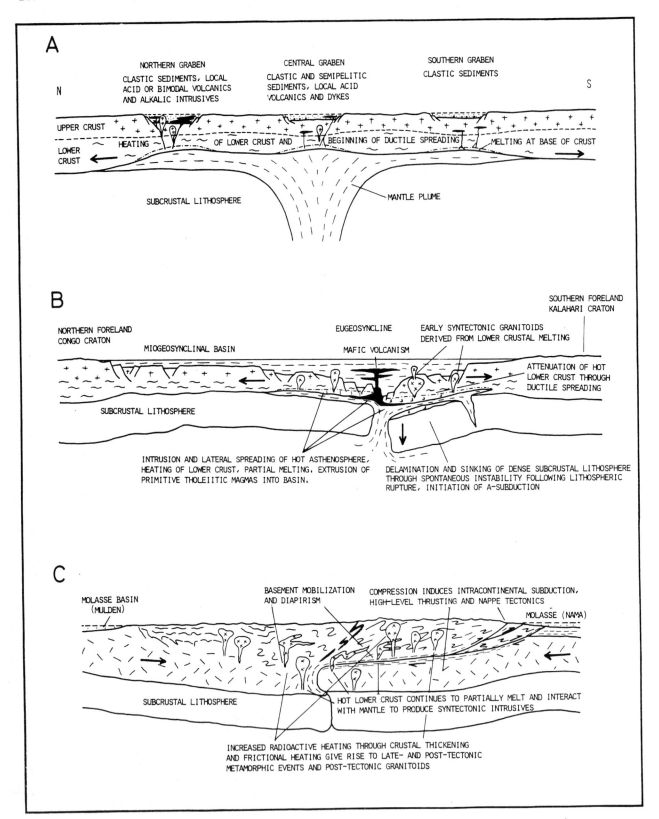

10.22. Schematic sections for the evolution of late Proterozoic, pre-Wilson-cycle tectonics (from Kröner, 1981).

proposed by Kröner (1981). This model is based on the assumption that buoyancy-driven subduction did not begin before about 1000 m.y. ago. The model involves convective upwelling beneath continental lithosphere, thinning of the lithosphere, and initiation of rift systems (fig. 10.21A). The spreading leads to underthrusting of ductile lithosphere beneath more rigid lithosphere on the flanks of the convective upwelling, with the net result of bringing lower crustal rocks to the surface. Continued sinking and widening of the rift is accompanied by accumulation of immature sediments and bimodal volcanism (B). However, because the lithosphere is not as rigid as that of today, complete splitting of the lithosphere and creation of an ocean basin does not occur. Instead, the convective upwelling or plume dies out and subsides (or the lithosphere moves away), leading to a compressive stress regime which produces folding, thrusting, ductile shearing, and thickening of the crust (C). Considered in conjunction with Proterozoic supracrustal Assemblage 2 which characterizes continental rifts, this type of model would begin about 2 b.y. ago.

Continued cooling of the Earth accompanied by thickening and increased rigidity of the lithosphere leads into the present tectonic regime, which begins at about 1 b.y. Buoyancy-driven subduction becomes widespread and the increased rigidity of the continental lithosphere results in complete fragmentation in some instances, and hence the onset of Wilson-cycle tectonics. Not all late Proterozoic and early Paleozoic crustal provinces, however, can readily be interpreted within a modern plate-tectonics framework. For instance, most of the Pan-African mobile belts in southern Africa are ensialic belts, indicating that the late Proterozoic and early Paleozoic was a time of transition between dominantly ensialic tectonics and modern plate-tectonics.

An idealized sequence of events during this transition period is illustrated in fig. 10.22. As with the ensialic model, convective upcurrents and/or plumes thin the continental lithosphere and produce a continental rift system (A). This process causes gravitational instability and eventual detachment of the dense mantle lithosphere from the crust (B). Injection of hot asthenosphere beneath the lower crust causes continued spreading of the rift system, magmatism, and mixing of crustal and mantle materials. This process may continue until the continent is completely ruptured and result in the opening of a small ocean basin, illustrated earlier by the Pan-African systems in northern Africa. Alternately, the downward pull of the delaminated slab may cause

continental convergence. This results in crustal thickening and complex mixing and deformation of supracrustal rocks and older sialic crust (C). The entire crustal section is intruded by syn- and post-tectonic granites and uplifted such that sediments are shed into marginal cratonic basins. In instances like this, where complete continental separation does not occur, ophiolites and arc volcanics are not formed as in the Pan-African belts of southern and eastern Africa. When complete separation occurs, Wilson-cycle tectonics begins, as in the Pan-African belts of northwestern and northeastern Africa and in the North Atlantic 500–700 m.y. ago.

PRECAMBRIAN PALEOMAGNETIC STUDIES

The problems discussed in Chapter 8 in relation to defining ancient pole positions in the geologic past by paleomagnetic methods become even more critical in Precambrian terranes. Accurate radiometric dates that represent the age of magnetization are difficult to obtain in most Precambrian rocks due to deformation and metamorphism. Metamorphism may partially or completely reset either or both the radiometric date and the magnetization direction, and it is often difficult to correlate the correct age with a measured pole direction. It is not clear yet which radiometric dating method most closely records the time at which magnetization is acquired by rocks that have been altered or metamorphosed. Recent experimental studies, however, suggest that little if any remanent magnetization can survive upper greenschist-facies metamorphism. Structural orientation of samples becomes more uncertain in deformed and metamorphosed terranes, and it is difficult to remove soft magnetization without affecting hard magnetization. In many successions, for instance basalts in the Belcher Island volcanics and sediments of the Huronian Supergroup in Canada, two periods of magnetization are recorded. One is pre- and the other post-folding. Also, as was mentioned in Chapter 8, it is not possible to identify longitudinal motions by the paleomagnetic method and it is commonly difficult to decide whether individual pole positions represent north or south poles. Paleomagnetic reversals have now been documented back at least as far as 3 b.y.

Because of these problems, Precambrian pole positions from a given locality are commonly more scattered than from comparable Phanerozoic sites. Investigators commonly reject and accept data based on sources of error that are difficult to evaluate, and

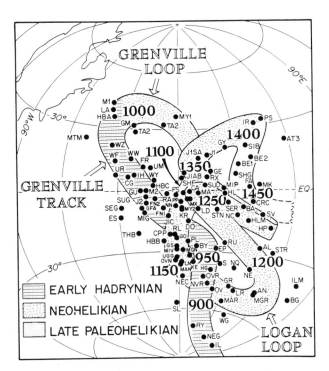

10.23. Suggested APW path for North America in the middle to late Proterozoic (from Irving, 1979).

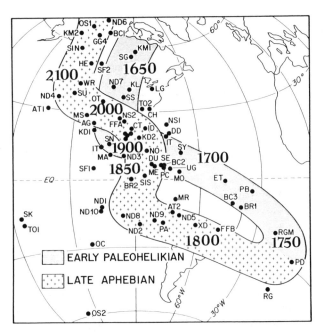

10.24. Suggested APW path for North America in the early Proterozoic (from Irving, 1979).

this leads to diverse and subjective apparent polar wandering curves. Some investigators connect individual poles to produce complex multiloop paths (Burke et al., 1976) while others draw broad bands encompassing many pole positions in an attempt to partially compensate for errors and sparsity of data. APW bands are most widely accepted and define curvilinear paths with approximately 10^8 y between loops (figs. 10.23, 10.24). Data indicate that APW bands can be defined back to at least 3.5 b.y.

The problem of whether loops exist in Precambrian polar wandering paths appears to be dependent upon the length of the increments of time over which pole positions are averaged. When data from North America are averaged for time intervals less than 200 m.y., the Logan Loop in fig. 10.23 is resolved. Many pole positions from Keweenawan basalts in the Lake Superior area define the descending limb of this loop. The fact that pole positions from stratigraphically lower horizons in the Keweenawan section define the upper part of the loop and those from higher horizons define the middle part of the descending limb indicate rapid polar wandering in a southwesterly direction or, alternately, rapid motion of the North American Plate in a northeasterly direction during a relatively short period of time (< 20 m.y.). Recent detailed paleo-

magnetic studies from sedimentary and volcanic rocks of the Grand Canyon Supergroup in northern Arizona (approx. 1000–1200 m.y. in age) indicate significant apparent polar wandering along a complex multiloop path as a function of stratigraphic height (Elston et al., 1973). Existing data suggest that loops in polar wandering curves with periods ≲ 200 m.y. are characteristic features for at least the last 2000 m.y. (fig. 10.23, 10.24).

Precambrian paleomagnetic data, despite inherent errors in interpretation, clearly preclude convergence of widely separated crustal provinces to explain mobile belts, and in this respect they support geologic data. Data constrain the amount of independent motion between adjacent crustal provinces, illustrated in fig. 10.25 for African crustal provinces. It is possible, allowing for the various sources of error described above, to open and close small ocean basins less than 1000 km across, provided adjacent crustal provinces always return to their approximate starting positions (McElhinny and McWilliams, 1977). Such limitations on the amount of interprovince movement have now been well documented in North America, Africa, and Australia. Results from North America and western Europe, furthermore, suggest that these continents were part of the same supercontinent in an approximate Laurussia configuration (fig. 10.8) during most of the Precambrian (Piper, 1976). The Southern Hemisphere continents appear to have comprised two

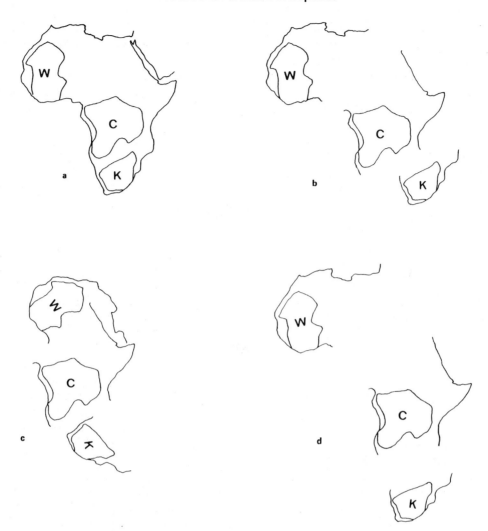

10.25. Diagrammatic illustration of motions between African crustal provinces which are allowed (a, b) and which are not allowed (c, d) by the tolerance levels of paleomagnetic dates (from Briden, 1976).

supercontinents, East and West Gondwana, until their collision during the early Cambrian (fig. 10.8) (McWilliams, 1981). How the three Precambrian supercontinents were located relative to each other is a topic of heated discussion.

Paleomagnetic data furthermore indicate that during most of the Proterozoic, Laurussia was located chiefly in low to middle latitudes, which is in good agreement with paleoclimatological data (Irving, 1979a). Implied rates of latitudinal continental drift during the Proterozoic of 4–5 deg/10^6 yr are more rapid than Phanerozoic rates of 1–2 deg/10^6 yr. Supporting the interpretation that only a few supercontinents existed during most of the Precambrian are the long segments ($\gtrsim 90°$) of APW paths between loops (fig. 10.23, 10.24). It would be difficult to

explain such long trajectories if the Earth were crowded with continents. It is notable that two loops (1750 and 1000 m.y.) and perhaps a third (600–650 m.y.) coincide with times of major orogeny. This implies rapidly changing plate motions at these times, which may reflect rapidly changing convection patterns in the Earth.

Although the Precambrian plate history of most of Asia remains largely unknown, data from other continents suggest that the Precambrian was a time of only a few supercontinents, and that major fragmentation of these continents did not begin until about 700 m.y. ago. The late Precambrian and early Paleozoic can be considered a time of continental breakup (Chapter 9), the middle and late Paleozoic as a time of continental reaggregation, and the

Mesozoic-Cenozoic, again, as dominantly a time of continental breakup. Clearly the Precambrian paleomagnetic data require a tectonic regime different from the modern plate regime characterized by numerous continents.

SUMMARY STATEMENTS

1. Cooling of the Earth with time favors a model in which buoyancy-driven subduction does not become important until after 1000 m.y. ago.

2. With increasing stratigraphic height, Archean greenstone successions exhibit decreasing amounts of ultramafic and mafic rocks, increasing amounts of calc-alkaline volcanics and clastic sediments, and increases in the proportion of volcaniclastic rocks to flows.

3. Two types of Archean greenstone belts are recognized: bimodal successions, comprised chiefly of ultramafic-mafic volcanics and minor felsic volcanics and calc-alkaline successions, comprised of mafic, intermediate, and felsic volcanics and clastic sediments. Bimodal successions may underlie calc-alkaline successions, but the reverse is not observed. Bimodal successions dominate in some provinces, calc-alkaline successions in others.

4. Archean greenstone belts appear to represent remnants of large volcanic-sediment basins.

5. Ultramafic and komatiite flows are common in Archean bimodal greenstones but are rare after 2.5 b.y. This distribution appears to reflect the falling geothermal gradient in the Earth.

6. Sediments in Archean greenstone belts are largely graywacke-argillite turbidites derived from dominantly volcanic sources. Conglomerate, quartzite, and arkose are of minor importance. Chert is the dominant nonclastic sediment.

7. Dominant Archean sedimentary environments are tectonically active basins, with slumping and turbidity currents related to associated volcanism and earthquake activity. Deltaic and tidal-flat sedimentation characterize the margin

of some basins. Cratonic sedimentary successions are of minor importance in the Archean.

8. Most data indicate that Archean low- and high-grade terranes evolve concurrently in different thermal-tectonic environments.

9. Archean greenstone belts may have developed in cratonic continental rifts or in back-arc basins, with most evidence favoring the former. Archean high-grade terranes may have developed over convective downcurrents where sialic crust is locally thickened.

10. Most Proterozoic supracrustal successions can be classified into one or a combination of four assemblages: (1) quartzite–carbonate–shale; (2) bimodal-volcanics–quartzite–arkose; (3) calc-alkaline-volcanics–graywacke; and (4) ophiolites and deep-sea sediments.

11. The vast amount of detrital quartz in the Proterozoic was probably derived from uplifted and eroded gneissic and granitic terranes. Prolonged weathering or a high-energy depositional environment seems necessary to provide feldspar-free detrital quartz.

12. Proterozoic supracrustal rocks appear to reflect the following tectonic settings: Assemblage 1—trailing continental edge, cratonic margin of back-arc basin, or intracratonic basin; Assemblage 2—chiefly cratonic rifts and aulacogens; Assemblage 3—"hot" cratonic basins (prior to 2.0 b.y. ago) and rift basins associated with convergent plate margins.

13. The anorthosite-granite suite, which defines a broad belt through Laurussia, appears to have developed in a continental rift environment at 1.4–1.5 b.y. in North America and at 1.7–1.9 b.y. in Europe.

14. Widespread mafic dike swarms emplaced chiefly during the Proterozoic indicate that the crust was cool enough by this time to sustain brittle fracture, and attest to the widespread occurrence of extensional stress regimes.

15. Geological and paleomagnetic data are consistent with the onset of subduction and aborted continental rifting at about 2000 m.y. ago, but

major Wilson-cycle tectonics did not begin until 1000–600 m.y. ago.

16. Proterozoic ensialic mobile belts probably developed over convective upwelling or chains of mantle plumes which thinned the crust and lithosphere. Movement of the lithosphere off the upwelling or subsidence of the upwelling leads to compressive deformation in the late stages of development.

17. Precambrian APW paths preclude convergence of widely separated crustal provinces to explain mobile-belt development. Small ocean basins (< 1000 km across) may open and close, provided adjacent crustal provinces always return to their starting positions.

18. Paleomagnetic data suggest that only a few supercontinents existed on the Earth during most of the Precambrian and that rates of plate mo-

tion were about twice as rapid as Phanerozoic rates. Major fragmentation of supercontinents began about 700 m.y. ago.

19. Some APW loops coincide with major orogenies, implying rapidly changing plate motions and convection patterns at these times.

SUGGESTIONS FOR FURTHER READING

Condie, K.C. (1981) *Archean Greenstone Belts*. Amsterdam: Elsevier. 434 pp.

Kröner, A., editor (1981) *Precambrian Plate Tectonics*. Amsterdam: Elsevier. 781 pp.

Windley, B.F., editor (1976) *The Early History of the Earth*. New York: Wiley. 619 pp.

Chapter 11

Crustal Origin and Evolution

ORIGIN OF THE CRUST

Crustal origin can be considered in terms of several questions (Condie, 1981a):

1. Was the first crust of local or worldwide extent?
2. When and by what process did the first crust form?
3. At what rate and by what process did the early crust grow?
4. What was the composition of the early crust?
5. When and how did oceanic and continental crustal types develop?

A closely related question is that of why the Earth's crust is so different from the crusts of other terrestrial planets. These and related problems will be considered in the following sections.

The oldest preserved fragments of crust are about 3.8 b.y. in age and are comprised chiefly of tonalitic gneiss complexes. Model lead ages from the Earth and dates from the lunar crust and from meteorites, however, suggest that the earliest terrestrial crust may have formed between 4.2 and 4.5 b.y. ago. Although the areal extent of early crustal fragments is not well known, they appear to comprise less than 10 percent of Archean crustal provinces (Plate I). The sparsity of rocks older than 3.0 b.y. may be related to losses resulting from partial recycling of the early crust into the mantle. Alternately, the first stable terrestrial crust may not have formed until about 4 b.y. because of delayed upper-mantle convection, as will be discussed later. It may be possible to determine the extent of the first crust by comparison with the lunar crust and crusts of other terrestrial planets. The lunar highlands, for instance, appears to represent the remnants of the early lunar crust (4.5–4.3 b.y.), which covered most or all of the lunar surface (Taylor, 1975). Studies of topographic features on Mercury and Mars also suggest the preservation of widespread primitive crusts. If the early history of the Earth was similar to these planets, it may also have had an early crust that covered its surface.

Theories for the origin of the crust fall broadly into three categories: inhomogeneous Earth accretion, catastrophic models, and noncatastrophic models. Catastrophic models involve an early, rapid melting episode which results in formation of much or most of the crust. Noncatastrophic models, on the other hand, involve more gradual growth of the crust in response to widespread heating of the Earth from within.

Inhomogeneous Accretion Model

In the inhomogeneous accretion model for the origin of the planets (Chapter 2), the last compounds to condense from the solar nebula produce a thin veneer on planetary surfaces rich in the alkali and other volatile elements which form or evolve into the first crust. A major problem with this model is that many nonvolatile elements, such as U, Th, and REE, which would be concentrated in the core and lower mantle in an inhomogeneously accreted Earth, are today concentrated in the crust. This would seem to necessitate magmatic transfer from within the Earth, thus producing a crust of magmatic origin. Another line of evidence supporting a magmatic origin for the Earth's early crust is that the early crust of the moon (≥ 4.4 b.y.) appears to be of magmatic origin; if the Earth evolved in a similar manner, it should have developed a crust of similar origin.

Catastrophic Models

Three catastrophic events that may have triggered widespread or localized melting in the Earth are: (a) rapid formation of the Earth's core during or after accretion, (b) surface impact, and (c) lunar capture. If the core separated rapidly during or soon after accretion of the Earth, a considerable amount of gravitational energy would be released into the mantle, causing extensive melting. Such melting could produce magmas that rapidly ascend to the surface, producing either a widespread, thin crust or many small crustal nuclei.

Impact melting in the Earth during the early Archean has become more plausible as our understanding of the timing of large impact events on the moon has progressed. Existing data indicate that the moon underwent severe bombardment by both small and large objects up to and including the culminating period of maria impact at 3.9 b.y. The Earth, being an even larger target, should also have undergone extensive bombardment during this period of time. The tremendous energy given off during large impacts may have produced localized melting in the Earth, with the derivative magmas forming the first crust. The impact model involves multiple cratering whereby one or more large bodies hit the Earth's surface, transferring energy but not mass to the Earth (fig. 11.1). Cratering involves excavation approximately 10 km deep and a drop in pressure of > 2500 atm. This sudden release of pressure together with the production of a series of radiating fractures results in partial melting of the mantle beneath the crater (1). Erosion of the crater rim and intrusion and extrusion of dominantly mafic magmas from below jointly fill the crater (2). Fractional crystallization of mafic magmas produces granitic magmas which are intruded at shallow depths, and the crater of sediments and igneous rocks isostatically rises, forming a protocontinent (3). The impact may initiate a convection cell beneath the protocontinent, thus thickening the crust and causing it to grow by peripheral magmatic accretion. Impact melting may also result in the production of a mantle plume, which rises and provides a means for future growth of continental nuclei. It has been suggested by Goodwin (1974) that Archean continental nuclei were produced over rising mantle plumes that were triggered by impact on the Earth's

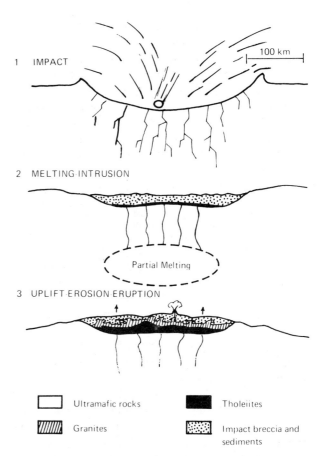

11.1. Possible sequence of events leading to the production of crust by impact melting (after Salisbury and Ronca, 1966).

surface. A major obstacle to the impact model, by analogy with early lunar cratering, is the fact that lunar cratering did not result in the production of significant amounts of magma.

Lunar capture may also have been a catastrophic event (Chapter 2). An early and close capture of the moon would result in the transfer of large amounts of angular momentum energy into the Earth's interior by tidal forces. This energy could produce widescale melting in the mantle and may result in the formation of a crust. Such a crust may not have been initially stable, however, in that tidal forces may have been strong enough to break it up. Only as the moon retreated from the Earth would a crust become stabilized.

Noncatastrophic Models

As was discussed later, it is likely that there was enough heat in the Earth at the time it formed or soon thereafter to melt the entire Earth. Complete melting probably did not occur because the Earth has retained volatile elements. If convection was widespread and served to homogenize the heat distribution in the Earth, partial melting should occur at the same depth throughout the upper mantle. This, in turn, should result in the formation of an unstable, thin, widespread crust. Such a crust may not have survived for long because surface impact or convective forces or both would serve to disrupt it and recycle it in the partially molten upper mantle.

Partial melting may also occur at inhomogeneities that form during or soon after accretion. Local concentrations of volatiles (principally H_2O and CO_2), structural discontinuities, or regions with high concentrations of heat-producing elements (U,Th,K) in the upper mantle may result in partial melting, rise of magmas, and production of crustal nuclei. Gradual core formation may also produce inhomogeneous melting in the mantle, and a corresponding inhomogeneous distribution of early crust. In the droplet model proposed by Elsasser (1963), crust forms over iron droplets which move toward the Earth's center under the influence of gravity. In this model, the number of crustal nuclei is equal to the number of droplets. Models involving the inhomogeneous distribution of magmas in the early Earth have in common the feature that the number and distribution of melting inhomogeneities controls the number and distribution of crustal nuclei.

Because of the large amount of heat available just after planetary accretion (see later section), inhomogeneous melting models are not as plausible as widespread homogeneous melting models. If, as is probable, the heat of core formation is distributed uniformly throughout the mantle, a world-wide magma "ocean" may develop in the upper mantle, and partial crystallization of this magma ocean would result in production of a thin crust covering the entire Earth. Some models for crustal formation appeal to early, viscous-drag plate tectonic processes (Condie, 1981a). The temperature of the Earth was greater in the early Precambrian than today, and hence numerous spreading centers may have existed to dissipate the heat. Plate tectonics may have begun in the thin crust overlying a magma ocean (fig. 11.2). Such crust is mafic to ultramafic in composition, and it is deformed and probably recycled through the upper mantle by convective forces. As the temperature falls, sialic crust forms at subduction zones and continents grow by continent-continent collisions, as will be discussed in a later section. If this process of crustal formation is valid, it has the advantage of providing a mechanism for production of both oceanic and continental crust early in the Earth's history.

COMPOSITION OF THE PRIMITIVE CRUST

Perhaps no other subject is more controversial at the present time than that of the composition of the primitive terrestrial crust. Most models fall into one or a combination of four categories: sialic, andesitic, anorthositic, or basaltic ($\pm$ultramafic). In part responsible for the diverging opinions are the different approaches to the subject. The most direct approach is to find and describe a relict of the primitive crust. Although some investigators have not given up on this approach, the chances that a remnant of this crust is preserved seem very small. Another approach is to deduce the composition of the early crust from studies of the preserved Archean crust. This approach is hazardous, in that the compositions and field relations of crustal rock types in the oldest preserved Archean terranes may not be representative of earlier crust. The oldest known supracrustal rocks in the Isua greenstone belt ($\sim$ 3.8 b.y.) contain a mixture of volcanic rocks (mafic and felsic), ultramafics, quartzites, iron formation, carbonates, and pelitic rocks. Another approach has been to assume that the Earth and moon have undergone similar early histories and

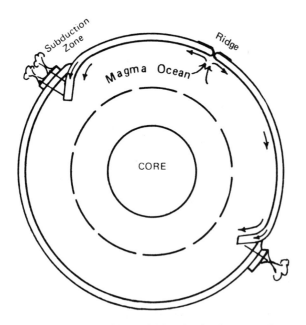

11.2. Non-catastrophic model for the development of a primitive terrestrial crust above a magma "ocean".

hence to go to the moon, where the early record is well preserved, to determine the composition of the Earth's primitive crust. Finally, a third approach is from geochemical models based on crystal-melt equilibria and a falling geothermal gradient with time.

Sialic Models

Some models for the production of a primitive sialic crust rely on the assumption that low degrees of partial melting in the mantle will be reached before high degrees, and hence granitic magmas should be produced before mafic ones. Other models call upon fractional crystallization of basalt to form sialic crust. Ramberg (1964) develops a model based on analogies to laboratory studies using various substances of different densities and viscosities; he concludes that granitic magmas, once formed, should rise to the surface buoyantly before basaltic magmas, which have lower densities.

Shaw (1976) presents a geochemical model for the formation of a widespread anorthositic-sialic crust. He proposes that the Earth heated soon after accretion, at which time the core formed and the mantle was molten. The mantle cools and crystallizes from the center outwards, concentrating incompatible elements into a near-surface basaltic magma layer. This layer undergoes fractional crystallization, resulting in the accumulation of an anorthositic scum in irregular patches. Impacting meteorites arrive frequently at the surface, breaking up the anorthositic veneer. Fractional crystallization of pyroxenes and olivine from the basaltic magma produces an ultramafic cumulate overlain by a residual granitic magma, which crystallizes to form the first stable crust by about 4 b.y.

Two main obstacles face the sialic crustal models. First, the high heat generation in the early Archean probably produced large degrees ($\geq 20\%$) of melting of the upper mantle, and hence it is unlikely that granitic melts could form directly. A sialic crust, however, could be produced by fractional crystallization of early basaltic magmas. If widespread sialic crust were produced in this manner, and it is essentially nondestructible because of its low density, why are there not remnants of the early sialic crust preserved today? The earliest preserved sialic crust in Greenland and other areas exhibits initial Sr, Pb, and Nd isotopic ratios that suggest that it was extracted from the mantle with little or no previous crustal history, and hence the oldest preserved remnants cannot represent reworked fragments of still older crust.

Andesitic Models

The average composition of the continents today is similar to calc-alkaline andesite (Chapter 4). Hence Taylor (1967) and others propose a model for the origin of the continental crust based on a modern island-arc setting where andesites are abundant. The model necessitates, by a strict modern plate-tectonic analog, that subduction zones began early in the Earth's history. The continents grow by arc development around the periphery of continents and by arc-continent collisions. Not supporting the andesite model, however, are (1) the sparsity of rocks of andesite composition in the Archean crust and (2) the high heat production in the early Archean, which would result in large degrees of melting of the mantle rather than the small degrees that are necessary for andesite production. It would appear that the andesitic composition of the continents reflects a mixture of rock types of which the average composition is similar to andesite, rather than necessitating an andesite-type origin for the continents.

Anorthosite Models

Geochemical and geochronological studies of lunar samples indicate that the oldest rocks on the lunar surface are gabbroic anorthosites and related high-alumina basalts of the lunar highlands (Taylor, 1975). These rocks are remnants of a widespread lunar crust formed bewteen 4.5 and 4.4 b.y. ago. This crust appears to have formed in response to one or more periods of catastrophic heating leading to widespread partial melting of the lunar interior and the production of voluminous basaltic magmas which ascend to the surface. Near the surface the magmas cool and undergo fractional crystallization, at which time pyroxenes and olivine sink and plagioclase (and some pyroxenes) floats, forming a crust of gabbroic anorthosites. Impact disrupts this crust and produces maria craters, and these craters are later filled with basaltic magmas (3.1–3.9 b.y.). The highland areas of the moon represent remnants of the primitive gabbroic anorthosite crust.

Windley (1970) has pointed out that early Archean anorthosites in Greenland are similar in composition (i.e., high An contents, associated chromite) to lunar anorthosites and not to younger terrestrial anorthosites. It is clear from the field relationships, however, that these anorthosites are not remnants of an early crust, since they commonly intrude tonalitic gneisses. If, however, the Earth did undergo an early melting history similar to the moon's, the Earth's first crust may have been composed dominantly of gabbroic anorthosites and the preserved early Archean anorthosites may represent the last stages of anorthosite production, which continued after both mafic and sialic magmas were also being produced.

The increased pressure gradient in the Earth limits the stability range of plagioclase to depths considerably shallower than on the moon. Available experimental data suggest that plagioclase would not be a stable phase at depths > 35 km in the Earth. Hence if such a model is applicable to the Earth, the anorthositic fraction, either as floating crystals or as magmas, must find its way to very shallow depths to be stable. The most serious problem with the anorthositic model comes from experimental data. Results indicate that plagioclase will readily float in an anhydrous lunar magmatic ocean but that even small amounts of water in the system causes plagioclase to sink (Taylor, 1979b). Hence in the terrestrial system, where water was probably abundant in the early mantle, an anorthositic scum on a magma ocean would not form.

Basaltic Models

A primitive crust of basaltic composition (±ultramafic and anorthositic components) is consistent with the probable large amounts of partial melting that would occur in the Earth soon after accretion. This crust may be similar compositionally to lower portions of Archean greenstone belts. A model for the development of a primitive mafic crust which appears to be most consistent with available geochemical and heat production data is discussed later in the chapter.

GROWTH OF THE EARLY CRUST

Mechanisms of Growth

Various mechanisms have been suggested for the growth of the crust and lithosphere. Crustal growth can occur by addition of new material from the mantle, by readdition of crustal material that has been cycled through the mantle, and by redistribution of crustal rocks due to sedimentary and tectonic processes. *Net crustal growth*, however, includes only additions to the crust of new (unrecycled) material from the mantle. Sr, Pb, and Nd isotopic systems are useful in distinguishing new from recycled crustal rocks. The major mechanisms of crustal-lithospheric growth are growth by magma ad-

ditions, interthrusting and stacking of crustal rocks, aggregation by microcontinental collisions, and welding of sedimentary prisms to continental margins.

As erosion removes material from the continents, it accumulates on the continental shelves and slopes and in oceanic basins near continental margins. Thick sedimentary prisms may accumulate in this manner. Burial leads to metamorphism and perhaps to partial melting, the net result being lateral accretion of the continents. How important this mechanism of growth was in the geologic past is difficult to evaluate, because ancient continental margins have been intensely reworked during later deformations. However, it may have been quite important adjacent to high-grade Precambrian provinces, where evidence indicates 15–30 km of uplift and erosion has occurred.

Magma from the mantle may be added to the crust and lithosphere by underplating involving the intrusion of sills and plutons and by overplating of volcanic rocks. Such additions may occur in a variety of tectonic environments. Oceanic crust and lithosphere are added at divergent plate boundaries. Growth of continents and arcs may occur by additions of magma from descending lithospheric slabs, and continents and oceanic islands may thicken by addition of magmas from mantle plumes. From investigations of Archean terranes in Greenland, it has been proposed that the early crust may have thickened by the interthrusting and stacking of thrust sheets and nappes of both oceanic and continental rocks. Such intimate intermixing of these rocks implies intense horizontal compression, which may be related to viscous-drag plate tectonics involving collisions of continental fragments. The simplest mechanism is by an arc-continent collision whereby an arc is added to the edge of a continent and the subduction zone relocates at the new continental margin.

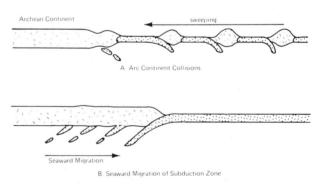

11.3. Possible plate-tectonic mechanisms for primitive continental growth during the Archean.

The descending plate may become the overriding plate if it carries the continent. Continual sweeping of arcs against a continent can result in substantial growth as can the seaward migration of a subduction zone (fig. 11.3). As will be discussed later, it is possible that such collisions were largely responsible for the aggregation of Precambrian supercontinents.

Thickness of the Early Continental Crust

Estimates of the thickness of the Archean continental crust can be made from three sources of data (Condie, 1973): minimum thickness of stratigraphic sections in greenstone belts, burial depths deduced from metamorphic mineral assemblages, and geochemical indices. Data from all three sources seem to indicate that Archean continents by 3.8 b.y. were equal to or greater than the thickness of the present continents (i.e., 30–40 km). If this is the case, little if any thickening has occurred in cratonic areas since their stabilization. Evidence at hand suggests that Archean sialic crust thickened to present-day thicknesses as it formed by one or a combination of the growth mechanisms described above.

Metamorphic mineral assemblages from Archean high-grade terranes imply crustal depths at the time of metamorphism of 30–40 km. Because these areas are underlain today by 30–40 km of continental crust, we are faced with three alternatives to explain the existence of such high P-T mineral assemblages at the Earth's surface: (1) sialic underplating has kept pace with uplift and erosion; (2) the Archean crust in these areas was originally on the order of 60–80 km thick; or (3) segments of the lower crust have been transported to the surface by low-angle thrust faults. The first mechanism would seem to necessitate a younging of sialic crust with depth, a feature which is not observed in radiometric dates from terranes of increasing metamorphic grade. It is also difficult to explain why the underplating magmatism does not have surface or near-surface manifestations. Regarding the second possibility, it is unlikely that Archean sialic crust could exceed about 25 km in thickness before its base began to melt (Fyfe, 1973). Davies (1979), however, has recently proposed that the Archean continental lithosphere was approximately 200 km thick and acted as a thermal buffer between the sialic crust and the hot upper mantle and thus prevented the bottom of the crust from being melted. Such thick lithosphere presents problems with alternative 3 above, in that low-angle thrusting of the lower crust would seem to necessitate decoupling of crust and mantle lithosphere.

Continental Growth Rates

The distribution of radiometric dates in the continents suggests that the volume of crust increases exponentially with decreasing age. Three models that can be considered to explain this observation are illustrated in fig. 11.4:

1. The rate of formation of continental crust has increased exponentially with time.
2. Rapid growth of the continental crust occurred early in the Earth's history with subsequent recycling of crustal material through the mantle.
3. Linear growth rate of continental crust is combined with recycling only through the crust.

Model 1 is based on the assumption that the volume of crust of a given age as deduced from present areal distribution reflects the actual amount of new crust extracted from the mantle over the given period of time. Any crustal material that is recycled through the mantle is also classified as new crust in this model. As discussed in Chapter 5, it is now clear that many relicts of older crust occur in Precambrian and Phanerozoic crustal provinces, and hence one cannot equate areal extent of "apparent" age provinces on geologic maps with volume of new crust produced during a given increment of time. Perhaps as much as 50 percent of Precambrian mobile belts

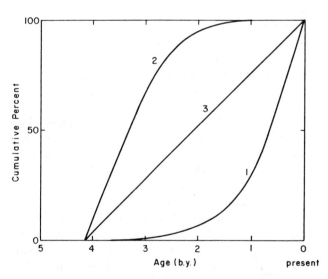

11.4. Three models for the growth of continental crust. Vertical axis is cumulative percent of present-day continental crust (modified after Veizer, 1976).

are composed of crustal rocks older than the dominant radiometric ages from the mobile belt. Sr and Nd isotopic studies of the crust-mantle system are also not compatible with continental growth as defined in Model 1 (Jacobsen and Wasserburg, 1979; Veizer, 1976).

Development of the second model was made by Armstong (1981). This model involves steady-state mixing of crust and mantle in subduction zones, with the mixing rate decreasing with time as thermal gradients in the Earth drop. The model, which is based on Sr, Nd, and Pb isotopic systems, indicates that most of the continental crust formed prior to 3.0 b.y. and that any growth after this time has involved chiefly recycling of early crust through a convecting upper mantle. Two lines of evidence, however, do not support this model. First, isotopic and trace-element studies of oceanic sediments and arc volcanics indicate that only a small amount of subducted continental sediment (<20%) has contributed to the production of calc-alkaline magmas. Furthermore, the low density of sialic crust should prevent it from being subducted or allow only a minor amount of subduction. Also, isotopic studies of oceanic basalts (both ridge and island basalts) indicate that mantle source areas are inhomogeneous, despite a long geologic history of convection, which will be discussed later in this chapter. The persistence of such heterogeneities in the mantle for long periods of time argues against widespread homogenization by convection in the upper mantle, which is a necessary condition for the second model.

Model 3, for linear growth of the continents, involves sialic crustal material that is recycled only in the crust and in which older continental crust is recycled faster than younger. Clearly, the simplest interpretation of the low initial $^{87}Sr/^{86}Sr$ ratios in many sialic rocks is that new sialic crust has been added to the continents throughout geologic time. Some high-grade terranes with high initial strontium isotope ratios, however, may have had appreciable crustal residence times (0.5–1.0 b.y.). This model allows for variable amounts of older crustal rock found in mobile belts. Moorbath (1977) suggests that continental crust has grown throughout geologic time, probably episodically, by processes involving essentially irreversible differentiation of the mantle. Existing Sr and Nd isotopic data indicate that the typical time interval of an episode in which new crustal material is extracted from the mantle is of the order of 50–100 m.y., although for some high-grade terranes it may have been longer ($\gtrsim 200$ m.y.). Both the upper and lower sialic crust appear to have grown

concurrently, with magmas emplaced in the lower crust crystallizing directly to granulite-facies mineral assemblages. Available Sr, Pb, and Nd isotopic data suggest that at least 50 percent of the continental crust was produced by 2.5 b.y.

THE EARTH'S THERMAL HISTORY

Models for the evolution of the crust and lithosphere and the role of plate tectonics in crustal evolution depend directly on the thermal history of the Earth. Data indicate that the Earth is cooling at the rate of about $200°C/10^9$ yr and that about 80 percent of the surface heat flow is due to the decay of radioactive isotopes in the Earth and 20 percent to cooling of the Earth (Turcotte, 1980). Important in any thermal model of the Earth is the distribution of the heat-producing isotopes of U, Th, and K (see Chapter 4) as a function of time and the initial temperature distribution in the Earth. Neither of these factors is well known. The initial temperature distribution in the Earth is dependent upon such poorly known parameters as the particle velocity distribution during accretion; the gravitational energy released by the Earth's interior as it grows; the abundance of short-lived radioactive isotopes such as ^{26}Al and ^{244}Pu; inductive heating related to early solar wind activity; and the process by which the core formed (Chapter 3). Order-of-magnitude estimates for these factors indicate that there was enough heat available if retained within the Earth to completely melt it. Because volatile elements are retained in the Earth, however, it is unlikely that it was ever completely melted. Hence it is necessary to remove much of this heat rapidly. Convection seems to be the only process capable of bringing heat from the deep interior of the Earth to surface in a short time (i.e., a few hundred million years). At this stage, mantle convection would have been more than twice as fast as today.

It is possible to estimate the heat productivity in the Earth as a function of time from a knowledge of the concentration of U, Th, and K in the Earth. Although such concentrations are not well known, constraints can be placed on heat production from U, Th, and K contents of ultramafic and mafic rocks and meteorites. Results indicate that heat production and surface heat flow were a factor of 3 to 5 greater in the early Archean than today (fig. 11.5). However, metamorphic mineral assemblages in the Archean terranes indicate geothermal gradients similar to present continental distributions in provinces with high heat flow such as the Rocky Mountains. Hence, the

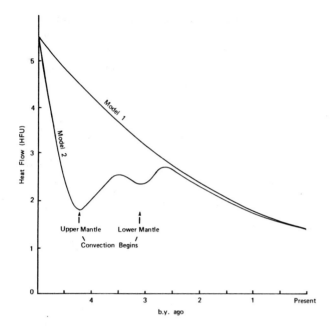

11.5. Heat flow variation at the Earth's surface as a function of time. Model 1 for an early convecting Earth (after McKenzie and Weiss, 1975) and model 2 for a delayed convecting Earth (after Lambert, 1980). $K/U \simeq 10^4$ in both models.

additional terrestrial heat in the Archean must have been lost from oceanic areas (Burke and Kidd, 1978). Whether or not oceanic Archean geothermal gradients were steeper than present gradients is a subject of considerable discussion. Evidence for steep Archean oceanic geotherms is ambiguous. For instance, the existence of abundant komatiites in the Archean may reflect steeper geotherms, or it may reflect abundant mantle plumes, in which komatiite magmas are produced by adiabatic melting as plumes move rapidly to the base of the lithosphere. It is not necessary to call upon steeper oceanic geotherms in the Archean, in that the additional heat may have been liberated by faster sea-floor spreading and/or by a greater total length of oceanic ridge systems.

The thickness of the lithosphere in the Archean and how it changes with time is a subject of disagreement. If steep geothermal gradients were present beneath the Archean crust, the lithosphere must have been thinner than the present lithosphere (fig. 10.1). If, however, geothermal gradients beneath continents were similar to present gradients, a lithosphere at least as thick as the present continental lithosphere is needed to prevent extensive melting of the lower crust (Davies, 1979).

Two models for the heat flow distribution in the Earth with time are shown in fig. 11.5. In Model 1,

the Earth is assumed to have an initial temperature sufficiently high to permit worldwide convection from the onset. In Model 2, it is assumed that upper mantle convection begins soon after accretion (~ 4.2 b.y.) and lower mantle convection at about 3.0 b.y. In this model, little of the heat generated in the first 1000 m.y. reaches the surface, due to delayed convection. It is noteworthy that the two maxima in the heat-flow curve (~ 3.5 and ~ 2.7 b.y.) correspond to the earliest periods of continent formation, suggesting major differentiation of the Earth at these times.

A MODEL OF EARLY CRUSTAL DEVELOPMENT

A tentative model for the development of the crust and lithosphere is based on the following assumptions (after Condie, 1980): (1) the geothermal gradient in the Earth prior to 4.0 b.y. was adiabatic; (2) heat production decreases with time; (3) mantle convection begins early; (4) the first stable crust is basaltic ($\pm$ anorthosite) in composition; (5) the Earth is subjected to intense impact cratering until 3.9 b.y. ago; and (6) continents are aggregated into one or a few supercontinents by 2.7 b.y., probably by collisions of microcontinent fragments.

Extensive melting occurs in the outer part of the Earth during the first 100 to 300 m.y. of Earth history. During this time, the core forms and an adiabatic geotherm is established. Loss of heat by radiation and volatile escape cools the surface of the Earth and a thin unstable ultramafic lithosphere is formed. As cooling continues, voluminous basaltic magmas segregate and rise to the surface, displacing the ultramafic veneer. Convection may start early

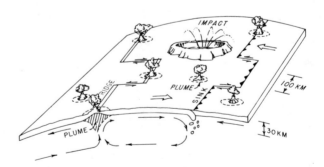

11.6. Schematic diagram of the early development of a primitive basaltic crust (after Condie, 1980). Tholeiitic and ultramafic magmas are produced at ridges and in plumes and the crust and lithosphere are completely melted and recycled in sinks.

11.7. Schematic diagram for the development of early tonalitic arcs at convergent plate boundaries (after Condie, 1980).

(Model I, fig. 11.5) or it may be delayed (Model 2). The basaltic crust and lithosphere by 4.0 b.y. are comprised of an array of ridges and, perhaps, sinks and are driven by viscous-drag associated with mantle convection (fig. 11.6). This early lithosphere is rapidly recycled through the mantle. Major impact cratering up until 3.9 b.y. serves to breakup the lithosphere and aids in recycling. Sometime between 3.8 and 4.0 b.y. geotherms at subduction zones decrease such that descending lithosphere is only partially melted, producing tonalite magmas which rise to form the earliest continental crust in proto-arc systems (fig. 11.7). In the delayed convection model (fig. 11.5), this would correspond to 3.8–3.5 b.y. The onset of deep convection at 2.7–2.8 b.y. is accompanied by renewed and intense tonalitic magmatism, during which time a large part of the continental crust is formed. This crust is aggregated into a few supercontinents (fig. 10.8) by 2.5 b.y., principally by microcontinent collisions. Partial melting of the lower part of this crust at 2.4–2.6 b.y. produces granites which are intruded into the dominantly tonalitic crust. Archean greenstone belts could form in either back-arc basins or in hot unstable cratonic basins, as was discussed in Chapter 10. During this time of extensive crustal growth, the upper mantle is largely devolatilized as the atmosphere and oceans rapidly grow.

COMPARATIVE EVOLUTION OF THE TERRESTRIAL PLANETS

It has long been known that the Earth is unique among the terrestrial planets. Not only is it a planet with oceans, an oxygen-bearing atmosphere, and living organisms, but it appears to be the only planet in which plate-tectonic processes are active. Why has the Earth undergone such a unique geologic history

among the terrestrial planets? Although this is a complicated question which has no single answer, we can consider many of the variables involved.

The terrestrial planets, including the moon, have similar densities (table 11.1) and probably similar bulk compositions. Each of them is probably evolving toward a stage of thermal and tectonic stability and quiescence as they cool. The rate of which each planet evolves is dependent upon a variety of factors which directly or indirectly control the loss of heat (Lowman, 1976; Schubert, 1979). First of all, the position of a planet in the solar system is important because it reflects the condensation sequence of elements from the cooling solar nebula (Chapter 2). Of particular importance are the abundances of elements with short- and long-lived radionuclides (such as Al, U, Th, and K) which contribute to heating planetary bodies. The moon, for instance, contains considerably smaller amounts of U, Th, and K than the Earth, as is evidenced by geochemical studies of lunar samples. Analyses of fine materials from the Viking landing sites suggest that Mars is significantly depleted in potassium relative to the Earth. Hence these bodies would have less radiogenic heating than the Earth. Planetary mass is important, in that the amount of accretional and gravitational energy is directly dependent upon mass. Planetary size is also important, in that greater area/mass ratios result in more rapid heat loss from the planetary surfaces (table 11.1). Of particular significance is the size of the iron core, in that much of the initial heat is provided during core formation (Chapter 3). Mercury, for instance, has the largest core/mantle ratio of any of the terrestrial planets, and hence would have had the largest contribution of core-forming heat. The volatile contents and especially the water content of planetary mantles and the rate of volatile release are important in controlling the amount of melting, the fractional crystallization trend, and the viscosity of the planetary interior which, in turn, affects rate of convection. Convection appears to be the primary mechanism by which heat has been lost during planetary evolution, and hence the rate of convection is important in terms of evolutionary state. It should be pointed out, however, that the only observational evidence that requires subsolidus convection today is from the Earth (Chapter 6). Because the surfaces of the other terrestrial planets do not show evidences of plate tectonics, convection in their interiors is inferred from geometrical and dynamic figures, such as the bulge on the Earth-facing side of the moon and the offsets in the center of mass and geometric center observed in most planets.

Table 11.1. Physical Properties of the Terrestrial Planets

	Mass (Earth = 1)	Radius (km)	Area/Mass[+]	Core/Mantle[*]	Density (gm/cm^3)	Distance from Sun (A.U.)
Mercury	0.0543	2440	2.5	~12	5.44	0.389
Venus	0.8137	6054	1.1	0.9	5.24	0.725
Earth	1.	6370	1	1	5.52	1
Mars	0.1077	3390	2.5	0.8	3.94	1.53
Moon	0.0123	1740	6.1	0.12	3.34	1

[*]Relative to Earth's core/mantle volume ratio = 1
[+]Relative to Earth's area/mass ratio = 1

Before proposing a model for the evolution of the terrestrial planets, it is important to review relevant features of each planet, which have been determined in large part from the Apollo, Mariner, Viking, and Pioneer space missions of the last decade.

The Moon

A great deal has been learned about the geochemistry and geophysics of the moon from the Apollo landings (Taylor, 1975, 1979b). Only a few of the findings will be reviewed here. Models for the interior of the moon indicate the presence of a crust 60–100 km thick comprised chiefly of anorthositic gabbros and anorthosites as represented by exposures in the lunar highlands. This crust overlies a mantle comprised of two layers. The upper layer or lithosphere extends to a depth of 400–500 km and is probably comprised of cumulate ultramafic rocks. The second layer extends to about 1100 km, where a sharp break in S-wave velocities occurs. Evidence for a lunar core is inconclusive and seismic and heat-flow data limit convection to the deeper mantle and core regions. Radiometric dates from the lunar crust indicate an age of 4.6 b.y., which is only 100–150 m.y. after the condensation of the solar system. Impact continued until about 3.9 b.y., when most of the large maria basins formed. Mare basalts cover about 17 percent of the lunar surface but extend up to depths of only one kilometer. Dated flows range from 3.9 to 3.16 b.y. and represent very fluid flows, probably erupted chiefly from fissures. The impacts which formed the large maria basins did not initiate the melting that produced the basalts, which were erupted up to hundreds of millions of years later. The youngest eruptions appear to be about 2.6 b.y. in age.

The currently most popular model for lunar evolution involves the production of an ultramafic "magmatic ocean" which covers the entire moon to a depth of 500 km and crystallizes between 4.46 and about 4.40 b.y. (fig. 11.8). Plagioclase floats, producing an anorthositic crust, and pyroxenes and olivine largely sink, producing an incompatible-element depleted upper mantle. Later, partial melting of this mantle produces the mare basalts. Detailed models for crystallization of the magma ocean indicate the process was complex, involving floating "rockbergs," cycles of assimilation and mixing, and trapping of residual liquids. The quenched surface and anorthositic rafts are also continually broken up by impact. It is possible, if not probable, that viscous-drag plate tectonics was operative.

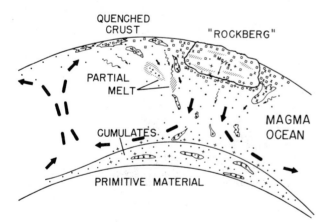

11.8. Schematic diagram illustrating the crystallization of a magma ocean on the moon (after Longhi, 1978). Arrows represent flow patterns and rockbergs are principally anorthosites.

Mercury

The high density (5.45 g/cm^3) of Mercury implies an Fe-Ni core that comprises about 66 percent of the planet's mass. Results from Mariner 10 show that the surface of Mercury is strikingly similar to that of the moon in terms of crater distribution. Mercury differs from the moon, however, in that it displays a global network of large lobate scarps. These scarps appear to represent high-angle reverse faults that developed during a period of global contraction related to continued cooling of the interior, and perhaps to core formation after crustal solidification. Although there is no unequivocal evidence for volcanism, the smooth plains are thought to represent widespread basaltic flows. The similarity of the highland-cratered terranes on Mercury to those of the moon suggest a similar origin and age (Lowman, 1976). The faulting that produced the lobate scarps occurred both before and after the volcanism. In terms of timing, it is clear that the major magmatic events on Mercury occurred prior to the basin-forming events, which by analogy with the moon probably formed at about 3.9 b.y.

Mars

The Mariner and Viking missions to Mars indicate that it is quite different from the moon and Mercury. The Martian surface includes major shield volcanoes, fracture zones, and rifts as well as large canyons that appear to have been cut by running water. Also, much of the planet is covered with wind-blown deposits. More than one half of the planet is covered by a variably cratered terrane similar to the surfaces of the moon and Mercury. Large near-circular basins are similar to lunar maria and probably formed at about 3.9 b.y. The Tharsis region includes gigantic shield volcanoes and volcanic plains, and the Valles Marineris represents an enormous rift valley that spans one-quarter of the equator. The size of these volcanoes implies a very thick lithosphere. Studies of the surface morphology of lava flows on Mars indicate a total volume much greater than on the Earth and low viscosities of eruption. Major fracture systems on Mars and perhaps some volcanism appear to have remained active to between 1000 to 100 m.y. ago. Polar caps consisting of H_2O and CO_2 ice and dust exhibit seasonal changes. Chemical analyses from the Viking landing sites suggest that the dominant volcanic rocks on Mars are Fe-rich basalts and that the weathering of the basalts occurred in a hydrous, oxidizing environment.

The geologic history of Mars was probably similar to the moon and Mercury for the first few hundred million years, with major impact terminating with the basin-forming event at 3.9 b.y. Crystallization of a planet-wide magmatic ocean resulted in the formation of crust, mantle, and core prior to 4.4 b.y.

Continued mantle convection results in extensive fracturing and rifting of the Martian lithosphere and in widespread volcanism. The Tharsis uplift developed at this time, perhaps in response to a large mantle plume (Mutch et al., 1976). The early atmosphere was very dense and rich in CO_2 and H_2O, and cooling led to extensive rainfall and erosion. With a drop in surface temperature permafrost formed, which was locally and perhaps catastrophically melted and caused massive floods that cut some of the large canyons. The major erosional events probably occurred during or just before formation of the cratered plains of volcanic origin (i.e., prior to 3.9 b.y.). Continued fracturing and volcanism on Mars extended to at least 1000 m.y. (and perhaps 100 m.y.), yet plate tectonics was not initiated, due perhaps to the presence of a thick Martian lithosphere.

Venus

Venus is similar to Earth in terms of its size (6054 and 6370 km, respectively) and mean density (5.24 and 5.52 g/cm^3, respectively). Recent mapping of the Venusian surface by radar imagery sent back by the Pioneer spacecraft has allowed identification of major topographic features for the first time. Also, analysis of material from one site on the surface sent back by Venera 8 indicates K, U, and Th contents similar to terrestrial granitic rocks. Not counting the Earth's oceans, ranges in elevation on Venus are somewhat greater than on Earth. Of the total surface, 84 percent is comprised of flat rolling plains, some of which are

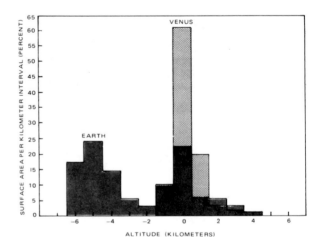

11.9. Comparison of relief on Venus and the Earth (after Pettengill, et al., 1980). Surface height is plotted in 1km intervals as a function of surface area. Height is measured from a sphere of average planetary radius for Venus and sea level for the Earth.

over 1 km above the average plain elevation. Only 8 percent of the surface comprises true highlands, and the remainder (16 percent) lies below the average radius, forming broad shallow basins. The large plateau-like areas might be similar to Earth's continents and the basins similar to oceanic basins. The topographic relief on Venus, however, is unimodal compared to the strikingly bimodal distribution on Earth (fig. 11.9). As with Mars, the Venusian surface appears to be characterized by large shield volcanoes and rift systems. Although resolution of surface features on Venus are not as good as those on other terrestrial planets, most of the surface appears to be ancient, as indicated by the distribution of impact craters. Whether plate tectonics actually began on Venus and was rapidly aborted due to a thick lithosphere or never started is a major question that must await better resolution of surface features.

An Evolutionary Model

Models for the evolution of the terrestrial planets fall into two categories: (1) those that propose that all planets evolve through the same stages at different rates, and (2) those models that require different histories for each planet. Although in the long run the second group of models may turn out to be more correct, existing data suggest that differences in the evolutionary histories of the terrestrial planets can be explained primarily by differences in heat productivities, volatile-element contents, and cooling rates. One possible scheme of events is illustrated in fig. 11.10 as a function of average planetary temperature. The temperature scale is schematic, in that the threshold temperatures for various processes vary from planet to planet depending on planetary mass and radius. All terrestrial planets undergo rapid heating during

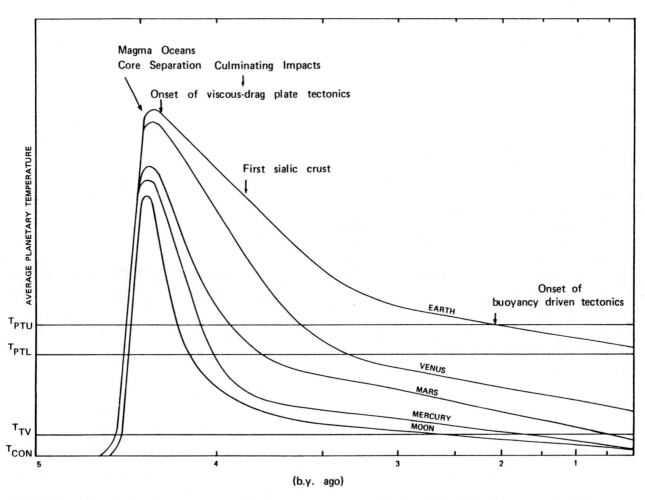

11.10. Schematic variation of average terrestrial planetary temperature with time. Threshold temperatures for major processes are indicated as follows: T_{PTU} and T_{PTL}, upper and lower temperatures for plate tectonics, respectively; T_{TV}, terminal volcanic temperature; T_{CON}, terminal subsolidus convection temperature.

the last stages of accretion, reaching maxima between 4.5 and 4.4 b.y. At this time, heat liberated primarily from core formation produces a widespread magmatic ocean and extensive degassing of the upper mantles. A thin mafic crust with variable numbers of anorthositic rafts is formed at this time, and viscous-drag plate tectonics begins. Impact is continually breaking up the crust. Major events in the cooling history of the Earth, discussed earlier in this chapter and in Chapter 10, are noted on the figure. Cooling curves of the other terrestrial planets are steeper than that of the Earth for the first 500–1000 m.y. with the moon cooling most rapidly. At or very soon after 4.0 b.y. a thick lithosphere has developed on all of the terrestrial planets except the Earth, which is cooling more slowly and, except for Venus, has a larger contribution of radiogenic heat than the other planets. Supercontinents grow on the Earth between 4.0 and 2.7 b.y. A supercontinent may also have grown on Venus, although not by plate-tectonic processes, and was probably completed before 3.9 b.y., evidenced by the crater distribution on Venus. Although all of the terrestrial planets passed through the upper thermal threshold for buoyancy-driven plate tectonics, only the Earth cooled slowly enough during this time and had a thin enough lithosphere for buoyancy-driven plate tectonics to become established. It is noteworthy that this model predicts that the Earth will pass through the lower threshold temperature for plate tectonics in about 500 m.y. The temperature of terminal volcanism was reached first by the moon at about 2.6 b.y. followed shortly thereafter by Mercury and, sometime after 1 b.y., by Mars. Whether or not Venus has passed this point is unknown. Only the moon and Mercury are nearing their thresholds for subsolidus convection.

This model suggests that all of the terrestrial planets went through the same evolutionary sequence, although not at the same rate. Only the Earth cools slowly enough so as not to form a thick lithosphere early in its history. Mars, Mercury, and the moon cool rapidly due to their small masses and relatively low radiogenic heat productivities. The relatively rapid cooling rate of Venus, however, remains unexplained.

EVOLUTION OF THE EARTH'S MANTLE

Compositional variation in the Earth's mantle can be evaluated from the chemical and isotopic composition of derivative magmas, discussed in Chapter 7.

Ratios of incompatible elements and isotopic ratios in magmas are not significantly affected by the amount of melting or fractional crystallization and hence should be approximately the same as source ratios. As Chapter 7 discussed, results from mantle-derived magmas indicate the presence of a heterogeneous mantle on scales ranging from hundreds to thousands of kilometers. Also, at least two different types of mantle sources are required today, and these sources have been separate for 1 to 2 b.y.

Geochemical studies of incompatible elements and Sr, Nd, and Pb isotope ratios indicate Archean mantle was also heterogeneous and that the source area of oceanic ridge tholeiites is more depleted in incompatible elements and more homogeneous than Archean mantle sources (Sun and Nesbitt, 1977). It is also known that Archean volcanic rocks are enriched in Ni, Co, and other transition metals relative to modern volcanics. Model studies require either that the early mantle be enriched in these elements and become depleted with time, or that higher Archean mantle temperatures result in greater partitioning of transition metals into the liquid phase than occurs today.

The growth of radiogenic ^{87}Sr in the mantle varies with the Rb/Sr ratio. Three models have been proposed (Jahn and Nyquist, 1976): (1) the Rb/Sr is constant with time; (2) Rb is extracted from the mantle preferentially to Sr and thus the Rb/Sr ratio in the upper mantle decreases with time; and (3) the Rb/Sr ratio in the upper mantle increases with time. Existing data from Archean basalts favor models 1 or 2. Possible methods by which a constant Rb/Sr could be maintained in the upper mantle are remixing of crust and upper mantle in proportions necessary to maintain a constant Rb/Sr, or replenishment of Rb from the lower mantle as it is preferentially extracted from the upper mantle in magmas. Sr isotope distributions in basalts of all ages, including early Archean basalts, indicate that the upper mantle has been inhomogeneous since 3.8 b.y. ago. Lead and Nd isotope results also confirm heterogeneity in U/Pb, Th/Pb, and Sm/Nd ratios which date back to the Archean.

It is not difficult to propose processes that can produce heterogeneity in the upper mantle during the Archean, but is is difficult to see how such heterogeneities have been maintained throughout geologic time. Among those processes capable of producing heterogeneities are extraction of magmas from the mantle, metasomatic additions from the lower mantle, and recycling of crust into the upper mantle. Because of the greater heat productivity in the early mantle, convection should have been more vigorous, resulting in rapid mixing and homogenization of compositional heterogeneities. As attested to by geochemical and isotopic results, however, mantle convection has not homogenized the mantle. Large segments of the upper mantle beneath continents, with their own geochemical peculiarities, have remained intact since the Archean. Although the oceanic upper mantle from which ocean ridge basalts are derived is more homogeneous, it also may contain ancient remnants of unmixed mantle (Chapter 7).

SECULAR COMPOSITIONAL CHANGES IN THE CONTINENTS

Secular compositional variations in the continental crust are difficult to document because of problems in sampling, lack of accurate dates, and changes in composition of specific rock types caused by metasomatism or regional metamorphism. To verify the existence of systematic compositional variations with time on the continents, it is necessary to have data available for large numbers of rocks of different ages and from many widespread locations. Data may represent a specific rock type or an average of many rock types. If compositional changes of averages are to be recognized, one is further faced with the prob-

lem of accurately weighting the contribution of each individual rock type as a function of time. The problem of adequate dating of rock units is particularly critical in the Precambrian, where younger periods of metamorphism may have reset radiometric dates. Also important in Precambrian terranes is the effect that regional metamorphism and/or metasomatism may have on the original compositions of rocks. One way to partially avoid this problem is to compare rocks of different ages that exhibit the same degree of metamorphism or metasomatism.

Two general approaches have been used to evaluate changes in the composition of the continental crust with time. One approach is to infer the composition of the crust from the composition of clastic sediments, and especially shales and some sandstones that appear to represent well-mixed samples of large segments of the continents. A more direct approach is to use geochemical data from widespread sampling in crustal provinces of various ages.

Sediment Results

As shown by Garrels and MacKenzie (1971), the preservation of sedimentary rocks varies with their age (fig. 11.11). Such a distribution can be interpreted in terms of recycling of sediments. Results indicate

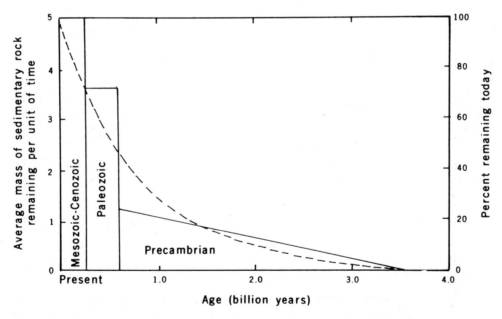

11.11. Distribution of sediment mass as a function of geologic time (after Garrels and MacKenzie, 1971). Dashed line is for a model assuming the total sediment mass has been recycled five times throughout geologic history. Model applies to either a constant or linearly growing mass of sediment.

that a mass of sediment equivalent to five times the present sedimentary mass has been recycled through erosion and deposition throughout geologic time. Recycling may produce changes in the proportions of sediments originally deposited, since some rocks are more resistant to weathering and erosion than others. Source-area composition, tectonic setting, atmosphere-ocean composition, and heat generation in the Earth also influence the proportions of sediments as a function of time. General trends are as follows (fig. 11.12):

1. The Archean is characterized by abundant volcanic rocks and graywackes, whereas shales, quartizites and arkoses dominate in the Proterozoic; carbonates, shales, and quartizites seem to dominate in the Phanerozoic. The importance of tectonic setting in explaining these distributions was discussed in Chapters 8 and 10.

2. The ratio of limestone to dolomite has increased with time, as will be discussed in a later section.

3. Occurrences of evaporites older than 600 m.y. are uncommon and extremely rare prior to 2000 m.y. Banded iron formation is abundant in the early Proterozoic, minor in the Archean, and absent in the Phanerozoic. Red beds make their first appearance in the Proterozoic just after the peak in iron-formation deposition. All three of these distributions appear to be related to the liberation of significant amounts of free oxygen into the atmosphere which will be discussed later.

4. Major organic-carbon bearing sediments appear in the Phanerozoic in response to the rapid development of living organisms. Coal appears at about 350 m.y. when land plants first become abundant.

Compositional changes in sediments also occur throughout the geologic record (Veizer, 1973). Some of the more important trends are summarized in fig. 11.13 and 11.14. SiO_2/Al_2O_3 increases with time and K_2O/Na_2O, total REE, and Th exhibit rapid increases between the Archean and Proterozoic.

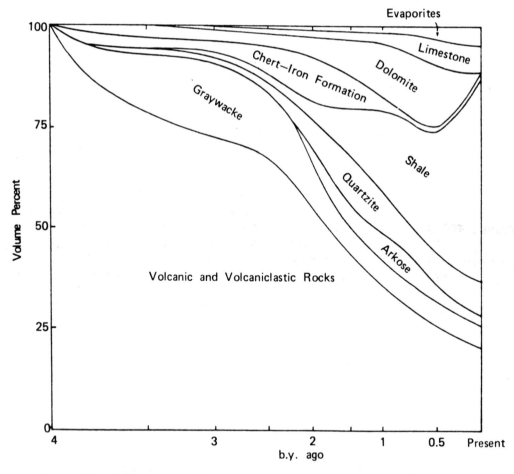

11.12. Relative abundances of supracrustal rocks as a function of age. Modified after Ronov and Migdisov (1971).

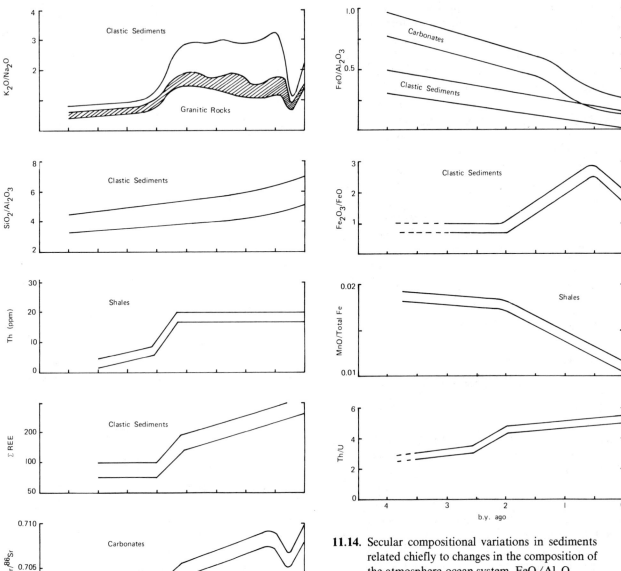

11.13. Secular compositional variations in sediments related chiefly to changes in composition of the continental crust. K_2O/Na_2O after Engel, *et al.* (1974); SiO_2/Al_2O_3 after Schwab (1978); Th and REE after McLennan and Taylor (1980); $^{87}Sr/^{86}Sr$ after Veizer and Compston (1976).

11.14. Secular compositional variations in sediments related chiefly to changes in the composition of the atmosphere-ocean system. FeO/Al_2O_3, Fe_2O_3/FeO, and MnO/total Fe after Ronov and Migdisov (1971); Th/U after McLennan and Taylor (1980).

K_2O/Al_2O_3, U, La/Yb and a negative Eu anomaly show similar variations (Ronov and Migdisov, 1971; Taylor, 1972). Noteworthy is the decrease in K_2O/Na_2O at about 600 m.y. followed by an increase at 200 m.y. Other similar trends are the decreases in FeO/Al_2O_3, MgO/CaO, and MnO/total

Fe and increases in Fe_2O_3/FeO, MgO/Al_2O_3, and Th/U with time (fig. 11.14). The variation of $^{87}Sr/^{86}Sr$ ratio in sedimentary carbonates with time is generally equated with the variation in the Sr isotope composition of sea water (Veizer, 1976). Its variation is similar to the K_2O/Na_2O ratio (fig. 11.13).

Secular geochemical variations in sediments are controlled chiefly by one or combination of five variables: (1) selective preservation of more resistant sediments during recycling; (2) diagenesis and low-grade metamorphism; (3) changing tectonic settings; (4) changing composition of the upper continental crust; and (5) changing composition of the atmosphere-ocean system. If recycling were important in controlling the sediment distributions shown in fig. 11.12, rocks more resistant to weathering and erosion (such as quartzite and chert) should increase in abundance with age. The fact that this is not observed suggests that recycling is not an important contributing factor to geochemical trends. Because there is not a consistent relationship between increasing diagenetic or metamorphic grade and the geochemical variations, and because samples of varying grade are represented throughout geologic time, diagenesis and metamorphism also do not appear to be major factors contributing to the trends. Changing tectonic setting is clearly important in controlling proportions of sedimentary rocks (see Chapters 8 and 10) and hence may also influence compositional trends. Changing composition of continental source areas, however, is commonly cited to best explain the trends shown in fig. 11.13 (Veizer, 1973; Schwab, 1978). Intrusion of voluminous high-K granite into the Archean crust at about 2.6 b.y. followed by uplift and erosion can account for the increases in K_2O/Na_2O, SiO_2/Al_2O_3, REE, La/Yb, Th and U in derivative sediments. The parallel trend to K_2O/Na_2O in granitic rocks and gneisses supports this interpretation. The negative Eu anomaly characteristic of post-Archean clastic sediments appears to be inherited from high-K granites intruded at or after 2.6 b.y. The increase in the $^{87}Sr/^{86}Sr$ ratio is probably caused by increased input of high Rb/Sr material into the oceans. The cause of the minima in the K_2O/Na_2O and $^{87}Sr/^{86}Sr$ curves is not understood, but it is noteworthy that the decrease in these parameters corresponds to the breakup of the Proterozoic supercontinent (~ 600 m.y.) and the renewed increase with the breakup of Pangaea (~ 200 m.y.)

Parameter variations in fig. 11.14 may be related to changes in the composition of the ocean-atmosphere system. The increase in Fe_2O_3/FeO and decrease in FeO/Al_2O_3 and MnO/total Fe correspond to the growth of free oxygen in the atmosphere and oceans. The MnO/total Fe trend reflects the fact that Mn is more readily oxidized than Fe at a given oxygen pressure. The decrease in the Fe_2O_3/FeO ratio during the last 600 m.y. probably reflects loss of oxygen accompanying the burial and decay of organic matter. The increase in the Th/U ratio with time appears to reflect the greater mobility of U which, when oxidized to U^{+6}, readily forms the very soluble $(UO_2)^{-2}$ ion. Organic carbon and sulfur also increase rapidly in Phanerozoic sediments, reflecting the rapid development of living organisms.

Average Continental Crust

Estimates of the composition of the Archean and Proterozoic crust based on widespread statistical sampling of Precambrian shields are summarized in table 11.2. Although the overall major-element composition of both ages of Precambrian crust are similar, the Proterozoic crust is clearly enriched in K_2O, U, and Th and depleted in Ni and Cr, and exhibits a higher K_2O/Na_2O ratio and lower Th/U and K/U ratios than the Archean crust. A comparison of Archean and Proterozoic granitic rocks results in similar differences.

The most adequate explanation for these trends is that of progressively increasing vertical compositional zonation of the continental crust with time (Eade and Fahrig, 1971). This zonation is produced by metamorphism and melting in the lower crust, which results in upward migration of elements like K, U, and Th either in magmas or in an aqueous phase, leaving a depleted lower crust composed of granulites. This zonation is more pronounced in the Proterozoic than in the Archean. Perhaps the intense heating in Precambrian mobile belts was responsible for developing such pronounced zonation.

Mineral Deposits

The occurrence and importance of various mineral deposits also varies with time. Archean greenstone belts contain a variety of important deposits of economic importance, such as Ni, Au, Cu, and Cr. The most remarkable feature of Archean ore deposits is their similarity in space and time. Massive sulfide deposits are of two types (Condie, 1981a): Zn-Cu deposits, which are associated with andesitic to felsic volcanic rocks, and Ni-Cu deposits, associated with ultramafic and komatiitic volcanics and intrusives. Pb is notably lacking in Archean sulfide deposits. Banded-iron formation is also of minor importance in many greenstone belts. Gold deposits occur in greenstone belts in quartz veins and lodes and in carbonated volcanic rocks, and associated with iron formation. Chromite occurs principally as cumulus ores in ultramafic bodies in greenstone belts. Nonmetallic deposits include asbestos and talc in greenstone belts and various Li, REE, Ta, and Sn-bearing minerals in pegmatites.

During the Proterozoic, several important types of sedimentary ore deposits appeared. Detrital gold and uraninite were deposited in early Proterozoic basins on most of the continents. Manganese formation occurs in some quartzite-shale successions in India and South America. The oldest continental rifts

Table 11.2. Average Compositions of Archean and Proterozoic Shield Areas

Shield	Archean ($\gtrsim 2.5$ b. yrs.)		Proterozoic (1.0–2.5 b. yrs.)	
	Canadian	Baltic-Ukrainian	Canadian	Baltic-Ukrainian
SiO_2	65.1	65.4	65.0	64.4
Al_2O_3	16.0	15.8	16.0	14.1
Fe_2O_3	1.5	2.1	1.2	2.5
FeO	3.0	3.3	3.4	4.9
MgO	2.3	2.2	2.1	2.5
CaO	3.4	3.4	3.3	2.2
Na_2O	4.1	3.1	3.5	2.0
K_2O	2.7	2.6	3.5	3.7
TiO_2	0.50	0.50	0.62	0.50
P_2O_5	0.15	0.10	0.19	0.10
MnO	0.08	0.08	0.09	0.07
Ba	790		810	
Sr	410		310	
Rb	100		125	
Ni	26		11	
Cr	88		45	
Th	9.7	10.7	13.6	27
U	1.2	1.6	2.2	4.4
Rb/Sr	0.24		0.40	
Th/U	8.1	6.7	6.2	6.1
$K/U \times 10^{-4}$	1.9		1.3	
K_2O/Na_2O	0.66	0.84	1.0	1.9
Fe_2O_3/FeO	0.50	0.64	0.35	0.51

Note: Major elements in oxide weight percent; trace elements in ppm; blank space indicates data not available. After Eade and Fahrig (1971); Ronov and Midgisov (1971)

at about 2 b.y. are the sites Pb-Zn-Ag deposits in carbonate rocks. Banded-iron formation, as was previously discussed, is also widely deposited in cratonic basins at this time. Copper deposits also appear in continental-margin and rift sediments. U, Be, Sn, and related rare elements are common in late pegmatites in mobile belts, and titaniferous-magnetite deposits are associated with massive anorthosites. Tin-bearing granites and pegmatites are widespread in Pan-African mobile belts in Africa and South America. Also associated with alkaline complexes in Proterozoic rift systems are deposits of U, Th, Nb, Zr, and REE.

Phanerozoic mineral deposits, discussed in Chapter 8, are characterized by a great variety of metallic and nonmetallic deposits associated with the various plate tectonic settings.

Major Discontinuities

Times of major change in crustal evolution occur at the end of the Archean (~ 2.5 b.y.), at about 2.0 b.y., near the Precambrian-Cambrian boundary (700–500 m.y.), and at the Paleozoic-Mesozoic boundary

(~ 200 m.y.). The end of the Archean is marked by a decrease in greenstone volcanism, development of widespread stable cratons, emplacement of voluminous, high-K granites (~ 2.6 b.y.), and relatively rapid uplift and erosion, producing thick successions of dominantly rift and cratonic sediments (Chapter 10). Geochemical trends in sediments and igneous rocks reflect these changes. Between 2.0 and 2.3 b.y., major amounts of banded-iron formation are deposited, as ferrous iron in seawater is oxidized by O_2 given off by photosynthetic microorganisms (see later section). When oxygen is liberated into the atmosphere at about 2 b.y., the first red beds and widespread evaporites form. The lithosphere becomes rigid enough to sustain brittle fracture, and continental rifting and buoyancy-driven subduction begins.

At 600–800 m.y. ago, widespread rifting results in fragmentation of Proterozoic supercontinents and onset of Wilson-cycle tectonics. Metazoans develop as oxygen rapidly increases in the atmosphere. After regrouping of the supercontinents, fragmentation occurs again at ~ 200 m.y. Beginning just before this time and increasing rapidly thereafter, animals and plants expand into continental habitats.

THE ATMOSPHERE

The Early Atmosphere

A discussion of the origin and development of the crust would not be complete without some mention of the origin of the atmosphere and oceans. Two models for the composition of the Earth's early atmosphere have been proposed. The Oparin-Urey model (Oparin, 1953) suggests that the first atmosphere was composed dominantly of CH_4 with smaller amounts of NH_3, H_2, He, and H_2O, while the Abelson model (Abelson, 1966) is based on an early atmosphere composed of CO_2, CO, H_2O, H_2, N_2, and HCl. Neither atmosphere allows significant amounts of free oxygen. Experimental studies indicate that reactions may occur in either atmosphere that could produce the first living organisms.

Early terrestrial eruptions may have expelled CH_4, H_2, and NH_3, producing an Oparin-Urey type atmosphere. CH_4 has been found in some kimberlite pipes which tap levels $\gtrsim 200$ km deep in the mantle and could represent trapped gas pockets dating back to the earliest periods of the Earth's history. The presence of significant amounts of CH_4 in the early atmosphere depends critically on the escape rate of H_2, which controls the reaction $CH_4 \rightleftharpoons C + 2H_2$. Considerable uncertainty exists in estimating an escape rate for H_2, with some models indicating that the rate may have been small enough for CH_4 to persist for on the order of 10^9 years while other models indicate a rapid escape rate, allowing CH_4 to collect for only 10^5–10^8 years. The general abundance of H and He in the solar system today suggests that they may have been abundant in the primitive Earth; if so, CH_4 and NH_3 may have been the most important gases in the early atmosphere. There are several problems encountered, however, if the early atmosphere were reducing and persisted for more than a few hundred million years. NH_3, for instance, would be rapidly degraded by ultraviolet radiation (in $< 50,000$ years). Also, relatively little carbon is found in Archean sedimentary rocks as it should be if the atmosphere were rich in CH_4, since ultraviolet radiation produces organic molecules by interaction with CH_4 and these should be absorbed by clays (Abelson, 1966).

Consideration of the depletion in heavy rare gases (Kr, Ne, Ar) in the Earth when compared to carbonaceous chondrites and the sun suggests that if a CH_4-rich atmosphere did exist on the Earth, it was lost in the first 200 m.y. after accretion. Also, the abundance of sedimentary rocks in Archean green-stone belts suggests that the Earth had an Abelson-type atmosphere and a hydrosphere by 3.8 b.y. An early reducing atmosphere, if such ever existed, may have been lost by a catastrophic event such as an intense solar wind as the sun evolved through a T-Tauri stage (figure 2.4), lunar capture, or by internal heating of the Earth and rapid degassing, in which H_2O, CO_2, and related gases are expelled into the atmosphere and CH_4 and NH_3 are lost (Fanale, 1971). Life was probably produced in this atmosphere sometime before 3.8 b.y. by the combination of HCN and H_2O in the presence of ultraviolet radiation (see next section).

The Secondary Atmosphere

It was first suggested by Rubey (1951) that the simplest explanation for excess volatiles (i.e., volatiles including H_2O, CO_2, and N_2 which cannot be accounted for by weathering of the crust) in the atmosphere-ocean system is by degassing of the Earth over geologic time. During core formation, it is likely the mantle contained free iron and that gases in equilibrium with mantle mineral assemblages were comprised dominantly of CO_2, H_2O, CO, H_2S, and N_2 (Holland, 1962). The rate at which this secondary atmosphere is outgassed from the Earth is a topic of considerable discussion. Two models are generally considered: the *big burp model*, in which the atmosphere is rapidly degassed early in the Earth's history (Fanale, 1971), and the *steady state model*, in which the atmosphere grows gradually over geologic time (Rubey, 1951). One way of evaluating atmospheric growth rate is to monitor the buildup of ^{40}Ar, ^{129}Xe, and ^4He in supracrustal rocks that have equilibrated with the atmosphere-ocean system. ^{40}Ar is produced by the radioactive decay of ^{40}K in the Earth, and as it escapes from the Earth's interior, it collects in the atmosphere. Since ^{36}Ar is nonradiogenic, the $^{40}Ar/^{36}Ar$ ratio should record distinct evolutionary paths for Earth degassing. The steady state model is characterized by a gradual increase in the $^{40}Ar/^{36}Ar$ ratio, while the big burp model is characterized by initially small changes in this ratio followed by rapid increases (fig. 11.15). The method, however, is complicated by problems related to estimating the amount of ^{40}Ar added to rocks by *in situ* decay of ^{40}K (which must be subtracted) and problems related to argon mobility after rock solidification.

Most rare gas data support the big burp model for the formation of the atmosphere, although large amounts of the deep mantle may be largely unde-

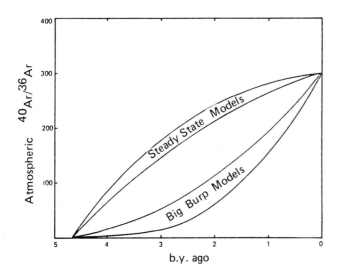

11.15. Evolution of atmospheric $^{40}Ar/^{36}Ar$ ratio in terms of the steady state and big burp models.

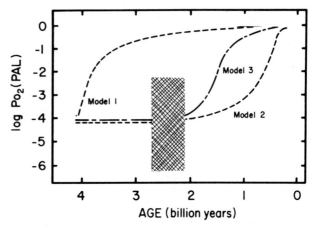

11.16. Models for the evolution of oxygen in the Earth's atmosphere (after Grandstaff, 1980). Shaded area shows range of maximum oxygen pressures which allow preservation of detrital uraninite. PAL = present atmospheric level.

gassed. Also, consistent with the big burp model is the high heat productivity in the early Archean discussed previously. It is likely that the Archean atmosphere was much denser than the present atmosphere and may have been similar to the atmosphere of Venus in composition (i.e., CO_2-rich). It is noteworthy in this respect that models for the evolution of the sun require an increase in solar luminosity of 25 percent since the formation of the solar system (Owen et al., 1979). Because there is no evidence for a corresponding warming trend in the Earth's surface temperature with time, a CO_2-rich atmosphere during the Archean provides a means of warming the Earth to, or in excess of, present surface temperatures by the greenhouse effect.

The Origin of Oxygen

The presence of free oxygen in the atmosphere is generally attributed to photodissociation of H_2O in the upper atmosphere or to O_2-producing photosynthetic reactions (Berkner and Marshall, 1965). Oxygen produced by these mechanisms is lost by oxidation of surface rocks, decay of organic material, respiration, and solution in the hydrosphere. For the growth of free oxygen in the atmosphere, it is necessary that the production rate exceed the removal rate. Three models for the growth of oxygen in the atmosphere are illustrated in fig. 11.16. Model 1 involves rapid growth of oxygen to near-present levels during the first 500 m.y. of the Earth's history. In Model 2, the O_2 level remains low until the early Phanerozoic, when it reaches 0.01 of the present level (Berkner and

Marshall, 1965). Oxygen level remains low until about 2 b.y. ago in Model 3, when the disappearance of banded-iron formation and widespread development of red beds and evaporites occur (Cloud, 1968a). Experimental studies on the stability of uraninite at various oxygen pressures indicate that detrital uraninite in Archean conglomerates (such as in the Witwatersrand in South Africa) can survive to oxygen pressures as high as 0.01 of present levels (Grandstaff, 1980), which eliminates Model 1. Also suggestive that free oxygen was not abundant in the Archean atmosphere is the fact that oxygen is lethal to the anaerobic microorganisms which developed during this period of time. The presence of banded-iron formation and oxidized margins on some pillows from Archean basalts, however, indicates that localized concentrations of oxygen were present in the Archean atmosphere-ocean system. Although photosynthesizing microorganisms were present by 3.8 b.y., they may not have produced free oxygen, and hence photodissociation of water may have been the principal source of Archean oxygen (Towe, 1978). Although Model 2 cannot be eliminated, the widespread occurrence of photosynthesizing microorganisms in Proterozoic cherts and iron formation, and the first major occurrences of red beds and sulfates at about 1.8–2.0 b.y. (which require an oxidizing atmosphere), strongly supports Model 3. Also, the presence of positive Eu anomalies but absence of Ce anomalies in pre-2 b.y. nonclastic sediments is consistent with a chiefly nonoxidizing atmosphere prior to 2 b.y. Banded-iron formation was widely deposited between 2.2 and 2.0

b.y., just prior to red-bed and sulfate deposition. It is possible that free oxygen was liberated into the atmosphere in significant amounts from photosynthesis of marine microorganisms just after the period of banded-iron formation deposition. This results from depletion of iron in the oceans, and hence sea water becomes rapidly saturated with O_2. As significant amounts of oxygen invade the atmosphere, red-bed and sulfate deposition begin. The increase in the Fe_2O_3/FeO ratio in shales during the Proterozoic may monitor the gradual build-up of oxygen in the atmosphere (fig. 11.14).

Atmospheres of Other Terrestrial Planets

The fact that the moon and Mercury do not have atmospheres probably reflects their condensation from a volatile-depleted portion of the solar nebula (Chapter 2). Mars has a very thin atmosphere (5–10 mbar surface pressure), comprised chiefly of CO_2 with minor amounts of N_2 and Ar and traces of O_2, CO, H_2O, and rare gases. Data indicate that, unlike the Earth, which may be largely degassed, Mars is only 27 percent degassed, a feature probably related to its rapid cooling rate and thick lithosphere. It is probable that the early secondary Martian atmosphere was quite dense and that the CO_2 greenhouse effect warmed the surface sufficiently to allow rainstorms and erosion. This early atmosphere was largely lost or incorporated into a frozen regolith as the planetary surface cooled. Changes in the atmosphere and climate of Mars may have occurred in response to changes in the energy output of the sun or to changes in the obliquity and eccentricity of the Martian orbit.

The very dense atmosphere on Venus (~ 90 bar surface pressure) is also comprised chiefly of CO_2 ($> 90\%$), N_2, and Ar, with traces of H_2O, CO, He, H_2SO_4, HCl, HF, and O_2. Complex photodissociation processes are generally invoked to explain the distribution of volatile species. The CO_2 content of the Venusian atmosphere produces a major greenhouse effect resulting in surface temperatures of the order 450°C. Venus, like the Earth, is probably largely degassed. The Earth contains about the same amount of CO_2 as Venus, but unlike Venus it is largely locked up in carbonates. Although causes of the differences in the Venusian and terrestrial atmospheres are not well understood, it is clear that the occurrence of living organisms on the Earth played a significant role in the evolution of the terrestrial atmosphere. Had it not been for photosynthesis, the present Earth's atmosphere may have been very much like that on Venus.

THE OCEANS

The Evolution of Seawater

As the early CO_2-rich atmosphere cools, water condenses and the oceans begin to form. By the time the Earth's early crust cooled below 100°C, much of the atmospheric water should have condensed into oceans. The acid components in the atmosphere (CO_2, HCl, H_2S, HF) rapidly react with exposed crustal rocks. Although the pH of the first seawater was probably low (< 7) due to the dissolved acid components, it must increase rapidly as Na, Ca, Mg, and Fe are leached from exposed crustal rocks and delivered to the oceans. Because of the buffering action caused by silicate-seawater reactions, the pH was probably maintained between 8 and 9 and the Eh, prior to 2 b.y., was probably neutral or negative. As early seawater is neutralized, Al and Fe oxides precipitate while alkalies increase in abundance. Because the composition of the early crust was probably mafic, as was previously discussed, more Mg, Ca, and Fe would be delivered to the early oceans than to the present oceans.

The study of the composition of seawater as a function of time relies upon indirect data. Most evidence from the sedimentary rocks suggests that by 3.8 b.y. ago, seawater had approached a composition similar to modern seawater, and that with exception of localized areas (such as evaporite basins), its composition, pH, and Eh have not greatly deviated from that of modern seawater (Holland, 1976). The near absence of sepiolite in marine sedimentary rocks indicates the silica content of seawater cannot have exceeded 25 ppm (at pH 8) in the geologic past; also the sparsity of marine brucite indicates that the pH of seawater did not exceed 10. After about 2.7 b.y. ago the average composition of continental crust appears to have been about the same, and hence the composition of river waters entering the sea has probably been approximately constant. Similar chemical and biochemical sediments are found in Precambrian and Phanerozoic successions. Precambrian evaporites show the same upward transition from gypsum-anhydrite to halite as do Phanerozoic evaporites. Bedded chert, siderite, sulfide-rich sediments, and carbonates occur in both Phanerozoic and Precambrian sections. All of these features suggest that the composition of seawater has not greatly changed in the last 3.8 b.y.

Oceanic Growth Rate

The rate of growth of the oceans probably paralleled that of the continents and the atmosphere. It is possible to consider oceanic growth rate independent of atmospheric growth rate by considering the freeboard of continents through time. There is considerable sedimentological evidence to support the idea that the *freeboard of continents*, which is defined as the relative elevation of continents with respect to sea level, has not varied greatly for 3.8 b.y. The presence of near-shore marine sediments throughout the preserved sedimentary record supports this conclusion. A constant freeboard model is important in that it provides a means of evaluating other variables. Wise (1973) proposes an internally consistent model of continent-ocean evolution based on the following assumptions: (1) The continents have been at least 90 percent of their present thickness for 2.5 b.y. Evidence supporting this assumption has been previously discussed. (2) The freeboard of continents and the area of the Earth have been constant with time. (3) There has been a net decrease in the volume of continental rocks and ocean waters recycled back into the mantle over this period of time. This limits changes in area or volume of continents or oceans to $\lesssim$ 20 percent of their present values. If the continents were 80% of their present thickness 2.5 b.y. ago, their total area would be 60% of the present area and the corresponding volume of ocean water would be 90% of present.

Carbonate Deposition

Archean carbonates are rare. Among the possible causes for such rarity are the following: (1) The pH of Archean seawater was too low for carbonate deposition. (2) Carbonates were deposited on stable shelves during the Archean and later eroded. (3) Carbonates were deposited in deep ocean basins during the Archean and later destroyed by plate-tectonic processes.

A higher P_{CO_2} in the Archean atmosphere has been appealed to by some to explain the sparsity of Archean carbonates (Cloud, 1968). The reasoning is that a high P_{CO_2} in the atmosphere results in more CO_2 being dissolved in seawater, thus lowering its pH and allowing Ca^{+2} to accumulate because of its increased solubility. Although the P_{CO_2} in the Pre-

cambrian atmosphere probably was higher than at present, the CO_2 content is not the only factor controlling pH. Recent evidence suggests that although the $CO_2 - CO_3^{-2}$ equilibria may have short-term control of pH, silicate equilibria have long-term control. In seawater held at a constant pH by silicate buffering reactions, the solubility of Ca^{+2} actually decreases with increasing P_{CO_2}, and thus the CO_2 mechanism does not seem capable of explaining the sparsity of Archean carbonates. As for the second explanation, carbonate rocks in Archean cratonic successions are rare, and hence erosion of such successions will not solve the missing Archean carbonate problem. The third possibility tentatively holds most promise. Because of the near absence of stable cratons in the Archean, carbonate sedimentation may have been confined to deep ocean basins, controlled perhaps by the deposition of planktonic algae that played a role similar to that played by foraminifera today. If so, the deep-sea carbonates would be largely destroyed by later plate-tectonic processes, thus accounting for their absence in the geologic record.

It has long been recognized that the ratio of dolomite to calcite in sedimentary carbonates decreases with increasing age (fig. 11.12). This is also reflected by the MgO/CaO ratio of carbonates (fig. 11.17). Most data suggest that these trends are

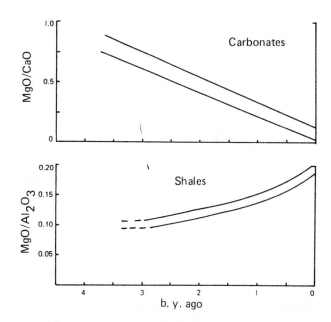

11.17. Secular variations in MgO/CaO in sedimentary carbonates and of MgO/Al$_2$O$_3$ in shales from the Russian platform. After Ronov and Migdisov (1971).

primary and have not been acquired during later diagenesis, metasomatism, or alteration. The residence times of CO_2, Mg^{+2}, and Ca^{+2} in seawater are short and are delicately controlled by input and removal rates of diagenetic silicates and carbonates (Holland, 1976). A decrease in the amount of dolomite precipitated with time may reflect one or, more likely, a combination of both of the following mechanisms:

1. Removal of Mg^{+2} from seawater changed from dominantly a carbonate depository in the Proterozoic to dominantly a silicate depository in the late Phanerozoic. There is a tendency for the MgO/Al_2O_3 in shales to increase over the same time period (fig. 11.17), providing some support for the mechanism, in that Mg appears to be transferred from carbonate to silicate reservoirs with time.

2. Silicate-carbonate equilibria are dependent upon the availability of CO_2 in the ocean-atmosphere system. Where only small amounts of CO_2 are available in the ocean, silicate-carbonate reactions shift such that Ca^{+2} is precipitated chiefly as calcite rather than as silicates. If the rate of CO_2 input into the ocean exceeds the release rate of Ca^{+2} during weathering, both Mg^{+2} and Ca^{+2} are precipitated as dolomite (Holland, 1976). Hence the greater amount of dolomite in the Precambrian may reflect an atmosphere-ocean system with greater amounts of CO_2, which is consistent with models for the development of the atmosphere (see previous section).

Oxygen Isotope Results

^{18}O is fractionated from ^{16}O in the ocean-atmosphere system during evaporation, with the vapor being preferentially enriched in ^{16}O. Fractionation also occurs as a function of temperature, with the amount of fractionation decreasing with increasing temperature and as a function of salinity (^{18}O increasing with salinity). The fractionation of the two isotopes is measured by the deviation of $^{18}O/^{16}O$ from standard mean ocean water (SMOW) in terms of $\delta^{18}O$ where,

$$\delta^{18}O(^0/_{00}) = \left[\frac{(^{18}O/^{16}O)_{sample}}{(^{18}O/^{16}O)_{SMOW}} - 1 \right] \times 1000.$$

$$(11.1)$$

Because marine cherts are resistant to secondary changes in their $\delta^{18}O$ values, they may be useful in monitoring the composition of seawater with geologic time (Knauth and Lowe, 1978). Available data suggest that cherts increase in $\delta^{18}O$ from the Archean to the present (fig. 11.18). Modern cherts have δ values

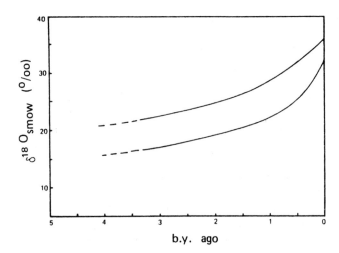

11.18. Variation in $\delta^{18}O$ in marine cherts as a function of geologic age. Modified after Knauth and Lowe (1978).

of 20–35$^0/_{00}$ while Archean cherts have values of 15-20$^0/_{00}$. Similar differences exist for sedimentary carbonates.

Available data suggest these trends are not due to secondary processes such as diagenesis and alteration. Two models have been suggested to explain increasing $\delta^{18}O$ in cherts, and hence in seawater, with geologic time:

1. The mantle-crust interaction model (Perry, Ahmad, and Swulius, 1978). If the Archean ocean was about the same temperature as the modern ocean, it would have a $\delta^{18}O$ value 34$^0/_{00}$ less than the cherts precipitated from it (fig. 11.18), or about $-14^0/_{00}$. For seawater to evolve to a δ value of zero per mil with time, it is necessary for it to interact with either (or both) mantle-derived magmas or water ($\delta^{18}O = +5$ to $+8^0/_{00}$) or continent-derived sediments ($\delta^{18}O = +7$ to $+20^0/_{00}$). One possible mechanism of interaction is the recycling of seawater through the upper mantle by subduction and convection.

2. The surface-cooling model (Knauth and Lowe, 1978). The observed $\delta^{18}O$ trend in marine cherts may be explained by a gradual cooling of the Earth's surface temperature from about 80°C at 3.5 b.y. ago to present temperatures. The $\delta^{18}O$ values of seawater may or may not be constant over this time; however, if they were not constant they must have changed in such a way so as not to affect the temperature control. Consistent with a cooling surface temperature is the probability that the CO_2 content, and hence the greenhouse effect, has decreased in the Earth's atmosphere with time.

PALEOCLIMATES

Climatic Evolution

As was discussed in Chapter 8, many approaches are available to reconstruct ancient climates. No single lithologic indicator, however, provides unambiguous paleoclimatic data, and it is only when many indicators are used together with fossil distributions and oxygen isotope results that ancient climatic regimes can be deduced with some degree of confidence. Available data allow reconstruction of average surface temperature and precipitation trends with geologic time, summarized in fig. 11.19. These curves represent proposed departures from global mean values, and both long- and short-term variations are apparent. The Precambrian portions of the curves are considerably more uncertain than the Phanerozoic portions. Five major glaciations are recognized in the temperature curves: ~ 2300 m.y., 600-800 m.y., 500 m.y., ~ 300 m.y., and the Pleistocene glaciations of the last 1.5 m.y. Minor glaciations are also recorded in the rock record in the Cambrian, Silurian, and Devonian. Times of widespread warm climates are the late Mesozoic and the Precambrian. The precipitation curve reflects the buildup of precipitation just before each major glaciation.

The existence of clastic sediments by 3.8 b.y. indicates that running water existed on the Earth's surface by this time. The abundance of CO_2 in the early atmosphere resulted in a greenhouse effect, discussed in the previous section, and hence surface temperatures were probably greater than at present. For instance, doubling the amount of CO_2 of the present atmosphere would result in an increase of about 2.5°C in the mean global temperature. Few climatic indicators are preserved in Archean supracrustal successions. Archean banded-iron formation is commonly thought to have been deposited in shallow marine waters, and hence implies low wind velocities to preserve the remarkable planar stratification. This is consistent with a CO_2-rich atmosphere, in which differences in temperature gradient with latitude, and hence wind velocities, are small (Frakes, 1979). The earliest evaporites and stromatolitic carbonates appear between 3.5 and 2.5 b.y. and also favor a generally warm climate during this time.

Cooling of the Earth's surface during the latter part of the Archean (fig. 11.19) leads to the first widespread glaciation at about 2.3 b.y. This glaciation is recorded by tillites at numerous localities in North America, Australia, and Africa. Glacial pavements and dropstones in varied marine sediments are

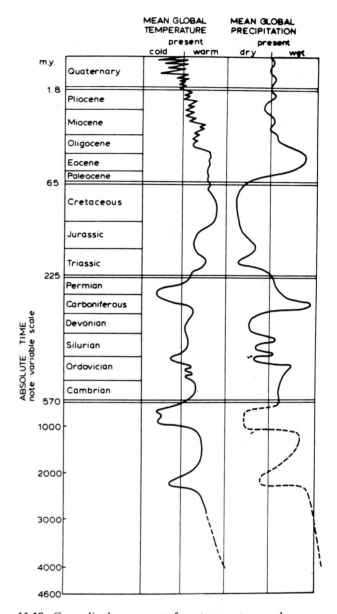

11.19. Generalized average surface temperature and precipitation history of the Earth (after Frakes, 1979).

preserved at several locations (Frakes, 1979). Existing radiometric dates indicate that this early glaciation was diachronous over ≳ 100 m.y. period. Shallow marine stromatolitic carbonates and evaporites become widespread between 1 b.y. and 2 b.y. ago, again reflecting a widespread warm climate. The appearance of red beds at about 2 b.y. reflects a more oxidizing atmosphere developing in response to rapid expansion of photosynthetic microorganisms. The existence of kaolinite and other alumina-rich minerals in a few Proterozoic formations (such as the Lorraine

Formation in North America) and numerous occurrences of evaporites indicate that both subtropical and semi-arid climates existed on the Earth during the Proterozoic.

Late Proterozoic tillites are widespread on most continents and appear to reflect several widespread glaciations in the interval of 500–800 m.y. ago. Most deposits are glacial marine deposits formed around continental margins (such as western North America) or in large intracratonic basins (such as those in Africa and Australia). Data suggest that major continental ice sheets were located in central Australia, west-central Africa, the Baltic shield area in western Europe, Greenland and Spitsbergen, western North America, South America, and Siberia. It is unlikely that glaciations were contemporaneous in all of these regions, however. Widespread shallow-water carbonates and paleomagnetic data (Chapter 10) indicate that the Proterozoic supercontinents were located dominantly at low latitudes during the late Precambrian and that generally warm climates were locally punctuated by continental glaciation.

Most of the middle and lower Paleozoic are characterized by climates that were, on the whole, warmer than present climates, while late Paleozoic climates show wide variations (fig. 11.19). Paleoclimatic data indicate that Europe and North America (Laurasia) underwent relatively small changes in climate during the Paleozoic, while Gondwana underwent several major changes. This appears to reflect chiefly the fact that the continental fragments comprising Laurasia remained at low latitudes during the Paleozoic, while Gondwana migrated over large latitudinal distances (figs. 9.2 to 9.5). Major glaciations occurred in Gondwana in late Ordovician, late Carboniferous, and early Permian and minor glaciations in the Cambrian and Devonian.

The Mesozoic is characterized by warm, dry climates that extend far north and south of equitorial regions. Paleontologic and paleomagnetic data also support this interpretation (see Chapter 8). The initial Triassic climates were cool and humid like the late Paleozoic. These were followed by a warm, drying trend until late Jurassic. Most of the Cretaceous is characterized by warm, commonly humid climates over most of the continents, regardless of their latitude, with tropical climates extending from 45°N to perhaps 75°S. Glacial climates are not reflected in polar regions, and mean annual surface temperatures, as determined from oxygen isotope data, are 10–15°C higher than today. The late Cretaceous is a time of maximum transgression of the continents by the oceans. The early Tertiary was a time of declining temperatures, significant increased precipitation, and falling sea level. Middle and late Tertiary are characterized by lower and more variable temperature and precipitation regimes. These variations continued and became more pronounced in the Quaternary, leading to the alternating glacial and interglacial epochs. Four major Quaternary glaciations are recorded, with the intensity of each glaciation decreasing with time: the Nebraskan (1.3–1.6 m.y.), the Kansan (0.7–0.9 m.y.), the Illinoian (400,000–550,000 y.), and the Wisconsin (10,000–80,000 y.). Each glaciation lasted 100 to 200 thousand years, with interglacial periods of 200 to 400 thousand years. Smaller glacial cycles with periods of the order of 20 to 40 thousand years are superimposed on the larger cycles.

Causes of Glaciation

Commonly cited causes for widespread glaciation on the Earth fall into one or a combination of the following categories: (1) a weakened greenhouse effect; (2) episodic decreases in solar luminosity; (3) periodic variation in the obliquity of the ecliptic with the Earth's equatorial plane; and (4) changes in the position and relative elevation of continents and corresponding changes in oceanic and atmospheric circulation patterns caused by continental drift. A decrease in the atmospheric and oceanic CO_2 can result in a worldwide cooling trend that could lead to widespread glaciation. It is possible, for instance, that the early Proterozoic glaciations at about 2300 m.y. resulted in withdrawal of substantial amounts of CO_2 from the atmosphere-ocean system by photosynthesizing stromatolites. The widespread distribution of stromatolitic carbonates at this time is consistent with such an interpretation. The obliquity of the Earth's orbit may have changed with time. When the equator is at high angles to the ecliptic, mean annual insolation is at a minimum, and hence widespread glaciation is favored, even in low latitudes. This is one of the few models that seems capable of explaining the widespread, non-latitude-dependent glaciations in the late Precambrian. Decreasing solar luminosity at this time could also have accounted for widespread glaciation.

The simplest explanation for the Gondwana glaciations during the Paleozoic is the positioning of the supercontinent over the South Pole. Glaciation terminated in the Silurian and Permian when Gondwana drifted away from the pole. The reassembly of Pangaea in the Permian, which resulted in a continent that reached from pole to pole, may also have been important in terminating the Permian glaciation. If poleward heat transport is more efficient in a single large ocean than in several small ones, global warming would have accompanied the reassembly of Pangaea. The late Cenozoic glaciations are most readily explained by various plate-tectonic events which lead to changing oceanic and atmospheric circulation patterns. Perhaps contributing to the Cenozoic glaciations is an increasing albedo of the Earth as the continents become more emergent, since continents reflect more than three times the solar radiation than oceans do. Perhaps most important in initiating the Pleistocene glaciations was the formation of the Panama arc, which forced the warm Gulf Stream northward into the Arctic Basin, and the final closing of the Tethys Ocean, which forced warm equatorial currents south toward Antarctica. These oceanic currents brought with them warm, moist air which, upon cooling, provided precipitation for growing ice sheets.

THE EVOLUTION OF LIFE

Origin of Life

Perhaps no other subject in geology has been the subject of more investigation and general inquiry than that of the origin of life (Kvenvolden, 1974). It has been approached from many points of view. Geologists have searched painstakingly for fossil evidences of the earliest life. Biologists and biochemists have provided a variety of evidence from experiments and models that must be incorporated into any model for the origin of life. One major question that has not been completely answered to everyone's satisfaction is whether life began on the Earth or on some other body, to be later carried to the Earth. The interest in the nature and origin of carbonaceous compounds in Type I carbonaceous chondrites has stemmed in part from this possibility.

Two factors seem to be necessary for the production of living cells (i.e., cells capable of reproducing themselves) in any model (Rutten, 1971):

1. Free oxygen must not be present in the atmosphere-hydrosphere system. Such oxygen has two deleterious effects on the production of life. First, any organic molecules formed in the presence of free oxygen would be immediately oxidized. Second, oxygen in the atmosphere would produce an ozone layer which prevents ultraviolet (UV) radiation from hitting the Earth's surface, and UV radiation appears to be a necessary catalyst to form organic molecules.

2. The elements (H, C, O, S, etc.) and catalysts necessary for the production of organic molecules must be present.

Most models for the origin of life on Earth call upon a primordial "soup" rich in carbon-bearing compounds which form by inorganic processes. Reactions in this "soup" result in increasingly larger and more complex "organic" molecules formed by inorganic reactions. These molecules grow at the expense of smaller molecules with similar structures, with clays or sulfides providing suitable sites for such reactions to begin. During the early growth stages, some molecules grow into polymers, such as peptides, which join to form amino acids. These, in turn, combine into particles of protein. The proteins must develop in such a way to form membranes, which allow living matter to maintain compositional and energy differences from their surroundings. Once membranes have formed, it is possible for metabolism to begin. The ultimate stage in biogenesis is the development of the ability to duplicate, such that a living cell can perpetuate itself. Mutation leads to molecules capable of organic photosynthesis. Only after sufficient oxygen accumulates in the atmosphere can mutation lead to forms capable of respiration.

The earliest experiments dealing with the production of life were carried out by Miller (1953), who sparked a hydrous mixture of H_2, NH_3, and CH_4 to form amino acids. Similar results have been reported using UV radiation at various temperatures. These experiments show that it is possible to produce life in an early reducing atmosphere on the Earth. However, as discussed previously, if such an atmosphere existed it is likely it was lost very soon after accretion, and if life were formed in such an atmosphere it is likely that it did not survive the catastrophic loss.

Mechanisms for the formation of amino acids have also been sought in a nonreducing atmosphere composed chiefly of H_2O, N_2, CO, CO_2, and H_2. Such mechanisms are more in line with the probable composition of the first stable atmosphere on the Earth. It is possible to build peptides by dehydrating smaller molecules if HCN is present in the aqueous solution of "soup" in which life forms (beneath a nonreducing atmosphere). Such reactions occur at room temperature in very dilute basic solutions.

"Organic" compounds and possible organic remains have been formed in carbonaceous chondrites, leading to the idea that life exists elsewhere in the solar system and perhaps was created elsewhere and brought to Earth. Claus and Nagy (1961) were the first to describe "organic" compounds and "organic" structures from meteorites. Subsequent studies have shown that such materials are common in carbonaceous chondrites and that many of them are contaminants picked up after the meteorite fell on the Earth. Some of the structures known as "organized elements" appear to be indigenous to the meteorites, however. Whether they represent remains of living organisms or not, however, is a subject of disagreement. Recent experimental studies clearly show that it is possible, if not likely, that the "organic" compounds in meteorites were produced by inorganic reactions at low temperatures, perhaps in the solar nebula. Although recent data seem to cast doubt on the life-in-meteorites hypothesis, the question is not closed.

The First Evidences of Life on Earth

Two lines of evidence are available for the recognition of living organisms in the earliest preserved Archean rocks: microfossils and organic geochemical evidence (Schopf, 1976). The microfossil-like objects that have been described under the microscope are readily subject to contamination during their preparation in the laboratory. Also, many structures are preserved in rocks that can be mistaken for cell-like objects (viz., inclusions, bubbles, microfolds, etc.), and progressive low-grade metamorphism can produce structures that look remarkably organic and at the same time can destroy real microfossils. For this reason, caution must be exercised in accepting such microstructures as totally biologic.

Carbonaceous compounds in rocks are of two types, extractable (viz., amino acids, hydrocarbons, sugars, etc.) and nonextractable (viz., kerogen). Although the extractable components from Archean sediments are similar (or indistinguishable) from those in modern organisms, it is not always clear when they were introduced into the rock system. Some may date to the time of sedimentation, while others were introduced by secondary processes in the recent geologic past. The insoluble carbonaceous compounds in kerogen, although very likely formed at the time a rock is deposited, may be of biologic or nonbiologic origin, or both. To date, analytical methods are unable to distinguish between these alternatives.

The most convincing, unambiguous evidence for early Archean life comes from the carbon isotopic analysis of kerogen. The $\delta^{13}C$ values of Archean kerogen are similar to modern values from living organisms, a feature which is difficult to interpret if living organisms were not present in the Archean. Such evidence, when considered together with the presence of microfossil-like structures and extractable

biologic compounds, indicates that primitive living organisms date back to at least 3.5 b.y.

The oldest well-described assemblage of microfossil-like structures comes from cherts in the Barberton greenstone belt in South Africa ($\gtrsim 3.5$ b.y.). The oldest unambiguous structures of organic origin are 3.5 b.y. old stromatolites from the Pilbara region of Western Australia (Walter et al., 1980). Three types of microstructures, ranging in size from $< 1 \mu m$ to $\sim 20 \mu m$, have been reported from the Barberton sequence and from other Archean sections. These are rodshaped bodies, filamentous structures, and spheroidal bodies. The spheroidal bodies are similar to alga-like bodies from Proterozoic assemblages and are generally interpreted as such. Of the two known types of cells, procaryotic and eucaryotic, only procaryotic types appear to be represented among Archean microfossils. *Procaryotic* cells are primitive cells which lack a cell wall around the nucleus and are not capable of cell division; *eucaryotic* cells possess these features and hence are capable of transmitting genetic coding material to various cells and to descendants of the organism. The first cells are interpreted as *heterotrophs*, in that they must have developed in an O_2-free environment and fed on organic substances of nonbiologic origin or on one another. *Autotrophs*, i.e., cells capable of manufacturing their own food by photosynthesis, probably developed sometime between 3.8 and 3.5 b.y.

Refinement of gas chromatographic and most spectrometric techniques has made it possible to detect and identify microquantities of complex organic compounds in rocks and minerals. It is possible from such results to develop a set of criteria to distinguish between primary and secondary sources for such compounds. Hydrocarbons (principally alkanes) and very small amounts of amino acids have been extracted from Archean cherts. Because of the mobility of these organic components, however, it is not certain at what time they entered the cherts. Analyses of the insoluble kerogen in Barberton cherts indicate that the Onverwacht samples contain chiefly aromatic degradation products of kerogen while the Fig Tree samples contain an abundance of n-alkanes. These components are generally thought to come from degraded unicellular organisms.

Organic compounds which have survived diagenesis (with little alteration) are sometimes called *chemical fossils*. Such compounds may be useful as biological markers. Branched alkanes, for instance, are thought to reflect bacterial activity, and branched fatty acids and porphyrins reflect evidence of photosynthesis. Nitriles and furans have been extracted from Archean cherts and are thought to represent the breakdown products of amino acids, peptides, porphyrins, and carbohydrates.

Organic photosynthesis fractionates carbon isotopes in that $^{12}CO_2$ is preferred over $^{13}CO_2$. Plants extract CO_2 principally from two sources: atmospheric CO_2 and aqueous carbonate or bicarbonate ions. Fractionations of carbon isotopes are generally expressed as deviations from a carbonate standard (usually the belemnite PDB-1) where

$$\delta^{13}C\%_{00} = \left[\frac{\left(^{13}C/^{12}C\right)_{sample}}{\left(^{13}C/^{12}C\right)_{standard}} - 1 \right] \times 1000.$$

(11.2)

$\delta^{13}C$ values are different for atmospheric, seawater, and freshwater CO_2, and carbonates and organisms obtaining their carbon from each of these sources reflects, in part, the $\delta^{13}C$ of the source. Many analyses of Precambrian organic carbon show low $\delta^{13}C$ values like those observed in modern plants, consistent with photosynthesis occurring in the Precambrian. Studies of Precambrian carbonates (Schidlowski et al., 1975) show that the average $\delta^{13}C$ value of sedimentary carbonates has remained approximately constant for at least the last 3.3 b.y. This implies that a ratio of organic to carbonate carbon of 1:4 has been maintained for this period of time and that O_2-producing photosynthetic organisms were in existence by 3.3 b.y.

Proterozoic Microbiotic Communities

With decreasing age, microfossil remains become increasingly more complex, increase in size, and contain more biogenic types of polymer degradation products. Such microbiotic communities are known from at least 28 localities and range from about 2.3 b.y. to 0.65 b.y. in age. Only a few have been studied in detail, but all are known to be dominated by procaryotic microorganisms (chiefly blue-green algae and bacterial and possibly aquatic fungi). Eucaryotic cells appear to have developed soon after 2 b.y. when significant amounts of oxygen are liberated into the atmosphere-ocean system (fig. 11.20). Some algae have built widespread stromatolitic mats, described in the next section. Three major categories of microfossils have been described: spheroidal bodies which appear to represent individual cells, cylindrical tube-like sheaths, and cellular filaments. Many of these forms are widespread enough and distinctive enough to use as index fossils.

One of the first assemblages described and one of the best known assemblages of Precambrian microorganisms occurs in the Gunflint Iron Formation (~ 2.0 b.y.) in the Lake Superior region. The number and diversity of forms in this assemblage represents an advanced level of evolution over the Archean assemblages. Only procaryotic forms have been described, however. Sixteen taxa have been recognized, falling into one of three categories (Awramik and Barghoorn, 1977): blue-green algae, budding bacteria, and 6 taxa of unknown affinities. Some of the best preserved Proterozoic microbiota come from black cherts of the Bitter Springs Formation (~ 1.0 b.y.) in central Australia. The Bitter Springs assemblage, although composed dominantly also of coccoid and filamentous algae, is even more varied and complex than the Gunflint assemblage, reflecting a more evolved assemblage. Of the 19 species of algae recognized, 14 have been classified into modern algae families. This suggests that these species have not evolved greatly in terms of surface morphology in the last 1 b.y. Other Proterozoic assemblages occur in the Skillogalee Dolomite (~ 1.0 b.y.) in southern Australia, the Beck Spring Dolomite in California (1.2 b.y.), and the Belcher Group (~ 1.8 b.y.) in Canada.

Stromatolites

Stromatolites are finely laminated sediments composed chiefly of carbonate minerals that have formed by the accretion of both detrital and biochemical precipitates on successive layers of microorganisms (commonly blue-green algae). They exhibit a variety of domical forms and range in age from about 3.5 b.y. in Western Australia to modern. Two parameters are especially important in stromatolite growth: water currents and sunlight. There are serious limitations to interpreting ancient stromatolites in terms of modern ones, however. First of all, modern stromatolites are not well understood and occur in a great variety of aqueous environments (Walter, 1977). Their distribution in the past is also controlled by the availability of shallow stable shelf environments, the types of organisms producing the stromatolites (i.e., bacteria or algae), the composition of the atmosphere, and the importance of burrowing animals. It is possible to use stromatolites to distinguish deep-water from shallow-water deposition, since reef morphologies are different in these environments.

One of the potentially most valuable uses of stromatolites is in establishing a Proterozoic biostratigraphy. Stromatolites increased in number and complexity from 3.0 b.y. ago to about 700 m.y. ago, when they decreased rapidly. It is likely that the abrupt decrease in numbers of stromatolites in the late Precambrian reflects an increase in burrowing organisms at this time. Such organisms ate the algae and destroyed the stromatolites by burrowing. The extent to which stromatolites can be used to establish a worldwide Proterozoic biostratigraphy is a subject of controversy at the present time (Walter, 1977). The controversy revolves around (1) the role of environment and diagenesis in determining stromatolite shape, and (2) the development of an acceptable taxonomy. Because the growth of stromatolites is at least partly controlled by organisms, it should in theory be possible to construct a worldwide biostratigraphic column. Another controversial subject is that of how stromatolite height is related to tidal range. Cloud (1968b) suggests that the height of intertidal stromatolites at maturity reflects the tidal range, whereas Walter (1977) suggests that the situation is much more complex. The distribution of laminations in stromatolites has been suggested as a means of studying secular variations, as was discussed in Chapter 2.

Metazoan Evolution

Metazoa are multicellular organisms which require oxygen for their metabolism and are capable of sexual reproduction. Before metazoa can evolve, eucaryotic cells must exist. This, in turn, means that the level of O_2 dissolved in seawater must be sufficient to support metazoan-type respiration. Although hard shells and hard skeletons did not evolve until the Phanerozoic, fossil imprints indicate that soft-bodied metazoans were present on the Earth by about 700 m.y. One of the basic problems regarding the development of metazoa is the fact that many phyla appear rather suddenly in the lower Cambrian and that these phyla must have had precursors in the late Precambrian. Late Precambrian soft-bodied faunal assemblages are known from South Australia, the Charnwood Forest in England, the Nama System in southwest Africa, northern Russia and Siberia, and perhaps in California. The best documented ancient metazoan assemblage is the *Ediacaran assemblage*. This assemblage is primarily a planktonic or pelagic association 600–800 m.y. in age. It represents at least 25 species in 15 genera and three phyla. Approximately two-thirds of the group are coelenterates, 25 percent annelids, 5 percent arthropods, and the remainder of unknown affinity.

Major events and suggested phylogenetic patterns of major groups of metazoans (including plants) are summarized in fig. 11.20. As has been pointed out in earlier sections, the evolution of life is closely tied with the evolution of the atmosphere-ocean system and of the crust. Life appears to have formed sometime prior to 3.8 b.y. in a nonoxidizing atmosphere composed chiefly of CO_2, CO, H_2O, H, and N_2. Interaction of these compounds with UV radiation produced a variety of "organic-like" compounds, concentrated perhaps in ponds on the Earth's surface. In this environment, heterotropic procaryotes formed. These evolved into photosynthetic autotrophs. Photosynthetically generated O_2 is poisonous to these early cells and must be disposed of. Cloud (1968a) suggests that ferrous iron in the primitive oceans provided a ready-made sink for such oxygen. Oxidization of $Fe^{+2} \rightarrow Fe^{+3}$ results in precipitation of ferric oxides and hydroxides, forming banded-iron formation. The early procaryotes must have been rather resistant to further breakdown by UV radiation, since the Earth lacked an ozone screen at this time. As time went on, some procaryotes developed cytochromes which absorb free oxygen, and hence it became possible for the organisms to survive in the presence of free oxygen and eventually use it in their metabolism. This period of time appears to have been reached just after the large amounts of banded-iron formation were deposited between 2.0 and 2.3 b.y. Oxygen begins to accumulate in the atmosphere as O_2-producing organisms (chiefly blue-green algae) expand, and UV radiation produces ozone, which begins to collect in the upper atmosphere and form a radiation screen. It is also very reactive and initially may have been partly responsible for the large amounts of red beds that appeared in the geologic record at about 1.8 b.y.

At about 1.8 b.y., oxygen in the atmosphere reached levels high enough (~ 1 percent of the present atmospheric level) for the first eucaryotes to appear. At this time, organisms changed from dominantly anaerobic types to aerobic types. By late Precambrian time (600–800 m.y.), the ozone screen in the upper atmosphere develops to such an extent that most UV radiation is prevented from reaching the Earth's surface, and thus the shallow-water environments are rapidly invaded by planktonic organisms. At about the same time, metazoa begin to develop and to occupy previously unoccupied ecological niches. It is possible that an increase in the atmospheric O_2 at the expense of CO_2 at this time was responsible for cooling the Earth and triggering the worldwide glaciations of late Precambrian age. Con-

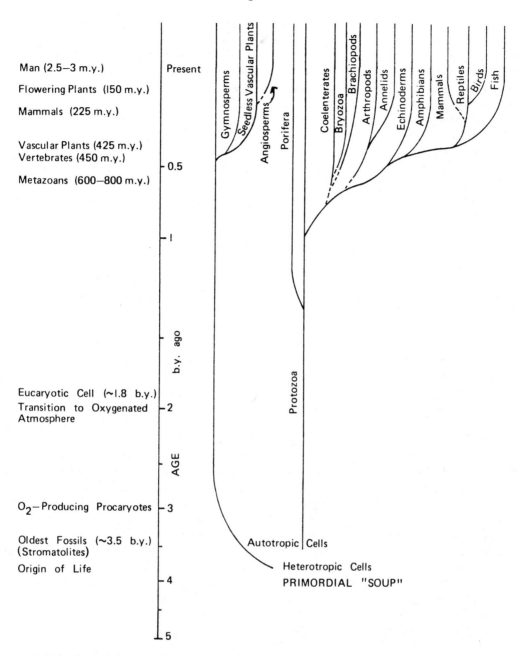

Man (2.5–3 m.y.)

Flowering Plants (150 m.y.)

Mammals (225 m.y.)

Vascular Plants (425 m.y.)
Vertebrates (450 m.y.)

Metazoans (600–800 m.y.)

Eucaryotic Cell (~1.8 b.y.)
Transition to Oxygenated
Atmosphere

O_2–Producing Procaryotes

Oldest Fossils (~3.5 b.y.)
(Stromatolites)
Origin of Life

11.20. Major events and possible phylogeny of major groups of living organisms.

tinuing increases in atmospheric O_2 (up to about 10 percent of the present level) led to protection of the land surface from UV radiation and thus expansion of both animal and plant populations on the continents during the Paleozoic.

All of the major invertebrate phyla (except the Protozoa) made their appearance in the early Paleozoic. This is primarily a time of primitive, shallow-marine life in which attached forms predominate. Cambrian faunas are dominated by trilobites ($\sim 60\%$) and brachiopods ($\sim 30\%$). By late Ordovician, most

of the common invertebrate classes that occur in modern oceans are well established. Trilobites reached the peak of their development during the Ordovician, with a great variety of shapes, sizes, and shell ornamentation. Bryozoans, which represent the first attached communal organisms, first appear in the Ordovician. Graptolites, cephalopods, crinoids, echinoderms, molluscs, and corals also begin to increase in numbers at this time. Vertebrates first appear during the Ordovician and Silurian as primitive fish-like forms without jaws. Marine algae and bacteria continue to be the important plant forms during the early Paleozoic

The late Paleozoic is a time of increasing diversification of plants and vertebrates and of decline in many invertebrate groups. Brachiopods, coelenterates, and crinoids all increase in abundance in the late Paleozoic seas, followed by a rapid decrease in numbers at the end of the Permian. The end of the Paleozoic was also a time of widespread extinction, with trilobites, eurypterids, fusulines, and many corals and bryozoa becoming extinct at this time. Insects appeared in the late Devonian. Fish greatly increased in abundance during the Devonian and Mississippian, amphibians appeared in the Mississippian, and reptiles in the Pennsylvanian. Plants greatly increased in numbers during the late Paleozoic as they moved into terrestrial environments. Psilopsids are most important during the Devonian, with lycopsids, ferns, and conifers becoming important thereafter. Perhaps the most important evolutionary event in the Paleozoic was the development of vascular tissue in plants, which made it possible for land plants to survive under extreme climatic conditions. Seed plants also began to become more important relative to spore-bearing plants in the late Paleozoic and early Mesozoic. The appearance and rapid evolution of amphibians in the late Paleozoic was closely related to the development of forests, which provided protection for these animals. The appearance for the first time of shell-covered eggs and of scales (in the reptiles) allowed vertebrates to adapt to a greater variety of climatic regimes.

During the Mesozoic gymnosperms rapidly increase in numbers, with cycads, ginkoes, and conifers being most important. During the lower Cretaceous, angiosperms (flowering plants) made their first appearance and rapidly grew in numbers thereafter. The evolutionary success of flowering plants is due to the development of a flower and enclosed seeds. Flowers attract birds and insects that provide pollination, and seeds may develop fleshy fruits which, when eaten by animals, can serve to disperse the seeds. Marine invertebrates, which decrease in numbers at the end of the Permian, make a comeback in the Mesozoic (such as bryozoans, molluscs, echinoderms, and cephalopods). Gastropods, pelecypods, foraminifera, and coiled cephalopods are particularly important Mesozoic invertebrate groups. Arthropods in the form of insects, shrimp, crayfish, and crabs also rapidly expand during the Mesozoic. Mesozoic reptiles are represented by a great variety of groups, including dinosaurs. Dinosaurs include both herbiferous and carniverous types, as well as marine and flying forms.

Birds and mammals evolve from reptilian ancestors in the lower Jurassic. The development of mammals is a major evolutionary breakthrough, in that their warm-blooded nature allowed them to adapt to a great variety of natural environments (including marine), and their increased brain size allowed them to learn more rapidly than other vertebrates. During the Cenozoic, mammals evolve into large numbers of groups filling numerous ecological niches. Man evolves in the anthropoid group about 3 m.y. ago. The vertebrate groups characteristic of the Mesozoic continue to increase in numbers and angiosperms expand exponentially.

Extinctions

Extinctions and diversification of organisms have alternated with time, as if new organisms developed to fill vacated niches following major extinctions (fig. 11.21). The most important times of widespread extinction occur at the end of the Cambrian, Ordovician, Devonian, Permian, Triassic, and Cretaceous periods. Extinction episodes affect a great variety of organisms, marine and terrestrial, stationary and swimming forms, carnivorous and herbivorous, protozoans and metazoans. Hence the causal processes do not appear to be related to certain ecological, morphological, or taxonomic groups. Proposed causes of extinction fall into three groups: extraterrestrial, physical, and biological.

Among the extraterrestrial models are increased production of cosmic and X-radiation from nearby stars, increased radiation during reversals in the Earth's magnetic field, and climatic changes during supernova events or as a result of asteroid impact on the Earth. Of these, only two have been the subjects

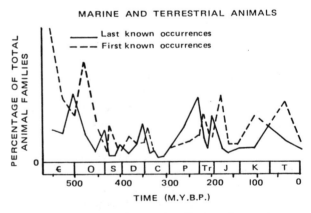

11.21. Animals extinctions and radiations during the Phanerozoic. After Newell (1967).

of detailed study. A recent correlation analysis between times of microfossil extinctions in deep-sea sediments and polarity reversals of the Earth's magnetic field does not support a relationship between the two (Plotnick, 1980). It has recently been proposed that the extinctions at the Cretaceous-Tertiary boundary have resulted from asteroidal impact (Alverez et al., 1980). Clay horizons at this boundary in various stratigraphic sections from around the world contain an anomalously large amount of Ir and other platinum metals. The proposed model involves injection of a large amount of pulverized rock into the atmosphere upon asteroid impact. This dust is rapidly distributed worldwide and remains in the atmosphere for several years. Resulting darkness suppresses photosynthesis and upsets the food chain, leading directly or indirectly to the extinction of many groups of animals, including the dinosaurs.

Among the physical environmental changes that have been proposed to explain extinctions are rapid climatic changes, reduction in oceanic salinity, fluctuations in atmospheric oxygen level, and changes in sea level. None of these has been rigorously evaluated. As was discussed in Chapter 8, collision of continents also leads to extinction of those groups least able to compete for the same ecological niches. Biological changes such as predators, disease, and competition may also lead to extinction of specialized groups of organisms. Rapid changes in environmental factors lead to widespread extinctions, while gradual changes permit organisms to adapt and may lead to diversification. It is clear that no single cause is responsible for all major extinctions, and it is probable that most extinctions have various contributing causes.

SUMMARY STATEMENTS

1. The earliest terrestrial crust appears to have formed between 4.2 and 4.5 billion years ago, and the earliest continental crust between 3.8 and 4.0 b.y. ago.

2. Catastrophic models for the origin of the Earth's crust call upon such events as rapid core formation, impact-generated melting, or lunar capture to provide energy for early rapid melting and eruption which produce a primitive crust.

3. Noncatastrophic models for the production of the early crust call upon homogeneous, or less likely inhomogeneous, heating and partial melting of the mantle soon after accretion. Because large amounts of heat are available at this time, it is likely that a worldwide magma "ocean" formed in the upper mantle and that partial crystallization gives rise to a widespread thin crust, which is perhaps subjected to viscous-drag plate tectonics.

4. Models for the composition of the primitive crust fall into one or a combination of four categories: sialic, andesitic, anorthositic, and basaltic. Existing data favor a basaltic composition, including perhaps ultramafic and anorthositic fractions.

5. Growth of the early crust occurs by a combination of magma additions, interthrusting of crustal segments, microcontinental collisions, and welding of sedimentary prisms to continental margins. Archean continents by 3.8 b.y. were equal to or greater in thickness than present-day continents. A linear model of continental growth in which at least 50 percent of the continental crust is produced by 2.5 b.y. is consistent with most isotopic and geologic data.

6. Heat production in the Archean was a factor of 3 to 5 higher than at present. Because geothermal gradients in Archean continents, as inferred from metamorphic mineral assemblages, are similar to present continental geotherms, this excess heat must have been lost from the oceanic crust, probably by increased sea-floor spreading rates and/or increased total length of Archean oceanic ridge systems.

7. The first sialic (tonalitic) crust may have formed in subduction-related arc systems by 3.8–4.0 b.y. and became aggregated into one or more supercontinents by microcontinent collisions by 2.7 b.y.

8. Differences in evolutionary patterns among the terrestrial planets can be accounted for by differences in heat productivities and cooling rates, which are in turn controlled chiefly by planetary mass and position in the solar system. Only the Earth cooled slowly enough to maintain a thin lithosphere (in oceanic areas) and hence to sustain buoyancy-driven plate tectonics after 2 b.y. Other planets cooled rapidly and developed thick lithospheres, which resist disruption by mantle convection.

9. Lead, Nd, and Sr isotopic data indicate that the Earth's mantle has been inhomogeneous for at least 3.8 b.y. There is no evidence that convection has been successful in mantle homogenization.

10. Secular compositional changes in sediments that appear to be chiefly related to changes in continental composition are changes in K_2O/Na_2O, K_2O/Al_2O_3, SiO_2/Al_2O_3, REE, Eu anomaly, La/Yb, $^{87}Sr/^{86}Sr$, Th, and U. Changes which appear to reflect chiefly increases in free oxygen in the atmosphere-ocean system are changes in FeO/Al_2O_3, MgO/CaO, Mn/total Fe, Fe_2O_3/FeO, MgO/Al_2O_3, and Th/U. Increases in K, Th, and U and decreases in Ni and Cr between average Archean and Proterozoic continental crust reflect increasing vertical compositional zonation of the crust with decreasing age.

11. Times of rapid crustal evolution occur at 2500, 2000, 600–800, and 200 m.y. ago.

12. If the Earth had a primitive reducing atmosphere, it must have been lost in response to some catastrophic event in the first 200 m.y. after accretion. Rapid degassing of the Earth during the Archean produced an atmosphere rich in CO_2, N_2, and H_2O. Oxygen is liberated into the atmosphere beginning about 2 b.y. ago and grows rapidly after 600 m.y., derived chiefly from photosynthesizing organisms.

13. Existing data suggest that the composition of seawater has not changed significantly in the last 3.8 b.y. and that the volume of the oceans at 2.5 b.y. was about 90 percent of the present volume.

14. Climatic evolution of the Earth is characterized by generally warm climates in the Precambrian and late Mesozoic and five worldwide glaciations at ~ 2300, 600–800, 500, and 300 m.y. ago and during the Pleistocene.

15. Life appears to have formed by 3.8 b.y. ago in aqueous ponds on the Earth's surface by reactions of carbon-bearing compounds in an oxygen-free environment, perhaps activated by ultraviolet radiation. The earliest fossil remains are organic microstructures and stromatolites which date back to about 3.5 b.y. Most evidence indicates that O_2-producing photosynthetic bacteria and algae were present by 3.3 b.y. ago.

16. The first eucaryotic cells appear about 1.8 b.y. ago in response to increasing amounts of oxygen in the atmosphere. Metazoans evolve by 800 m.y. ago and rapidly expand, together with plants, in the Phanerozoic. Major Phanerozoic evolutionary events are the origin of vertebrates (450 m.y.), vascular plants (425 m.y.), mammals (225 m.y.), flowering plants (150 m.y.), and man (3 m.y.).

SUGGESTIONS FOR FURTHER READING

Condie, K.C., (1981) *Archean Greenstone Belts.* Amsterdam: Elsevier., 434 pp.

Frakes, L.A. (1979) *Climates Throughout Geologic Time.* Amsterdam: Elsevier., 310 pp.

Philips, R.J., and Ivins, E.R. (1979) Geophysical observations pertaining to solid-state convection in the terrestrial planets. *Phys. Earth Planet. Interiors*, **19**, 107–148.

Smith, J.V. (1979) Mineralogy of the planets: A voyage in space and time. *Mineral. Mag.*, **43**, 1–89.

Walker, J.C.G. (1976) *Evolution of the Atmosphere.* New York: Hafner. 318 pp.

Windley, B.F., editor (1976) *The Early History of the Earth.* New York: Wiley. 619 pp.

Windley, B.F. (1977) *The Evolving Continents.* New York: Wiley. 385 pp.

Acknowledgments

We are grateful to the listed authors and publishers for permission to reprint the following figures.

Figure

1.1   B. Isacks *et al.*, *Jour. Geophys. Res.*, Vol. 73, p. 5857, 1968, copyright by American Geophysical Union.

1.3   Seismic data from Anderson *et al.*, *Science*, Vol. 171, p. 1104, 1971, copyright 1971 by the American Association for the Advancement of Science.

2.1   S.R. Taylor, *Geochim. et Cosmochim. Acta*, Vol. 28, p. 1995, copyright 1964, Pergamon Press, Ltd.

2.3   From *Composition and Petrology of the Earth's Mantle* by A.E. Ringwood. Copyright © 1975 by McGraw Hill, Inc. Used with permission of McGraw Hill Book Company.

3.1   C.B. Archambeau *et al.*, *Jour. Geophys. Res.*, Vol. 74, p. 5829, 1969, copyright by American Geophysical Union.

3.2   C.B. Archambeau *et al.*, *Jour. Geophys. Res.*, Vol. 74, p. 5839, 1969, copyright by American Geophysical Union, and A.L. Hales, *Tectonophys.*, Vol. 13, p. 474, 1972.

3.3   C.B. Archambeau *et al.*, *Jour. Geophys. Res.*, Vol. 74, p. 5860, 1969, copyright by American Geophysical Union.

3.4   E. Herrin and J. Taggart, *Bull. Seismol. Soc. America*, Vol. 58, 1968.

3.5   J.H. Whitcomb and D.L. Anderson, *Jour. Geophys. Res.*, Vol. 75, p. 5722, 1970, copyright by American Geophysical Union.

3.6   L.R. Johnson, *Bull. Seismol. Soc. America*, Vol. 59, 1969.

3.8   T. Rikitake, *Phys. Earth and Planet. Interiors*, Vol. 7, p. 249, 1973.

3.9   T. Rikitake, *Tectonophys.*, Vol. 7, p. 262, 1969.

3.10   D.I. Gough, *Phys. Earth and Planet. Interiors*, Vol. 7, p. 383, 1973.

3.11   From *The Nature of the Solid Earth* edited by E.C. Robertson. Copyright 1972 by McGraw-Hill, Inc. Used with permission of McGraw-Hill Book Company.

3.12   Adapted from D.H. Green and A.E. Ringwood, *Jour. Geol.*, Vol. 80, p. 286, 1972, and K. Ito and G.C. Kennedy, *Am. Geophys. Union Mon.*, No. 14, p. 309, 1971, copyright by American Geophysical Union.

3.13   A.E. Ringwood, *Am. Geophys. Union Mon.*, No. 13, p. 8, 1969, copyright by American Geophysical Union.

3.14   D.W. Forsyth and F. Press, *Jour. Geophys. Res.*, Vol. 76, p. 7968, 1971, copyright by American Geophysical Union.

3.15   P.J. Wylie, *Jour. Geophys. Res.*, Vol. 76, p. 1331, 1971, copyright by American Geophysical Union.

4.2   D.E. Karig, *Jour. Geophys. Res.*, Vol. 76, p.

2543, 1971, copyright by American Geophysical Union.

4.5   Reproduced by permission of the National Research Council of Canada from the *Canadian Journal of Earth Sciences*, Vol. 9, pp. 1100 and 1108, 1972.

4.7   G. Palmason, *Soc. Sci. Isl.*, Vol. 40, 1971.

4.8   Adapted from H.W. Menard, *Jour. Geophys. Res.*, Vol. 72, p. 3063, 1967, copyright by American Geophysical Union.

4.9   (a) M. Landisman *et al.*, *Am. Geophys. Union Mon.* No. 14, p. 20, 1971, copyright by American Geophysical Union; (b) B. Mitchell and M. Landisman, *Geophysics*, Vol. 36, p. 378, 1971.

4.10   K.C. Condie, *Geol. Soc. America Bull.*, Vol. 84, 1973.

4.14   D.D. Blackwell, *Am. Geophys. Union Mon.*, No. 14, p. 1970, 1971, copyright by American Geophysical Union.

4.15   R.D. Hyndman *et al.*, *Phys. Earth and Planet. Interiors*, Vol. 1, p. 134, 1968, and D.D. Blackwell, *Am. Geophys. Union Mon.*, No. 14, p. 178, 1971, copyright by American Geophysical Union.

4.16   J.G. Sclater, *Tectonophys.*, Vol. 13, pp. 262 and 267, 1972.

4.19   J. Talwani *et al.*, *Jour. Geophys. Res.*, Vol. 70, p. 343, 1965, copyright by American Geophysical Union.

4.20   J. Talwani *et al.*, *Jour. Geophys. Res.*, Vol. 70, p. 345, 1965, copyright by American Geophysical Union.

4.21   L.W. Morley *et al.*, *Magnetic Anomaly Map of Canada*, Map. No. 1255A, 1967. Used with permission of the Geological Survey of Canada.

4.22   A.D. Raff and R.D. Mason, *Geol. Soc. America Bull.*, Vol. 72, 1961.

4.23   Adapted from R.D. Hyndman and D.W. Hyndman, *Earth and Planet. Sci. Letters*, Vol. 4, p. 430, 1968.

5.1.   G.W. Wetherill and M.E. Bickford, *Jour. Geophys. Res.*, Vol. 70, pp. 3672 and 4673, 1965, copyright by American Geophysical Union.

5.2   T.W. Stern *et al.*, *Geol. Soc. America Bull.*, Vol. 82, 1971.

5.3   C.H. Stockwell, *Amer. Assoc. Petrol. Geol. Bull.*, Vol. 49, pp. 890–891, 1965.

5.7   F.A. Hills *et al.*, *Geol. Soc. America Bull.*, Vol. 79, 1968.

5.8   A.G. Fischer *et al.*, *Science*, Vol. 168, p. 1211, 1970, Copyright 1969 by the American Association for the Advancement of Science and R.G. Douglas and M. Moullade, *Geol. Soc. America Bull.*, Vol. 83, p. 1163, 1972.

5.9   C.R. Anhaeusser *et al.*, *Geol. Soc. America Bull.*, Vol. 80, 1969.

5.10   L.D. Ayres, unpublished Ph.D. thesis, Princeton Univ., 1969, and A.M. Goodwin, *Geol. Soc. Australia Spec.* Publ. No. 3, p. 164, 1971.

5.11   Adapted from K.G. Cox et al., *Phil. Trans. Roy. Soc. Lond A*, Vol. 275, pp. 71–218, 1965.

5.12   A.J. Baer, *Phil. Trans. Roy. Soc. Lond. A.*, Vol. 280, pp. 499–515, 1976.

6.1   A. Cox, *Science*, Vol. 163, p. 240, 1969, copyright 1969 by the American Association for the Advancement of Science.

6.3   F.J. Vine, *Science*, Vol. 154, p. 1409, 1966. copyright 1966 by the American Association for the Advancement of Science.

6.4   J.R. Heirtzler *et al.*, *Jour. Geophys. Res.*, Vol. 73, p. 2120, 1968, copyright by American Geophysical Union.

6.5   From "The Deep-Ocean Floor," by H.W. Menard. Copyright © 1969 by Scientific American, Inc. All rights reserved.

6.8   B. Isacks *et al.*, *Jour. Geophys. Res.*, Vol. 73, p. 5874, 1968, copyright by American Geophysical Union.

6.10   J. Oliver *et al.*, *Tectonophys.*, Vol. 19, p. 143, 1973.

6.11   B.P. Luyendyk, *Geol. Soc. America Bull.*, Vol. 81, 1970.

6.12   M. Barazangi and B. Isacks, *Jour. Geophys. Res.*, Vol. 76, p. 8511, 1971, copyright by American Geophysical Union, and I.S. Sacks and H.

Okada, *Carnegie Inst. Wash. Yearbook*, 1973, p. 232.

6.13 W.J. Morgan, *Jour. Geophys. Res.*, Vol. 73, p. 1962, 1968, copyright by American Geophysical Union.

6.14 J.F. Dewey *et al.*, *Geol. Soc. America Bull.*, Vol. 84, 1973.

6.15 T. Johnson and P. Molnar, *Jour. Geophys. Res.*, Vol. 77, p. 5001, 1972, copyright by American Geophysical Union.

6.16 R.L. Larson and W.C. Pitman, *Geol. Soc. America Bull.*, Vol. 83, 1972.

6.18 J.T. Wilson, *Science*, Vol. 150, p. 482, 1965, copyright 1965 by the American Association for the Advancement of Science.

6.19 D.P. McKenzie and W.J. Morgan, *Nature*, Vol. 224, 1969.

6.20 D.P. McKenzie and W.J. Morgan, *Nature*, Vol. 224, 1969.

6.21 Reproduced, with permission, from "Mantle Convection and Global Tectonics," *Annual Review of Fluid Mechanics*, Volume 4, page 59. Copyright © 1972 by Annual Reviews Inc. All rights reserved.

6.22 N.M. Toksoz *et al.*, *Jour. Geophys. Res.*, Vol. 76, p. 1124, 1971, copyright by American Geophysical Union.

6.23 Adapted from N. Sleep and M. Toksoz, *Nature*, Vol. 233, 1971.

6.24 M.C. Zoback and M. Zoback, *Jour. Geophys. Res.*, Vol. 85, pp. 6113–6156, 1980.

6.26 From "The Origin of the Oceanic Ridges," by E. Orowan. Copyright © 1969 by Scientific American, Inc. All rights reserved.

6.27 M.H.P. Bott, *The Interior of the Earth*, 1971, Edward Arnold Publ., Ltd.

6.28 W.J. Morgan, *Geol. Soc. America Memoir*, No. 132, 1972.

7.1 Adapted from A.E. Ringwood, *Am. Geophys. Union Mon.* No. 13, p. 8, 1969, copyright by American Geophysical Union.

7.6 K.C. Condie, *Geol. Soc. America Bull.*, Vol. 84, 1973.

7.9 R.T. Martin and A.J. Piwinskii, *Jour. Geophys. Res.*, Vol. 77, p. 4968, 1972, copyright by American Geophysical Union.

7.10 D.H. Green, *Tectonophys.*, Vol. 13, p. 56, 1972.

8.1 E.C. Bullard *et al.*, *Phil. Trans. Roy. Soc. London*, Vol. 258A, 1965. Used with permission of the Royal Society of London.

8.2 A.G. Smith and A. Hallam, *Nature*, Vol. 225, 1970.

8.3 From the "Confirmation of Continental Drift," by P.M. Hurley. Copyright © by Scientific American, Inc. All rights reserved.

8.4 J.W. Valentine and E.M. Moores, *Nature*, Vol. 228, 1970.

8.9 K.M. Creer, *Phil. Trans. Roy. Soc. Lond.*, Vol. 258A, 1965. Used with permission of the Royal Society of London.

8.11 A.G. Smith and A. Hallam, *Nature*, Vol. 225, 1970, and D.H. Tarling, *Nature*, Vol. 229, 1971.

8.12 Adapted from R.G. Gordon *et al.*, *Jour. Geophys. Res.*, Vol. 84, p. 5484, 1979. Copyright by the American Geophysical Union.

8.13 W.C. Pitman and M. Talwani, *Geol. Soc. America Bull.*, Vol. 83, 1972.

8.14 R.G. Coleman, *Jour. Geophys. Res.*, Vol. 76, p. 1214, 1971, copyright by American Geophysical Union.

8.15 J.F. Dewey and J.M. Bird, *Jour. Geophys. Res.*, Vol. 76, p. 3193, 1971, copyright by American Geophysical Union.

8.19 S.T. Crough, *Geology*, Vol. 9, p. 2, 1981.

8.21 J.F. Dewey and K. Burke, *Geology*, Vol. 2, p. 58, 1974.

8.22 & 8.23 D.W. Scholl et al., *Geology*, Vol. 8, p. 565, 1980.

8.24 W.G. Ernst, *Contr. Mineral. and Petrol.*, Vol. 34, pp. 45, 49, and 51, 1971.

8.26   Reprinted from *Journal of Geology*, Vol. 81, pp. 406–433, 1973, K. Burke and J.F. Dewey, by permission of the University of Chicago Press.

8.27   J. Dewey and J.M. Bird, *Tectonophys.*, Vol. 10., p. 627, 1970.

8.28   Adapted from AMC Sengor, *Geology*, Vol. 4, p. 780, 1976.

8.31   J.F. Dewey and J.M. Bird, *Jour. Geophys. Res.*, Vol. 75, p. 2627, 1970, copyright by American Geophysical Union.

8.32   J.F. Dewey and B. Horsfield, *Nature*, Vol. 225, 1970.

8.33   E. Irving *et al.*, *Jour. Geophys. Res.*, Vol. 79, p. 5497, 1974, copyright by American Geophysical Union.

8.34   J.F. Dewey and J.M. Bird, *Jour. Geophys. Res.*, Vol. 75, p. 2642, 1970, copyright by American Geophysical Union.

8.35   Reprinted from *Journal of Geology*, Vol. 82, 1973, J.F. Dewey and K. Burke, by permission of the University of Chicago Press.

8.36   E. R. Oxburgh, "Flake Tectonics and continental collision," *Nature*, Vol. 239, pp. 202–204, 1972.

8.37   J.F. Dewey and J.M. Bird, *Jour. Geophys. Res.*, Vol. 75, p. 2641, 1970, copyright by American Geophysical Union.

8.38   Adapted from A.H.G. Mitchell and M.S. Garson, *Mineral Science and Engineering*, Vol. 8, No. 2, April 1976. Copyright by the National Institute for Metallurgy.

8.39   P.A. Rona, *EOS*, Vol. 58, p. 632, 1977, copyright by American Geophysical Union.

9.2–9.7   Reprinted from *Journal of Geology*, Vol. 87, No. 3, 1979, Scotese *et al.*, by permission of the University of Chicago Press.

9.8–9.11   Scotese *et al.*, 1981, written communication.

9.12   Modified after J.M. Bird and J.F. Dewey, *Geol. Soc. America Bull.*, Vol. 81, 1970.

9.15   Adapted from S.P. Coleman-Sadd, *American Journal of Science*, Vol. 280, 1980.

9.17   B.C. Burchfiel and G.A. Davis, *Am. Jour. Science*, Vol. 272, pp. 100–101, 1972, and J.H. Stewart, *Geol. Soc. America Bull.*, Vol. 83, 1972.

9.18   W.G. Ernst, *Jour. Geophys. Res.*, Vol. 75, p. 894, 1970, copyright by American Geophysical Union.

9.19   R.L. Larson and W.C. Pitman, *Geol. Soc. America Bull.*, Vol. 83, 1972.

9.20   R.L. Christiansen and P.W. Lipman, *Phil. Trans. Roy. Soc. Lond*, Vol. 271A, 1972. Used with permission of the Royal Society of London.

9.21   D.H. Scholz *et al.*, *Geol. Soc. America Bull.*, Vol. 82, 1971.

9.23   Reprinted by permission from *Nature*, Vol. 288, p. 330. Copyright © 1980, MacMillan Journals Ltd.

9.24   G.W. Moore and L.D. Castillo, *Geol. Soc. America. Bull.*, Vol. 85, 1974 and B.T. Malfait and M.G. Dinkelman, *Geol. Soc. America Bull.*, 83, 1972.

9.25   Adapted from D.E. James, *Geol. Soc. America Bull.*, Vol. 82, p. 3341, 1971.

9.26   J.F. Dewey *et al.*, *Geol. Soc. America Bull.*, Vol. 84, 1973.

9.27   J.F. Dewey *et al.*, *Geol. Soc. America Bull.*, Vol. 84, 1973.

9.28   Adapted from J.F. Dewey et al., *Geol. Soc. America Bull.*, Vol. 84, p. 3166, 1973.

9.29   J.F. Dewey *et al.*, *Geol. Soc. America Bull.*, Vol. 84, 1973.

10.2   Modified after C.R. Anhaeusser, *Geol. Soc. Australia Spec. Pub.*, No. 3, p. 61, 1971.

10.5   A.Y. Glikson, *Geology*, Vol. 4, p. 202, 1976.

10.7   J. Tarney *et al.*, in *The Early History of the Earth*, B.F. Windley, ed. Copyright 1976, John Wiley and Sons. Reprinted by permission.

10.12   Reprinted by permission from *Nature*, Vol. 279, p. 490. Copyright © 1979, MacMillan Journals Ltd.

10.16   Adapted from Dimroth *et al.*, *Geol. Surv. of Canada*, Paper 70-40, p. 48, 1970.

10.19 & H.R. Wynne Edwards, *American Journal of*
10.20 *Science*, Vol. 276, 1976.

11.1 J.W. Salisbury and L.B. Ronca, *Nature*, Vol. 210, 1966.

11.4 Modified after J. Veizer in *The Early History of the Earth*, B. F. Windley, ed. Copyright 1976, John Wiley and Sons. Reprinted by permission.

11.5 D.P. McKenzie and M. Weiss, *Geophys. J. Roy. Astro. Soc.*, Vol. 42, 1975.

11.9 From "The surface of Venus" by G.H. Pettengill *et al*. Copyright © 1980 by Scientific American, Inc. All rights reserved.

References

Abelson, P.H. (1966) Chemical events on the primitive earth. *Proc. Nat. Acad. Sci.*, **55**, 1365–1372.

Ádám, A., editor (1976) Geoelectric and geothermal studies (East Central Europe and Soviet Asia). *KAPG Geophys. Mon.* Budapest: Adakémiai Kiadó.

Ahmad, F. (1973) Have there been major changes in the earth's axis of rotation through time? in D.H. Tarling and S.K. Runcorn, editors. *Implications of Continental Drift to the Earth Sciences*. Vol. 1. London: Academic Press, pp. 487–501.

Ahrens, T.J., and Peterson, C.F. (1969) Shock wave data and the study of the earth. In S.K. Runcorn, editor, *The Application of Modern Physics to Earth and Planetary Interiors*. New York: Wiley-Interscience, pp. 449–461.

Alvarez, L.W., Alvarez, W., Asaro, F., and Michel, H.V. (1980) Extraterrestrial cause for the Cretaceous-Tertiary extinction. *Science*, **208**, 1095–1108.

Anderson, D.L. (1966) Earth's viscosity. *Science*, **151**, 321–322.

Anderson, D.L. Sammis, C., and Jordan, T. (1971) Composition and evolution of the mantle and core. *Science*, **171**, 1103–1112.

Anhaeusser, C.R. (1971) Cyclic volcanicity and sedimentation in the evolutionary development of Archean greenstone belts of shield areas. *Geol. Soc. Australia Spec. Publ.* No. 3, 57–70.

Anhaeusser, C.R., Mason, R., Viljoen, M.J., and Viljoen, R.P. (1969) A reappraisal of some aspects of Precambrian shield geology. *Geol. Soc. America Bull.*, **80**, 2175–2200.

Archambeau, C.B., Flinn, E.A., and Lambert, D.G. (1969) Fine structure of the upper mantle. *Jour. Geophys. Res.*, **74**, 5825–5865.

Armstrong, R.L. (1981) Radiogenic isotopes: The case for crustal recycling on a near steady-state no-continental-growth Earth. (In press.)

Armstrong, R.L., and Cooper, J.A. (1971) Lead isotopes in island arcs. *Bull. Volcanol.*, **35**, 27–63.

Atwater, T. (1970) Implications of plate tectonics for the Cenozoic tectonic evolution of western North America. *Geol. Soc. America Bull.*, **81**, 3513–3536.

Ave'Lallemant, H.G., and Carter, N.L. (1970) Syntectonic recrystallization of olivine and modes of flow in the upper mantle. *Geol. Soc. America Bull.*, **81**, 2203–2220.

Awramik, S.M., and Barghoorn, E.S. (1977) The Gunflint microbiota. *Precamb. Res.*, **5**, 121–142.

Ayres, L.D. (1969) Early Precambrian stratigraphy of part of Lake Superior Provincial Park, Ontario, Canada, and its implications for the origin of the Superior Province, unpubl. Ph.D. thesis, Princeton University. 399 pp.

Baer, A.J. (1976) The Grenville Province in Helikian times: A possible model of evolution. *Phil. Trans. Roy. Soc. Lond. A*, **280**, 499–515.

Baer, A.J. (1977) Speculations on the evolution of the lithosphere. *Precamb. Res.*, **5**, 249–260.

Baragar, W.R.A. (1977) Volcanism of the stable crust. *Geol. Assoc. Canada, Spec. Paper 16*, 377–406.

Barazangi, M., and Isacks, B. (1971) Lateral variations of seismic-wave attenuation in the upper mantle above the inclined earthquake zone for the Tonga Island Arc, deep anomaly in the upper mantle. *Jour. Geophys. Res.*, **76**, 8493–8515.

Barker, P.F. (1972) A spreading center in the East Scotia Sea. *Earth and Planet. Sci. Letters*, **15**, 123–132.

Barr, M.W.C. (1976) Crustal shortening in the Zambezi belt. *Phil. Trans. Roy. Soc. Lond. A.*, **280**, 555–567.

Bass, M.N., Moberly, R., Rhodes, J.M., Shih, C., and Church, S.E. (1973) Volcanic rocks cored in the central Pacific, Leg 17,

Deepsea Drilling Project. *Trans. Am. Geophys. Union*, **54**, 991–995.

Bell, K., and Powell, J.L. (1970) Strontium isotopic studies of alkali rocks: The alkalic complexes of eastern Uganda. *Geol. Soc. America Bull.*, **81**, 3481–3490.

Berkner, L.V., and Marshall, L.C. (1965) On the origin and rise of oxygen concentration in the earth's atmosphere. *Jour. Atmos. Sci.*, **22**, 225–261.

Bertrand, J.M.L., and Caby, R. (1978) Geodynamic evolution of the Pan-African orogenic belt: A new interpretation of the Hoggar shield. *Geol. Rund.*, **67**, 357–388.

Birch, F. (1960) The velocity of compressional waves in rocks to 10 kilobars. *Jour. Geophys. Res.*, **65**, 1083–1102.

Bird, J.M., and Dewey, J.F. (1970) Lithosphere plate continental margin tectonics and the evolution of the Appalachian orogen. *Geol. Soc. America Bull.*, **81**, 1031–1060.

Blackwell, D.D. (1971) The thermal structure of the continental crust. *Am. Geophys. Union Mon.* No. 14, 169–191.

Boettcher, A.L. (1973) Volcanism and orogenic belts—the origin of andesites. *Tectonophys.*, **17**, 223–240.

Bonatti, E. (1967) Mechanisms of deep-sea volcanism in the South Pacific. In P.H. Abelson, editor, *Researches in Geochemistry*. New York: Wiley, pp. 453–491.

Bott, M.H.P. (1971) *The Interior of the Earth*. London: Edward Arnold. 316 pp.

Bott, M.H.P. (1979) Subsidence mechanisms at passive continental margins. *Am. Assoc. Petrol. Geol., Mem. 29*, 3–9.

Bottinga, Y., and Allegre, C.J. (1973) Thermal aspects of sea-floor spreading and the nature of the oceanic crust. *Tectonophys.*, **18**, 1–17.

Boyd, F.R. (1973) A pyroxene geotherm. *Geochim. Cosmochim. Acta*, **37**, 2533–2546.

Brett, R. (1976) The current status of speculations on the composition of the core of the Earth. *Rev. Geophys. Space Phys.*, **14**, 375–383.

Brewer, J.A., and Oliver, J.E. (1980) Seismic reflection studies of deep crustal structure. *Ann. Rev. Earth Planet. Sci.*, **8**, 205–230.

Briden, J.C. (1976) Application of paleomagnetism to Proterozoic tectonics. *Phil. Trans. Roy. Soc. Lond. A.*, **280**, 405–416.

Bridgewater, D., Escher, A., and Watterson, J. (1973) Tectonic displacements and thermal activity in two contrasting Proterozoic mobile belts from Greenland. *Phil. Trans. Roy. Soc. Lond. A.*, **273**, 513–533.

Brooks, C., James, D.E., and Hart, S.R. (1976) Ancient lithosphere: Its role in young continental volcanism. *Science*, **193**, 1086–1094.

Brown, L.D., Krumhansel, P.K., Chapin, C.E., Sanford, A.R., Cook, F.A., Kaufman, S., Oliver, J.E., and Schilt, F.S. (1979) COCORP seismic reflection studies in the Rio Grande rift. In R.E. Riecker, editor, *Rio Grande Rift: Tectonics and Magmatism*. Washington, D.C.: Am. Geophys. Union, pp. 169–184.

Brune, J.N. (1969) Surface waves and crustal structure. *Am. Geophys. Union Mon.* No. 13, 230–242.

Buffler, R.T., Schaub, F.J., Watkins, J.S., and Worzel, J.L. (1979) Anatomy of the Mexican Ridges, southwestern Gulf of Mexico. *Am. Assoc. Petrol. Geol., Mem. 29*, 319–327.

Bullard, E.C., Everett, J.E., and Smith, A.G. (1965) The fit of the continents around the Atlantic. *Phil. Trans. Roy. Soc. Lond. A.*, **258**, 41–51.

Bultitude, R.J., and Green, D.H. (1967) Experimental study at high pressures on the origin of olivine nephelinite and olivine melilite nephelinite magmas. *Earth and Planet. Sci. Letters*, **3**, 325–337.

Burchfiel, B.C., and Davis, G.A. (1972) Structural framework and evolution of the southern part of the Cordilleran orogen, western United States. *Amer. Jour. Sci.*, **272**, 97–118.

Burke, K. and Dewey, J.F. (1973) Plume-generated triple junctions: Key indicators in applying plate tectonics to old rocks. *Jour. Geol.*, **81**, 406–433.

Burke, K., Dewey, J.F., and Kidd, W.S.F. (1976) Precambrian paleomagnetic results compatible with contemporary operation of the Wilson Cycle. *Tectonophys.*, **33**, 287–299.

Burke, K., and Kidd, W.S.F. (1978) Were Archean continental geothermal gradients much steeper than those of today? *Nature*, **272**, 240–241.

Burke, K., Kidd, W.S.F., and Wilson, J.T. (1973) Relative and latitudinal motion of Atlantic hot spots. *Nature*, **245**, 133–137.

Burke, K., and Sengör, A.M.C. (1979) Review of plate tectonics. *Revs. Geophys. Space Phys.*, **17**, 1081–1090.

Burrett, C., and Richardson, R. (1980) Trilobite biogeography and Cambrian tectonic models. *Tectonophys.*, **63**, 155–192.

Burwash, R.A. (1969) Comparative Precambrian geochronology of the North American, European, and Siberian Shields. *Can. Jour. Earth Sci.*, **6**, 357–365.

Cahen, L., and Snelling, N.J. (1966) *Geochronology of Equatorial Africa*, Amsterdam: North-Holland. 162 pp.

Cameron, A.G.W. (1973) Accumulation processes in the primitive solar nebula. *Icarus*, **18**, 407–450.

Camfield, P.A., and Gough, D.I. (1977) A possible Proterozoic plate boundary in North America. *Can. Jour. Earth Sci.*, **14**, 1229–1238.

Cara, M. (1979) Lateral variations of S velocity in the upper mantle from higher Raleigh modes. *Geophys. Jour.*, **57**, 649–670.

Card, K.D., et al. (1972) The Southern Province. *Geol. Assoc. Canada, Spec. Paper 11*, 335–380.

Carey, S.W. (1958) The tectonic approach to continental drift. *Continental Drift Symposium*, Univ. Tasmania, pp. 177–355.

Carey, S.W. (1976) *The Expanding Earth*. Amsterdam: Elsevier. 488 pp.

Carr, M.J., Stoiber, R.E., and Drake, C.L. (1973) Discontinuities in the deep seismic zones under the Japanese arcs. *Geol. Soc. America Bull.*, **84**, 2917–2930.

Chandra, N.N., and Cumming, G.L. (1972) Seismic refraction studies in western Canada. *Can. Jour. Earth Sci.*, **9**, 1099–1109.

Chase, C.G., and Gilmer, T.H. (1973) Precambrian plate tectonics: the Midcontinent gravity high. *Earth Planet. Sci. Lett.*, **21**, 70–78.

Christensen, N.I. (1966) Elasticity of ultrabasic rocks. *Jour. Geophys. Res.*, **71**, 5921–5931.

Christensen, N.I. (1970) Composition and evolution of the oceanic crust. *Marine Geol.*, **8**, 139–154.

Christensen, N.I. (1972) The abundance of serpentinites in the oceanic crust. *Jour. Geol.*, **80**, 709–719.

Christensen, N.I., et al. (1973) Deep-Sea Drilling Project: Properties of igneous and metamorphic rocks of the oceanic crust. *Trans. Am. Geophys. Union*, **54**, 972–1035.

Christensen, N.I., and Salisbury, M.H. (1972) Sea floor spreading, progressive alteration of layer 2 basalts, and associated changes in seismic velocities. *Earth and Planet. Sci. Letters*, **15**, 367–375.

Christiansen, R.L., and Lipman, P.W. (1972) Cenozoic volcanism and plate-tectonic evolution of the western United States. II. Late Cenozoic. *Phil. Trans. Roy. Soc. Lond. A.*, **271**, 249–284.

Church, S.E. (1973) Limits of sediment involvement in the genesis of orogenic volcanic rocks. *Contr. Mineral. Petrol.*, **39**, 17–32.

Churkin, Jr., M., Carter, C., and Trexler, Jr., J.H. (1980) Collision-deformed Paleozoic continental margin of Alaska—foundation for microplate accretion. *Geol. Soc. Amer. Bull.*, **91**, 648–654.

Churkin, Jr., M., and Trexler, Jr., J.H. (1980) Circum-Arctic plate accretion—isolating part of a Pacific plate to form the nucleus of the Arctic basin. *Earth Planet. Sci. Lett.*, **48**, 356–362.

Clark, S.P., Jr., and Ringwood, A.E. (1964) Density distribution and constitution of the mantle. *Revs. Geophys.*, **2**, 35–88.

Clark, S.P., Turekian, K., and Grossman, L. (1972) Model for the early history of the Earth. In E.C. Robertson, editor, *The Nature of the Solid Earth*. New York: McGraw-Hill, pp. 3–18.

Claus, G., and Nagy, B. (1961) A microbiological examination of some carbonaceous chondrites. *Nature*, **192**, 594–596.

Cleary, J.R. and Anderssen, R.S. (1979) Seismology and the internal structure of the Earth. In M.W. McElhinny, editor, *The Earth: Its Origin, Structure, and Evolution*. London: Academic Press, pp. 137–170.

Clifford, T.N. (1970) The structural framework of Africa. In T.N. Clifford and I.G. Gass, editors, *African Magmatism and Tectonics*. Darien, Conn.: Hafner, pp. 1–26.

Cloud, P.E. (1968a) Atmospheric and hydrospheric evolution on the primitive earth. *Science*, **160**, 729–736.

Cloud, P.E. (1968b) Pre-metazoan evolution and the origins of the metazoa. In E.T. Drake, editor, *Evolution and Environment*. New Haven: Yale Univ. Press, pp. 1–72.

Coleman, R.G. (1971) Plate tectonic emplacement of upper mantle peridotites along continental edges. *Jour. Geophys. Res.*, **76**, 1212–1222.

Coleman, R.G. (1977) *Ophiolites*. Berlin: Springer. 229 pp.

Coleman-Sadd, S.P. (1980) Geology of south-central Newfoundland and evolution of the eastern margin of Iapetus. *Amer. Jour. Sci.*, **280**, 991–1017.

Condie, K.C. (1973) Archean magmatism and crustal thickening. *Geol. Soc. America Bull.*, **84**, 2981–2992.

Condie, K.C. (1978) Geochemistry of Proterozoic granitic plutons from New Mexico, U.S.A. *Chem. Geol.*, **21**, 131–149.

Condie, K.C. (1980) Origin and early development of the Earth's crust. *Precamb. Res.*, **11**, 183–197.

Condie, K.C. (1981a) *Archean Greenstone Belts*. Amsterdam: Elsevier. 434 pp.

Condie, K.C. (1981b) Early and middle Proterozoic supracrustal successions and their tectonic settings. *Am. J. Sci.*, (in press).

Condie, K.C., and Hayslip, D.L. (1975) Late bimodal volcanism at the Medicine Lake volcanic center, northern California, *Geochim. et Cosmochim. Acta*, **39**, 1165–1178.

Condie, K.C. and Hunter, D.R. (1976) Trace element geochemistry of Archean granitic rocks from the Barberton region, South Africa. *Earth and Planet. Sci. Lett.*, **29**, 389–400.

Condie, K.C., Macke, J.E., and Reimer, T.O. (1970) Petrology and geochemistry of early Precambrian graywackes from the Fig Tree Group, South Africa. *Geol. Soc. America Bull.*, **81**, 2759–2776.

Condie, K.C., and Swenson, D.H. (1973) Compositional variations in three Cascade volcanoes: Jefferson, Rainier, and Shasta. *Bull. Volcanol.*, **37**, 205–230.

Condie, K.C., Viljoen, M.J., and Kable, E.J.D. (1977) Effects of alteration on element distributions in Archean tholeiites from the Barberton greenstone belt, South Africa. *Contrib. Mineral. Petrol.*, **64**, 75–89.

Coney, P.J. (1970) The geotectonic cycle and the new global tectonics. *Geol. Soc. America Bull.*, **81**, 739–748.

Coney, P.J. (1973) Non-collision tectogenesis in western North America. In D.H. Tarling and S. K. Runcorn, editors, *Implications of Continental Drift to the Earth Sciences*, Vol. 2. London: Academic Press, pp. 713–730.

Coney, P.J. (1978) Mesozoic-Cenozoic Cordilleran plate tectonics. *Geol. Soc. American Mem. 152*, 33–50.

Coney, P.J., Jones, D.L., and Monger, J.W.H. (1980) Cordilleran suspect terranes. *Nature*, **288**, 329–333.

Cook, F.A., Albaugh, D.S., Brown, L.D., Kaufman, S., Oliver, J.E., and Hatcher, Jr., R.D. (1979) Thin-skinned tectonics in the crystalline southern Appalachians: COCORP seismic-reflection profiling of the Blue Ridge and Piedmont. *Geology*, **7**, 563–567.

Corliss, J.B., et al. (1979) Submarine thermal springs on the Galapagos rift. *Science*, **203**, 1073–1083.

Coward, M.P. (1976) Archean deformation patterns in southern Africa. *Phil. Trans. Roy. Soc. Lond. A*, **283**, 313–331.

Coward, M.P., James, P.R., and Wright, L. (1976) Northern margin of the Limpopo belt, southern Africa. *Geol. Soc. America Bull.*, **87**, 601–611.

Cox, A. (1969) Geomagnetic reversals. *Science*, **163**, 237–245.

Cox, A., and Cain, J.C. (1972) International conference on the core-mantle interface. *Trans. Am. Geophys. Union*, **53**, 591–597.

Cox, K.G., Johnson, R.L., Monkman, L.J., Stillman, C.J., Vail, J.R., and Wood, D.N. (1965) The geology of the Nuanetsi igneous province. *Phil. Trans. Roy. Soc. Lond. A*, **275**, 71–218.

Creer, K.M. (1965) Paleomagnetic investigations in Great Britain. V. The remnant magnetization of unstable Keuper Marls. *Phil. Trans. Roy. Soc. Long.*, **240**, 130–143.

Creer, K.M., editor (1980) Electromagnetic induction in the Earth and moon. *Geophys. Surveys*, **4**, Nos. 1 and 2, 1–185.

Cross, T.A., and Pilger, Jr., R.H. (1978) Constraints on absolute motion and plate interaction inferred from Cenozoic igneous activity in the western United States. *Am. Jour. Sci.*, **278**, 865–902.

Crough, S.T. (1981) Mesozoic hotspot epeirogeny in eastern North America. *Geology*, **9**, 2–6.

Crough, S.T., Morgan, W.J., and Hargraves, R.B. (1980) Kimberlites: Their relation to mantle hotspots. *Earth Planet. Sci. Lett.*, **50**, 260–274.

Dalrymple, G.B., and Lanphere, M.A. (1969) *Potassium-Argon Dating*. San Francisco: Freeman. 258 pp.

Darracott, B.W., Fairhead, J.D., and Girdler, R.W. (1972) Gravity and magnetic surveys in northern Tanzania and southern Kenya. *Tectonophys.*, **15**, 131–141.

Davidson, A., et al., (1972) The Churchill Province. *Geol. Assoc. Canada Spec. Paper.* No. 11, 382–433.

Davies, D. (1968) A comprehensive test ban. *Science Jour. Lond.*, Nov. 1968, 78–84.

Davies, G.F. (1977) Whole-mantle convection and plate tectonics. *Geophys. Jour.*, **49**, 459–486.

Davies, G.F. (1979) Thickness and thermal history of continental crust and root zones. *Earth Planet. Sci. Lett.*, **44**, 231–238.

Davies, G.F. (1980) Review of oceanic and global heat flow estimates. *Revs. Geophys. Space Phys.*, **18**, 718–722.

DeLong, S.E., and Fox, P.J. (1977) Geological consequences of ridge subduction. In M. Talwani and W.C. Pittman III, editors, *Island Arcs, Deep-sea Trenches, and Back-arc Basins*. Washington, D.C.: Am. Geophys. Union, Maurice Ewing Series 1, pp. 221–228.

Dewey, J.F., and Bird, J.M. (1970a) Mountain belts and the new global tectonics. *Jour. Geophys. Res.*, **75**, 2625–2647.

Dewey, J.F., and Bird, J.M. (1970b) Plate tectonics and geosynclines. *Tectonophys.*, **10**, 625–638.

Dewey, J.F., and Bird, J.M. (1971) Origin and emplacement of the ophiolite suite: Appalachian ophiolites in Newfoundland. *Jour. Geophys. Res.*, **76**, 3179–3206.

Dewey, J.F., and Burke, K. (1973) Tibetan, Variscan, and Precambrian reactivation: Products of continental collision. *Jour. Geol.*, **81**, 683–692.

Dewey, J.F., and Burke, K. (1974) Hotspots and continental break-up: Implications for collisional orogeny. *Geology*, **2**, 57–60.

Dewey, J.F., and Horsfield, B. (1970) Plate tectonics, orogeny and continental growth. *Nature*, **225**, 521–525.

Dewey, J.F., and Kidd, W.S.F. (1977) Geometry of plate accretion. *Geol. Soc. America Bull.*, **88**, 960–968.

Dewey, J.F., Pitman III, W.C., Ryan, W.B.F., and Bonnin, J. (1973) Plate tectonics and the evolution of the Alpine system. *Geol. Soc. America Bull.*, **84**, 3137–3180.

Dewey, J.F., and Sengör, A.M.C. (1979) Aegean and surrounding regions: Complex multiplate and continuum tectonics in a convergent zone. *Geol. Soc. America Bull.*, **90**, 84–92.

Dickinson, W.R. (1970) Relations of andesites, granites, and derivative sandstones to arctrench tectonics. *Rev. Geophys. Space Phys.*, **8**, 813–860.

Dickinson, W.R. (1971a) Plate tectonic models for orogeny at continental margins. *Nature*, **232**, 41–42.

Dickinson, W.R. (1971b) Plate tectonic models of geosynclines. *Earth and Planet. Sci. Letters*, **10**, 165–174.

Dickinson, W.R. (1971c) Plate tectonics in geologic history. *Science*, **174**, 107–113.

Dickinson, W.R. (1972) Evidence for plate tectonic regimes in the rock record. *Am. Jour. Sci.*, **272**, 551–576.

Dickinson, W.R. (1973) Widths of modern arc-trench gaps proportional to past duration of igneous activity in associated magmatic arcs. *Jour. Geophys. Res.*, **78**, 3376–3389.

Dickinson, W.R. (1974b) Subduction and oil migration. *Geology*, **2**, 421–424.

Dickinson, W.R., and Luth, W.C. (1971) A model for plate tectonic evolution of mantle layers. *Science*, **174**, 400–404.

Dickinson, W.R., and Suczek, C.A. (1979) Plate tectonics and sandstone compositions. *Am. Assoc. Petrol. Geol. Bull.*, **63**, 2164–2182.

Dietz, R.S. (1961) Continent and ocean basin evolution by spreading of the sea floor. *Nature*, **190**, 854–857.

Dietz, R.S. (1972) Geosynclines, mountains and continent-building. *Scient. American*, **226**, 30–38.

Dimroth, E. (1981) Labrador geosyncline: Type example of early Proterozoic cratonic reactivation. In A. Kröner, editor, *Precambrian Plate Tectonics*. Amsterdam: Elsevier. pp. 331–352.

Dimroth, E., Baragar, W.R.A., Bergeron, R., and Jackson, G.D. (1970) The filling of the Circum-Ungava geosyncline. *Geol. Surv. Canada* Paper 70:40, 45–142.

Douglas, R.G., and Moullade, M. (1972) Age of the basal sediments on the Shatsky Rise, western north Pacific Ocean. *Geol. Soc. America Bull.*, **83**, 1163–1168.

DuToit, A. (1937) *Our Wandering Continents*. London: Oliver and Boyd. 366 pp.

Eade, R.E., and Fahrig, W.G. (1971) Geochemical evolutionary trends of continental plates—a preliminary study of the Canadian Shield. *Geol. Surv. Canada Bull.*, **179**, 51.

Eggler, D.H. (1979) Experimental igneous petrology. *Revs. Geophys. Space Phys.*, **17**, 744–761.

Eggler, D.H., and Burnham, C.W. (1973) Crystallization and fractionation trends in the system andesite-H_2O-CO_2-O_2 at pressures to 10 kb. *Geol. Soc. America Bull.*, **84**, 2517–2532.

Egyed, L. (1956) Determination of changes in the dimensions of the earth from paleographic data. *Nature*, **178**, 534.

Eichelberger, J.C. (1975) Origin of andesite and dacite: Evidence of mixing at Glass Mountain in California and at other Circum-Pacific volcanoes. *Geol. Soc. America Bull.*, **86**, 1381–1391.

Eicher, D.L. (1968) *Geologic Time*. Englewood Cliffs, N.J.: Prentice-Hall. 150 pp.

Elsasser, W.M. (1963) Early history of the earth. In J. Geiss and E.D. Goldberg, editors, *Earth Science and Meteoritics*. Amsterdam: North-Holland; pp. 1–30.

Elsasser, W.M. (1971) Sea-floor spreading as thermal convection. *Jour. Geophys. Res.*, **76**, 1101–1112.

Elston, D.P., Gromme, C.S., and McKee, E.H. (1973) Precambrian polar wandering and behavior of earth's magnetic field from stratified rocks of the Grand Canyon Supergroup, Arizona. *Geol. Soc. Abst. with Prog.*, **5**, 611–612.

Emslie, R.F. (1978) Anorthosite massifs, rapakivi granites, and late Proterozoic rifting of North America. *Precamb. Res.*, **7**, 61–98.

Engel, A.E.J., Engel, C.G., and Havens, R.G. (1965) Chemical characteristics of oceanic basalts and the upper mantle. *Geol. Soc. America Bull.*, **76**, 719–734.

Engel, A.E.J., Itson, S.P., Engel, C.G., and Stickney, D.M. (1974) Crustal evolution and global tectonics: A petrogenetic view. *Geol. Soc. America Bull.*, **85**, 843–858.

England, P.C., Kennett, B.L.N., and Worthington, M.H. (1978) A comparison of the upper mantle structure beneath Eurasia and the North Atlantic and Arctic Oceans. *Geophys. Jour.*, **54**, 575–585.

Eriksson, K.A. (1977) Tidal deposits from Archean Moodies Group, Bar Mountain Land, South Africa. *Sediment. Geol.*, **18**, 257–281.

Ernst, W.G. (1970) Tectonic contact between the Franciscan Melange and the Great Valley Sequence-crustal expression of a late Mesozoic Benioff Zone. *Jour. Geophys. Res.*, **75**, 886–901.

Ernst, W.G. (1971) Metamorphic zonations of presumably subducted lithospheric plates from Japan, California and the Alps. *Contr. Mineral. and Petrol.*, **34**, 43–59.

Ernst, W.G. (1973) Interpretative synthesis of metamorphism in the Alps. *Geol. Soc. America Bull.*, **84**, 2053–2078.

Fairbridge, R.W. (1973) Glaciation and plate migration. In D.H. Tarling and S.K. Runcorn, editors, *Implications of Continental Drift to the Earth Sciences*, Vol. 1. London: Academic Press, pp. 503–515.

Fallow, W.C. (1977) Trends in trans-North Atlantic commonality among Phanerozoic invertebrates and plate tectonic events. *Geol. Soc. America Bull.*, **88**, 62–66.

Fanale, F.P. (1971) A case for catastrophic early degassing of the earth. *Chem. Geol.*, **8**, 79–105.

Faure, G. (1977) *Principles of Isotope Geology*. New York: Wiley. 464 pp.

Fisher, R.L., and Hess, H.H. (1963) Trenches. In M.N. Hill, editor, *The Sea*, Vol. 3. New York: Wiley-Interscience, pp. 411–436.

Fitton, J.G. (1971) The generation of magmas in island arcs. *Earth and Planet. Sci. Letters*, **11**, 63–67.

Forsyth, D.W., and Press, F. (1971) Geophysical tests of petrological models of the spreading lithosphere. *Jour. Geophys. Res.*, **76**, 7963–7979.

Fox, P.J., Schreiber, E., and Heezen, B.C. (1971) The geology of the Caribbean crust II: Tertiary sediments, granitic, and basic rocks from the Aves Ridge. *Tectonophys.*, **12**, 89–109.

Frakes, L.A. (1979) *Climates Throughout Geologic Time*. Amsterdam: Elsevier. 310 pp.

Frey, F.A. (1979) Trace element geochemistry: Applications to the igneous petrogenesis of terrestrial rocks. *Revs. Geophys. Space Phys.*, **17**, 803–823.

Frey. F.A., Haskin, M.A., Poetz, J., and Haskin, L.A. (1968) Rare earth abundances in some basic rocks. *Jour. Geophys. Res.*, **73**, 6085–6098.

Furumoto, A.S., Wiebenga, W.A., Webb, J.P., and Sutton, G.H. (1973) Crustal structure of the Hawaiian Archipelago, northern Melanesia, and the central Pacific Basin by seismic refraction methods. *Tectonophys.*, **20**, 153–164.

Fyfe, W.S. (1973) The granulite facies, partial melting and the Archean crust. *Phil. Trans. Roy. Soc. Lond.*, **273A**, 457–461.

Gainanov, A.G., et al. (1968) The crust and the upper mantle in the transition zone from the Pacific Ocean to the Asiatic continent. *Am. Geophys. Union Mon.* No. 12, 367–378.

Garland, G.D. (1979) *Introduction to Geophysics*, Second ed. Philadelphia: Saunders 494 pp.

Garrels, R.M., and MacKenzie, F.T. (1971) *Evolution of Sedimentary Rocks: A Geochemical Approach*. New York: Norton. 397 pp.

Gast, P.W. (1968) Trace element fractionation and the origin of tholeiitic and alkaline magma types. *Geochim. et Cosmochim. Acta*, **32**, 1057–1086.

Gibb, R.A., and Walcott, R.I. (1971) A Precambrian suture in the Canadian Shield. *Earth and Planet. Sci. Letters*, **10**, 417–422.

Gill, J.B. (1970) Geochemistry of Vita Levu, Fiji, and its evolution as an island arc. *Contr. Mineral and Petrol.*, **27**, 179–203.

Gilluly, J. (1969) Oceanic sediment volumes and continental drift. *Science*, **166**, 992–994.

Glikson, A.Y. (1970) Geosynclinal evolution and geochemical affinities of early Precambrian systems. *Tectonophys.*, **9**, 397–433.

Glikson, A.Y. (1976) Earliest Precambrian ultramafic-mafic volcanic rocks: Ancient oceanic crust or relict terrestrial maria? *Geology*, **4**, 201–206.

Glikson, A.Y. (1980) Precambrian sial-sima relations: Evidence for Earth expansion. *Tectonophys.*, **63**, 193–234.

Goldreich, P., and Ward, W.R. (1973) The formation of planetesimals. *Astrophys. Jour.*, **183**, 1051–1061.

Goldschmidt, V.M. (1954) *Geochemistry* (editor, A. Muir). London: Oxford Univ. Press. 730 pp.

Goodwin, A.M. (1968) Evolution of the Canadian Shield. *Geol. Assoc. Canada Proc.*, **19**, 1–14.

Goodwin, A.M. (1971) Metallogenic patterns and evolution of the Canadian Shield. *Geol. Soc. Australia Spec. Publ. 3*, 157–174.

Goodwin, A.M. (1974) Precambrian belts, plumes, and shield development. *Am. Jour. Sci.*, **274**, 987–1028.

Goodwin, A.M. et al. (1972) The Superior Province. *Geol. Assoc. Canada Spec. Paper 11*, 528–623.

Goodwin, A.M., and Ridler, R.H. (1970) The Abitibi orogenic belt. *Geol. Surv. Canada Paper 70–40*, 1–30.

Gordon, R. B., and Nelson, C.W. (1966) Anelastic properties of the earth. *Revs. Geophys.*, **70**, 457–474.

Gordon, R.G., McWilliams, M.O., and Cox, A. (1979) Pre-Tertiary velocities of the continents: A lower bound from paleomagnetic data. *Jour. Geophys. Res.*, **84**, 5480–5486.

Grandstaff, D.E. (1980) Origin of uraniferous conglomerates at Elliot Lake, Canada and Witwatersrand, South Africa: Implications for oxygen in the Precambrian atmosphere. *Precamb. Res.*, **13**, 1–26.

Green, D.H. (1972b) Magmatic activity as the major process in the chemical evolution of the earth's crust and mantle. *Tectonophys.*, **13**, 47–71.

Green, D.H. (1972a) Archean greenstone belts may include terrestrial equivalents of lunar maria. *Earth and Planet. Sci. Letters*, **15**, 263–270.

Green, D.H., Hibberson, W.O., and Jaques, A.L. (1979) Petrogenesis of mid-ocean ridge basalts. In M.W. McElhinny, editor, *The Earth: Its Origin, Structure and Evolution*. London: Academic Press, pp. 265–300.

Green, D.H., and Ringwood, A.E. (1967) The genesis of basaltic magmas. *Contr. Mineral. and Petrol.*, **15**, 103–190.

Green, D.H., and Ringwood, A.E. (1972) A comparison of recent experimental data on the gabbro-garnet granulite-eclogite transition. *Jour. Geol.*, **80**, 277–288.

Green, T.H. (1980) Island arc and continent-building magmatism —a review of petrogenetic models based on experimental petrology and geochemistry. *Tectonophys.*, **63**, 367–385.

Green, T.H., and Ringwood, A.E. (1968) Genesis of the calc-alkaline igneous rock suite. *Contr. Mineral and Petrol.*, **18**, 105–162.

Grossman, L. (1972) Condensation in the primitive solar nebula. *Geochim. Cosmochim. Acta*, **36**, 597–619.

Grushinsky, N.P. (1967) The earth's crust: Its gravity field and topography. *Jour. Geol. Soc. Aust.*, **14**, 31–37.

Hales, A.L. (1972) The travel times of P seismic waves and their relevance to the upper mantle velocity distributions. *Tectonophys.*, **13**, 447–482.

Hales, A.L., Muirhead, J., and Rynn, J.M.W. (1980) A compressional velocity distribution for the upper mantle. *Tectonophys.*, **63**, 309–348.

Hall, D.H. (1968) Regional magnetic anomalies, magnetic units, and crustal structure in the Kenora District of Ontario. *Can. Jour. Earth Sci.*, **5**, 1277–1294.

Hallam, A. (1973) Provinciality, diversity, and extinction of Mesozoic marine invertebrates in relation to plate movements. In D.H. Tarling and S.K. Runcorn, editors, *Implications of Continental Drift to the Earth Sciences*, Vol. 1. London: Academic Press, pp. 287–294.

Hallam, A. (1974) Changing patterns of provinciality and diversity of fossil animals in relation to plate tectonics. *Jour. Biogeography*, **1**, 213–225.

Hargraves, R.B. (1978) Punctuated evolution of tectonic style. *Nature*, **276**, 459–461.

Harris, P.G., Reay, A., and White, I.G. (1967) Chemical composition of the upper mantle. *Jour. Geophys. Res.*, **72**, 6359–6369.

Hart, S.R. et al. (1970) Ancient and modern volcanic rocks: A trace element model. *Earth and Planet. Sci. Letters*, **10**, 17–28.

Hart, S.R., Glassley, W.E., and Karig, D.E. (1972) Basalts and sea floor spreading behind the Mariana Island Arc. *Earth and Planet. Sci. Letters*, **15**, 12–18.

Hart, S.R., Schilling, J.G., and Powell, J.L. (1973) Basalts from Iceland and along the Reykjanes Ridge: Sr isotope geochemistry. *Nature Phys. Sci.*, **246**, 104–107.

Hatcher, Jr., R.D. (1972) Developmental model for the southern Appalachians. *Geol. Soc. Amer. Bull.*, **83**, 2735–2760.

Hatcher, R.D., and Odom, A.L. (1980) Timing of thrusting in the southern Appalachians, U.S.A.: Model for orogeny. *Jour. Geol. Soc. London*, **137**, 321–327.

Hatherton, T., and Dickinson, W.R. (1969) The relationship between andesitic volcanism and seismicity in Indonesia, the Lesser Antilles, and other island arcs. *Jour. Geophys. Res.*, **74**, 5301–5310.

Hayes, D.E., and Ringis, J. (1973) Seafloor spreading in Tasman Sea. *Nature*, **243**, 454–458.

Heirtzler, J.R., Dickson, G.O., Herron, E.M., Pitman, W.C., and LePichon, X. (1968) Marine magnetic anomalies, geomagnetic field reversals, and motions of the ocean floor and continents. *Jour. Geophys. Res.*, **73**, 2119–2135.

Helmke, P.A., et al. (1973) Major and trace elements in igneous rocks from Apollo 15. *Moon*, **8**, 129–148.

Helmke, P.A., and Haskin, L.A. (1973) Rare-earth elements, Co, Sc, and Hf in the Steens Mountain Basalts. *Geochim. Cosmochim. Acta*, **37**, 1513–1529.

Henderson, J.B. (1972) Sedimentology of Archean turbidites at Yellow Knife, Northwest Territories. *Can. J. Earth Sci.*, **9**, 882–902.

Herman, Y. (1981) Causes of massive biotic extinctions and explosive evolutionary diversification throughout Phanerozoic time. *Geology*, **9**, 104–108.

Herrin, E., and Taggart, J. (1968) Regional variations in P travel times. *Bull. Seismol. Soc. America*, **58**, 1325–1337.

Herron, E.M. (1972) Sea-floor spreading and the Cenozoic history of the east-central Pacific. *Geol. Soc. America Bull.*, **83**, 1671–1692.

Hess, H.H. (1962) History of ocean basins. In A.E.J. Engel, H.L. James, and B.F. Leonard, editors, *Petrologic Studies: A Volume to Honor A.F. Buddington*. New York: Geol. Soc. America, pp. 599–620.

Hey, R. (1977) A new class of pseudofaults and their bearing on plate tectonics: A propagating rift model. *Earth Planet. Sci. Lett.*, **37**, 321–325.

Hey, R., Duennebier, F.K., and Morgan, W.J. (1980) Propagating rifts on mid-ocean ridges. *Jour. Geophys. Res.*, **85**, 3647–3658.

Hietanen, A. (1975) Generation of potassium-poor magmas in the northern Sierra Nevada and the Svecofennian of Finland. *J. Research U.S. Geol. Surv.*, **3**, 631–645.

Hilde, T.W.C., Uyeda, S., and Kronenke, L. (1977) Evolution of the western Pacific and its margin. *Tectonophys.*, **38**, 145–165.

Hills, F.A., Gast, P.W., Houston, R.S., and Swainbank, I.G. (1968) Precambrian geochronology of the Medicine Bow Mountains, southeastern Wyoming. *Geol. Soc. America Bull.*, **79**, 1757–1784.

Hills, F.A., and Houston, R.S. (1979) Early Proterozoic tectonics of the central Rocky Mountains, North America. *Contrib. Geol.*, **17**, 89–109.

Hoffman, P.F. (1980) Wopmay orogen: A Wilson-cycle of early Proterozoic age in the northwest of the Canadian Shield. *Geol. Assoc. Canada, Spec. Paper 20*, 523–549.

Hofmann, A.W., White, W.M., and Whitford, D.J. (1978) Geochemical constraints on mantle models: The case for a layered mantle. *Carnegie Inst. Washington Yearbook* **77**, 548–562.

Holland, H.D. (1962) Model for the evolution of the Earth's atmosphere. In A.E.J. Engel, H.L. James, and B.F. Leonard, editors, *Petrologic Studies*. Boulder, Colo.: Geol. Soc. America, pp. 447–477.

Holland, H.D. (1976) The evolution of sea water. In B.F. Windley, editor, *The Early History of the Earth*. New York: Wiley, pp. 559–567.

Holland, J.G., and Lambert, R.S. (1972) Major element chemical composition of shields and the continental crust. *Geochim. Cosmochim. Acta*, **36**, 673–683.

Holloway, J.R. (1971) Composition of fluid phase solutes in a basalt-H_2O-CO_2 system. *Geol. Soc. America Bull.*, **82**, 233–238.

Honkura, Y. (1978) Electrical conductivity anomalies in the Earth. *Geophys. Surveys*, **3**, 225–253.

Hunter, D.R. (1978) The Bushveld Complex and its remarkable rocks. *Am. Scient.*, **66**, 551–559.

Hurley, P.M. (1968a) Absolute abundance and distribution of Rb, K, and Sr in the earth. *Geochim. Cosmochim. Acta*, **32**, 273–283.

Hurley, P.M. (1968b) Correction to: Absolute abundance and distribution of Rb, K, and Sr in the earth. *Geochim. Cosmochim. Acta*, **32**, 1025–1030.

Hurley, P.M. (1968c) The confirmation of continental drift. *Scient. Amer.*, **218**, 52–64.

Hyndman, R.D., and Hyndman, D.W. (1968) Water saturation and high electrical conductivity in the lower continental crust. *Earth and Planet. Sci. Letters*, **4**, 427–432.

Hyndman, R.D., Lambert, I.B., Heier, K.S., Jaeger, J.C., and Ringwood, A.E. (1968) Heat flow and surface radioactivity measurement in the Precambrian shield of western Australia. *Phys. Earth and Planet. Interiors*, **1**, 129–135.

Irving, E. (1979a) Paleopoles and paleolatitudes of North America and speculations about displaced terrains. *Can. J. Earth Sci.*, **16**, 669–694.

Irving, E. (1979b) Pole positions and continental drift since the Devonian. In M.W. McElhinny, editor, *The Earth: Its Origin, Structure and Evolution*. London: Academic Press, pp. 567–594.

Irving, E., Emslie, R.F., and Ueno, H. (1974) Upper Proterozoic paleomagnetic poles from Laurentia and the history of the Grenville structural province. *Jour. Geophys. Res.*, **79**, 5491–5502.

Irving, E., and Puelaiah, G. (1976) Reversals of the geomagnetic field, magnetostratigraphy and relative magnitude of paleosecular variation in the Phanerozoic. *Earth. Sci. Revs.*, **12**, 35–64.

Isacks, B.L., and Barazangi, M. (1977) Geometry of Benioff zones: Lateral segmentation and downwards bending of the subducted lithosphere. In M. Talwani and W.C. Pittman III, editors, *Island Arcs, Deep-sea Trenches, and Back-arc Basins*. Washington, D.C.: Am. Geophys. Union, Maurice Ewing Series 1, pp. 99–114.

Isacks, B., Oliver, J., and Sykes, L.R. (1968) Seismology and the new global tectonics. *Jour. Geophys. Res.*, **73**, 5855–5899.

Ito, K., and Kennedy, G.C. (1971) An experimental study of the basalt-garnet granulite-eclogite transition. In J.G. Heacock, editor, *Am. Geophys. Union Mon.* No. 14, 303–314.

Jacobsen, S.B., and Wasserburg, G.J. (1979) The mean age of mantle and crustal reservoirs. *Jour. Geophys. Res.*, **84**, 7411–7427.

Jacoby, W., Björnsson, A., and Möller, D., editors (1980) Iceland, active tectonics and structure. *Jour. Geophys.*, **47**.

Jahn, B.M., and Nyquist, L.E. (1976) Crustal evolution in the earth-moon system: Constraints from Rb-Sr studies. In B.F. Windley, editor, *The Early History of the Earth*. New York: Wiley, pp. 55–76.

Jakes, P., and Gill, J. (1970) Rare earth elements and the island arc tholeiite series, *Earth and Planet. Sci. Letters*, **9**, 17–28.

Jakes, P., and White, A.J.R. (1972) Major and trace element abundances in volcanic rocks of orogenic areas. *Geol. Soc. America Bull.*, **83**, 29–40.

James, D.E. (1971) Plate tectonic model for the evolution of the central Andes, *Geol. Soc. America Bull.*, **82**, 3325–3346.

Jarrard, R.D., and Clague, D.A. (1977) Implications of Pacific island sea-mount ages for the origin of volcanic chains. *Revs. Geophys. Space Phys.*, **15**, 57–76.

Jessop, A.M., and Lewis, T. (1978) Heat flow and heat generation in the Superior Province of the Canadian Shield. *Tectonophys.*, **50**, 55–77.

Johnson, J.G. (1971) Timing and coordination of orogenic, epeirogenic, and eustatic events, *Geol. Soc. America Bull.*, **82**, 3265–3298.

Johnson, L.R. (1969) Array measurements of P velocities in the lower mantle. *Bull. Seismol. Soc. America*, **59**, 973–1008.

Johnson, T., and Molnar, P. (1972) Focal mechanisms and plate tectonics of the Southwest Pacific. *Jour. Geophys. Res.*, **77**, 5000–5032.

Jones, M.R., VanderVoo, R., and Bonhommet, N. (1979) Late Devonian to early Carboniferous paleomagnetic poles from the Amoricain Massif, France. *Geophys. Jour.*, **58**, 287–308.

Kanamori, H. (1977) Seismic and aseismic slip along subduction zones and their tectonic implications. In M. Talwani and W.C. Pittman III, editors, *Island Arcs, Deep-sea Trenches, and Back-arc Basins*. Washington, D.C.: Am. Geophys. Union, Maurice Ewing Series 1, pp. 163–174.

Karig, D.E. (1971) Origin and development of marginal basins in the western Pacific. *Jour. Geophys. Res.*, **76**, 2542–2561.

Karig, D.E., and Sharman III, G.F. (1975) Subduction and accretion in trenches. *Geol. Soc. America Bull.*, **86**, 377–389.

Katsumata, M., and Sykes, L.R. (1969) Seismicity and tectonics of the western Pacific: Izu-Mariana-Caroline and Ryuku-Taiwan regions. *Jour. Geophys. Res.*, **74**, 5923–5948.

Kaula, W.M. (1972) Global gravity and tectonics. In E.C. Robertson, editor, *The Nature of the Solid Earth*. New York: McGraw-Hill, pp. 385–405.

Kaula, W.M., and Harris, A.W. (1975) Dynamics of lunar origin and orbital evolution. *Revs. Geophys. Space Phys.*, **13**, 363–371.

Kay, R., Hubbard, N.J., and Gast, P.W. (1970) Chemical characteristics and origin of oceanic ridge volcanic rocks. *Jour. Geophys. Res.*, **75**, 1585–1613.

Kelleher, J., and McCann, W. (1977) Bathymetric highs and development of convergent plate boundaries. In M. Talwani and W.C. Pittman III, editors, *Island Arcs, Deep-sea Trenches, and Back-arc Basins*, Washington, D.C.: Am. Geophys. Union, Maurice Ewing Series 1, pp. 115–122.

Keller, G.V. (1971) Electrical studies of the crust and upper mantle. *Am. Geophys. Union Mon.* No. 14, 107–121.

Kluth, C.F., and Coney, P.J. (1981) Plate tectonics of the Ancestral Rocky Mountains. *Geology*, **9**, 10–15.

Knauth, L.P., and Lowe, D.R. (1978) Oxygen isotope geochemistry of cherts from the Onverwacht Group, South Africa. *Earth Planet. Sci. Lett.*, **41**, 209–222.

Knopoff, L. (1969) The upper mantle of the earth. *Science*, **163**, 1277–1287.

Knopoff, L. (1972) Observation and inversion of surface-wave dispersion. *Tectonophys.*, **13**, 497–519.

Krogh, T.E., and Davis, G.L. (1971) Zircon U-Pb ages of Archean volcanic rocks in the Canadian Shield. *Carnegie Inst. Wash. Yearbook* **70**, 241–242.

Kröner, A. (1977) The Precambrian geotectonic evolution of Africa: Plate accretion versus plate destruction. *Precamb. Res.*, **4**, 163–213.

Kröner, A. (1979) Pan-African plate tectonics and its repercussions on the crust of northeast Africa. *Geol. Rund.*, **68**, 565–583.

Kröner, A. (1981) Precambrian plate tectonics. In A. Kröner, editor, *Precambrian Plate Tectonics*. Amsterdam: Elsevier.

Kumazawa, M., Helmstaedt, H., and Masaki, K. (1971) Elastic properties of eclogite xenoliths from diatremes of the east Colorado Plateau and their implication to the upper mantle structure. *Jour. Geophys. Res.*, **76**, 1231–1247.

Kumazawa, M., Sawamoto, H., Ohtani, E., and Masaki, K. (1974) Postspinel phase of forsterite and evolution of the earth's mantle. *Nature*, **247**, 356–358.

Kushiro, I. (1972) Effect of water on the composition of magmas formed at high pressure. *Jour. Petrol.*, **13**, part 2, 311–334.

Kvenvolden, K.A. (1974) Natural evidence for chemical and early biologic evolution. *Origins of Life*, **5**, 71–86.

Lachenbruch, A.H. (1968) Preliminary geothermal model of the Sierra Nevada. *Jour. Geophys. Res.*, **72**, 6977–6989.

Lachenbruch, A.H. (1970) Crustal temperature and heat production: Implications of the linear heat-flow relation. *Jour. Geophys. Res.*, **75**, 3291–3300.

Lambeck, K. (1980) *The Earth's Rotation: Geophysical Causes and Consequences*. Cambridge: Cambridge University Press. 449 pp.

Lambert, R.S.J. (1980) The thermal history of the earth in the Archean. *Precamb. Res.*, **11**, 199–213.

Landisman, M., Mueller, S., and Mitchell, B.J. (1971) Review of evidence for velocity inversions in the continental crust. *Am. Geophys. Union Mon.* No. 14, 11–34.

Larimer, J.W. (1967) Chemical fractionations in meteorites I. Condensation of the elements. *Geochim. Cosmochim. Acta*, **31**, 1215–1238.

Larimer, J.W. (1971) Composition of the earth: Chrondritic or achondritic? *Geochim. Cosmochim. Acta*, **35**, 769–786.

Larson, R.L., and Pitman, W.C. (1972) Worldwide correlation of Mesozoic magnetic anomalies, and its implications. *Geol. Soc. America Bull.*, **83**, 3645–3662.

Lee, T. (1979) New isotopic clues to solar system formation. *Revs. Geophys. Space Phys.*, **17**, 1591–1608.

Leeds, A.R., Knopoff, L., and Kausel, E.G. (1974) Variations of upper mantle structure under the Pacific Ocean. *Science*, **186**, 141–143.

Le Pichon, X., Houtz, R.E., Drake, C.L., and Nafe, J.E. (1965) Crustal structure of the mid-ocean ridges, Part I. *Jour. Geophys. Res.*, **70**, 319–339.

Lewis, B.T.R. (1978) Evolution of ocean crust seismic velocities. *Ann. Rev. Earth Planet. Sci.*, **6**, 377–404.

Lilley, F.E.M. (1979) Geomagnetism and the Earth's core. In M.W. McElhinny, editor, *The Earth: Its Origin, Structure and Evolution*. London: Academic Press, pp. 83–112.

Lipman, P.W. (1969) Alkalic and tholeiitic basaltic volcanism related to the Rio Grande Depression, southern Colorado and northern New Mexico. *Geol. Soc. America Bull.*, **80**, 1343–1354.

Lipman, P.W., Prostka, H.J., and Christiansen, R.L. (1972) Cenozoic volcanism and plate-tectonic evolution of the western United States. I., Early and Middle Cenozoic. *Phil. Trans. Roy. Soc. Lond. A.*, **271**, 217–248.

Liu, L. (1976) The post-spinel phase of forsterite. *Nature*, **262**, 770–772.

Liu, L. (1979) Phase transformations and the constitution of the deep mantle. In M.W. McElhinny, editor, *The Earth: Its Origin, Structure, and Evolution*. London: Academic Press, pp. 177–202.

Long, G.H., Brown, L.D., and Kaufman, S. (1978) A deep seismic reflection survey across the San Andreas fault near Parkfield, California. *EOS*, **59**, 385.

Longhi, J. (1978) Pyroxene stability and the composition of the lunar magma ocean. *Proc. Lunar Planet. Sci. Conf. 9th*, 285–306.

Lowe, D.R. (1980) Archean sedimentation. *Ann. Rev. Earth Planet. Sci.*, **8**, 145–167.

Lowman, Jr., P.D. (1976) Crustal evolution in silicate planets: Implications for the origin of continents. *Jour. Geol.*, **84**, 1–26.

Lubimova, E.A. (1969) Thermal history of the earth. *Am. Geophys. Union Mon.* No. 13, 63–77.

MacDonald, G.A., and Katsura, T. (1964) Chemical composition of Hawaiian lavas. *Jour. Petrol.*, **5**, 82–133.

MacDonald, G.J.F. (1964) The deep structure of continents. *Science*, **143**, 921–929.

MacGregor, I.D. (1974) The system $MgO-Al_2O_3-SiO_2$: Solubility of Al_2O_3 in enstatite for spinel and garnet peridotite compositions. *Am. Mineral.*, **59**, 110–119.

MacLaren, A.S., and Charbonneau, B.W. (1968) Characteristics of magnetic data over major subdivisions of the Canadian Shield. *Geol. Assoc. Canada Proc.*, **19**, 57–65.

Malfait, B.T., and Dinkelman, M.G. (1972) Circum-Caribbean tectonic and igneous activity and the evolution of the Caribbean plate. *Geol. Soc. America Bull.*, **83**, 251–272.

Marko, G.M. (1980) Velocity and attenuation in partially molten rocks. *Jour. Geophys. Res.*, **85**, 5173–5189.

Martin, R.T., and Piwinskii, A.J. (1972) Magmatism and tectonic settings. *Jour. Geophys. Res.*, **77**, 4966–4975.

Mason, B. (1966) Composition of the earth. *Nature*, **211**, 616–618.

Mason, R. (1973) The Limpopo mobile belt-Southern Africa. *Phil. Trans. Roy. Soc. Lond. A*, **273**, 463–485.

Masuda, A. (1966) Lanthanides in basalts of Japan with three distinct types. *Geoch. Jour.*, **1**, 11–26.

McBirney, A.R. (1969a) Andesitic and rhyolitic volcanism of orogenic belts. *Am. Geophys. Union Mon.* No. 13, 501–507.

McBirney, A.R. (1969b) Compositional variations in Cenozoic calc-alkaline suites of Central America. *Oreg. Dept. Geol. and Min. Industries*, Bull. 65, 185–189.

McConnell Jr., R.K., Gupta, R.N., and Wilson, J.I. (1966) Compilation of deep crustal seismic refraction profiles. *Revs. Geophys.*, **4**, 41–55.

McDougall, I. (1979) The present status of the geomagnetic polarity time scale. In M.W. McElhinny, editor, *The Earth: Its Origin, Structure and Evolution*. London: Academic Press, pp. 543–566.

McDougall, I., and Duncan, R.A. (1980) Linear volcanic chains—recording plate motions? *Tectonophys.*, **63**, 275–295.

McElhinny, M.W. (1971) Geomagnetic reversals during the Phanerozoic. *Science*, **172**, 157–159.

McElhinny, M.W. (1973a) *Paleomagnetism and Plate Tectonics*. Cambridge: Cambridge Univ. Press, 358 pp.

McElhinny, M.W. (1973b) Mantle plumes, paleomagnetism, and polar wandering. *Nature*, **241**, 523–524.

McElhinny, M.W., and McWilliams, M.O. (1977) Precambrian geodynamics—a paleomagnetic view. *Tectonophys.*, **40**, 137–159.

McElhinny, M.W., Taylor, S.R., and Stevenson, O.J. (1978) Limits to the expansion of Earth, moon, Mars, and Mercury and to changes in the gravitational constant. *Nature*, **271**, 316–321.

McIver, J.R., and Lenthall, D.H. (1973) Mafic and ultramafic extrusives of the Onverwacht Group in terms of the system $XO-YO-R_2O_3-ZO_2$. *Econ. Geol. Research Unit, Univ. Witwatersrand, Johannesburg, S. Africa, Inf. Cir.* No. 80. 8 pp.

McKee, E.H. (1971) Tertiary igneous chronology of the Great Basin of western United States—implications for tectonic models. *Geol. Soc. America Bull.*, **82**, 3497–3502.

McKenzie, D.P., and Morgan, W.J. (1969) Evolution of triple junctions. *Nature*, **224**, 125–133.

McKenzie, D., Watts, A., Parsons, B., and Roufosse, M. (1980) Planform of mantle convection beneath the Pacific Ocean. *Nature*, **288**, 442–446.

McKenzie, D.P., and Weiss, N. (1975) Speculations on the thermal and tectonic history of the Earth. *Geophys. J. Roy. Astro. Soc.*, **42**, 131–174.

McLennan, S.M., and Taylor, S.R. (1980) Th and U in sedimentary rocks: Crustal evolution and sedimentary recycling. *Nature*, **285**, 621–624.

McWilliams, M.O. (1981) Paleomagnetism and Precambrian tectonic evolution of Gondwana. In A. Kröner, editor, *Precambrian Plate Tectonics*. Amsterdam: Elsevier, pp. 649–688.

Melville, R. (1973) Continental drift and plant distribution. In D.H. Tarling and S.K. Runcorn, editors, *Implications of Continental Drift to the Earth Sciences*, Vol. 1. London: Academic Press, pp. 439–446.

Menard, H.W. (1967) Transitional types of crust under small ocean basins. *Jour. Geophys. Res.*, **72**, 3061–3073.

Menard, H.W. (1969) The deep ocean floor. *Scient. Amer.*, **221**, 127–142.

Meyerhoff, A.A. (1970) Continental drift: Implications of paleomagnetic studies, meteorology, physical oceanography, and climatology. *Jour. Geol.*, **78**, 1–51.

Miller, S.L. (1953) A production of amino acids under possible primitive earth conditions. *Science*, **117**, 528–529.

Minster, J.B., and Jordan, T.H. (1978) Present-day plate motions. *Jour. Geophys. Res.*, **83**, 5331–5354.

Mitchell, A.H.G., and Garson, M.S. (1976) Mineralization at plate boundaries. *Minerals Sci. Engineering*, **8**, 129–169.

Mitchell, B., and Landisman, M. (1971) Electrical and seismic properties of the earth's crust in the southwestern Great Plains of the U.S.A. *Geophys.*, **36**, 363–381.

Miyashiro, A. (1972) Pressure and temperature conditions and tectonic significance of regional and ocean-floor metamorphism. *Tectonophys.*, **13**, 141–159.

Miyashiro, A. (1973) Paired and unpaired metamorphic belts. *Tectonophys.*, **17**, 241–254.

Miyashiro, A. (1975) Classification, characteristics, and origin of ophiolites, *Jour. Geol.*, **83**, 249–281.

Mohr, R.E. (1975) Measured periodicities of the Biwbik stromatolites and their geophysical significance. In G.D. Rosenberg and S.K. Runcom, editors, *Growth Rhythms and the History of the Earth's Rotation*. New York: Wiley, pp. 43–55.

Molnar, P., and Atwater, T. (1973) Relative motion of hot spots in the mantle. *Nature*, **246**, 288–291.

Molnar, P., and Atwater, T. (1978) Interarc spreading and Cordilleran tectonics as alternates related to the age of subducted oceanic lithosphere. *Earth Planet. Sci. Lett.*, **41**, 330–340.

Molnar, P., and Gray, D. (1979) Subduction of continental lithosphere: Some constraints and uncertainties. *Geology*, **7**, 58–62.

Molnar, P., and Tapponnier, P. (1975) Cenozoic tectonics of Asia: Effects of a continental collision. *Science*, **189**, 419–426.

Monger, J.W.H., Souther, J.G., and Gabrielse, H. (1972) Evolution of the Canadian Cordillera: A plate tectonic model. *Amer. Jour. Sci.*, **272**, 577–602.

Moorbath, S. (1977) Ages, isotopes, and evolution of Precambrian continental crust. *Chem. Geol.*, **20**, 151–187.

Moore, G.W., and Castillo, L.D. (1974) Tectonic evolution of the southern Gulf of Mexico. *Geol. Soc. America Bull.*, **85**, 607–618.

Moores, E.M., and Vine, F.J. (1971) The Troodes Massif, Cyprus and other ophiolites as oceanic crust: Evaluation and implications. *Phil. Trans. Roy. Sci. Lond. A*, **268**, 443–466.

Morgan, W.J. (1968) Rises, trenches, great faults, and crustal blocks. *Jour. Geophys. Res.*, **73**, 1959–1982.

Morgan, W.J. (1972a) Deep mantle convection plumes and plate motions. *Amer. Assoc. Petrol. Geol. Bull.*, **56**, 203–213.

Morgan, W.J. (1972b) Plate motions and deep mantle convection. *Geol. Soc. America Mem.* **132**, 7–22.

Morley, L.W., MacLaren, A.S., and Charbonneau, B.W. (1967) Magnetic Anomaly Map of Canada, Geological Survey of Canada, Map no. 1255A.

Murauchi, S. et al. (1968) Crustal structure of the Philippine Sea. *Jour. Geophys. Res.*, **73**, 3143–3171.

Murthy, V.R., and Hall, H.T. (1970) The chemical composition of the earth's core: Possibility of sulfur in the core. *Phys. Earth and Planet. Interiors*, **2**, 276–282.

Mutch, T., Arndson, R.E., Head, J.W., Jones, K.L., and Saunders, R.S. (1976) A summary of Martian geologic history. In T. Mutch et al., editors, *The Geology of Mars*. Princeton: Princeton Univ. Press, pp. 316–319.

Naqvi, S.M., Divakara, V., and Narain, H. (1978) The primitive crust: Evidence from the Indian shield. *Precamb. Res.*, **6**, 323–345.

Neumann, E.R., and Ramberg, I.B. (1977) Paleorifts—concluding remarks. In E.R. Neumann and I.B. Ramberg, editors, *Tectonics and Geophysics of Continental Rifts*. Dordrecht, Holland: D. Reidel, pp. 409–424.

Newell, N.D. (1967) Revolutions in the history of life. *Geol. Soc. America*, *Spec. Paper 89*, 62–91.

Nicholls, I.A., and Ringwood, A.E. (1972) Production of silica-saturated tholeiitic magmas in island arcs. *Earth and Planet. Sci. Letters*, **17**, 243–246.

Nielson, D.R., and Stoiber, R.E. (1973) Relationship of potassium content in andesitic lavas and depth to the seismic plane. *Jour. Geophys. Res.*, **78**, 6887–6892.

Nilsen, T.H., and Stewart, J.H. (1980) The Antler orogeny—mid-Paleozoic tectonism in western North America. *Geology*, **8**, 298–302.

Nockolds, S.R. (1954) Average chemical compositions of some igneous rocks, *Geol. Soc. America Bull.*, **65**, 1007–1032.

Norton, I.O., and Sclater, J.G. (1979) A model for the evolution of the Indian Ocean and the breakup of Gondwanaland. *Jour. Geophys. Res.*, **84**, 6803–6830.

O'Hara, M.J. (1967) Mineral parageneses in ultramafic rocks. In P.J. Wyllie, editor, *Ultramafic and Related Rocks*. New York: Wiley, pp. 393–403.

O'Nions, R.K., Carter, S.R., Evensen, N.M., and Hamilton, P.J. (1979) Geochemical and cosmochemical applications of Nd isotope analysis. *Ann. Rev. Earth Planet. Sci.*, **7**, 11–38.

Oliver, J., Isacks, B., Baranzangi, M., and Tronovas, W. (1973) Dynamics of the downgoing lithosphere. *Tectonophys.*, **19**, 133–147.

Oparin, A.I. (1953) *The Origin of Life*. New York: Dover. 157 pp.

Orowan, E. (1969) The origin of the oceanic ridges. *Scient. American*, **221**, 103–119.

Oversby, V.M., and Ewart, A. (1972) Lead isotopic compositions of Tonga-Kermadec volcanics and their petrogenetic significance. *Contr. Mineral. and Petrol.*, **37**, 181–210.

Owen, T., Cess, R.D., and Ramanathan, V. (1979) Enhanced CO_2 greenhouse to compensate for reduced solar luminosity on the Earth. *Nature*, **277**, 640–642.

Oxburgh, E.R. (1969) The deep structure of orogenic belts—the root problems. In *Time and Place in Orogeny*. London: Geol. Soc. London, pp. 251–273.

Oxburgh, E.R. (1972) Flake tectonics and continental collision. *Nature*, **239**, 202–204.

Palmason, G. (1971) Crustal structure of Iceland from explosion seismology. *Soc. Sci. Isl.*, **40**, 187 pp.

Pannella, G. (1972) Paleontologic evidence on the Earth's rotational history since early Precambrian. *Astrophys. Space Sci.*, **16**, 212–237.

Patton, H. (1980) Crust and upper mantle structure of the Eurasian continent from phase velocity and Q of surface waves. *Revs. Geophys. Space Phys.*, **18**, 605–625.

Pearce, J.A., and Cann, J.R. (1973) Tectonic setting of basic volcanic rocks determined using trace element analysis. *Earth Planet. Sci. Lett.*, **19**, 290–300.

Pearce, T.H., Gorman, B.E., and Birkett, T.C. (1977) The relationship between major element chemistry and tectonic environment of basic and intermediate volcanic rocks. *Earth Planet. Sci. Lett.*, **36**, 121–132.

Perfit, M.R., Gust, D.A., Beuce, A.E., Arculus, R.J., and Taylor, S.R. (1980) Chemical characteristics of island arc basalts: Implications for mantle sources. *Chem. Geol.*, **30**, 227–256.

Perry, E.C., Jr., Ahmad, S.N., and Swulius, T.M. (1978) The oxygen isotope composition of 3800-million-year-old metamorphosed chert and iron formation from Isukasia, West Greenland. *Jour. Geol.*, **86**, 223–239.

Peterman, Z.E., and Hedge, C.E. (1971) Related strontium isotopic and chemical variations in oceanic basalts. *Geol. Soc. America Bull.*, **82**, 493–500.

Peterson, J.J., Fox, P.J., and Schrieber, E. (1974) Newfoundland ophiolites and the geology of the oceanic layer. *Nature*, **247**, 194–196.

Pettengill, G.H., Campbell, D.B., and Masursky, H. (1980) The Surface of Venus. *Scient. Amer.*, **243**, 54–65.

Phillips, R.J., and Ivins, E.R. (1979) Geophysical observations pertaining to solid state convection in the terrestrial planets. *Phys. Earth Planet. Interiors*, **19**, 107–148.

Phillips, R.J., and Lambeck, K. (1980) Gravity fields of the terrestrial planets: Long-wavelength anomalies and tectonics. *Revs. Geophys. Space Phys.*, **18**, 27–76.

Philpotts, J.A., and Schnetzler, C.C. (1968) Europium anomalies and the genesis of basalt. *Chem. Geol.*, **3**, 5–13.

Philpotts, J.A., and Schnetzler, C.C. (1970) Phenocryst-matrix partition coefficients for K, Rb, Sr, and Ba with applications to anorthosite and basalt genesis. *Geochim. Cosmochim. Acta*, **34**, 307–322.

Piper, J.D.A. (1976) Paleomagnetic evidence for a Proterozoic supercontinent. *Phil. Trans. Roy. Soc. Lond. A.*, **280**, 469–490.

Pitman, W.C., and Talwani, M. (1972) Sea-floor spreading in the North Atlantic. *Geol. Soc. America Bull.*, **83**, 619–646.

Pittman III, W.C. (1978) Relationship between eustacy and stratigraphic sequences of passive margins. *Geol. Soc. America Bull.*, **89**, 1389–1403.

Plotnick, R.E. (1980) Relationship between biological extinctions and geomagnetic reversals. *Geology*, **8**, 578–581.

Plumstead, E.P. (1973) The enigmatic *Glossopteris* flora and uniformitarianism. In D.H. Tarling and S.K. Runcorn, editors, *Implications of Continental Drift to the Earth Sciences*, Vol. 1. London: Academic Press, pp. 413–424.

Poldervaart, A. (1955) Chemistry of the earth's crust. *Geol. Soc. America Special Paper 62*, 119–144.

Porada, H. (1979) The Damara-Ribeira orogen of the Pan-African-Braziliano cycle in Namibia. *Tectonophys.*, **57**, 237–265.

Poster, C.K. (1973) Ultrasonic velocities in rocks from the Troodos Massif, Cyprus. *Nature Phys. Sci.*, **243**, 2–3.

Powell, C.M., and Conaghan, P.J. (1973) Plate tectonics and the Himalayas. *Earth Planet. Sci. Lett.*, **20**, 1–12.

Powell, C.M., Johnson, B.D., and Veevers, J.J. (1980) A revisited fit of east and west Gondwanaland. *Tectonophys.*, **63**, 13–29.

Presnall, D.C., and Bateman, P. (1973) Fusion relations in the system $Ab-An-Or-Q-H_2O$ and generation of granitic magmas in the Sierra Nevada Batholith. *Geol. Soc. America Bull.*, **84**, 3181–3202.

Press, F. (1972) The earth's interior as inferred from a family of models. In E.C. Robertson, editor, *The Nature of the Solid Earth*, New York: McGraw-Hill, pp. 147–171.

Pretorius, D.A. (1974) Gold in the Proterozoic sediments of South Africa: Systems, paradigms, and models. *Econ. Geol. Res. Unit, Univ. Witwatersrand, S. Africa, Inf. Circular 87*.

Raff, A.D., and Mason, R.G. (1961) Magnetic survey off the west coast of North America, 40°N to 52 1/2°N. *Geol. Soc. America Bull.*, **72**, 1259–1265.

Raitt, R.W., Shor, G.G., Francis, T.J.F., and Morris, G.B. (1969) Anisotropy of the Pacific upper mantle. *Jour. Geophys. Res.*, **74**, 3095–3109.

Ramberg, H. (1964) A model for the evolution of continents, oceans, and orogens. *Tectonophys.*, **2**, 159–174.

Raymond, L.A., and Swanson, S.E. (1980) Accretion and episodic plutonism. *Nature*, **285**, 317–319.

Richardson, R.M., Solomon, S.C., and Sleep, N.H. (1979) Tectonic stress in the plates. *Revs. Geophys. Space Phys.*, **17**, 981–1002.

Rikitake, T. (1969) The undulation of an electrically conductive layer beneath the islands of Japan. *Tectonophys.*, **7**, 257–264.

Rikitake, T. (1973) Global electrical conductivity of the earth. *Phys. Earth and Planet. Interiors*, **7**, 245–250.

Rikitake, T., Miyamura, S., Tsubakawa, I., Murauchi, S., Uyeda, S., Kuno, H., and Gorar, M. (1968) Geophysical and geological data in and around the Japan Arc. *Can. Jour. Earth Sci.*, **5**, 1101–1118.

Ringwood, A.E. (1966) The chemical composition and origin of the earth. In P.M. Hurley, editor, *Advances in Earth Sciences*. Cambridge: Massachusetts Institute of Technology Press. pp. 287–326.

Ringwood, A.E. (1969a) Composition and evolution of the upper mantle. *Am. Geophys. Union Mon.* No. 13, 1–17.

Ringwood, A.E. (1969b) Phase transformations in the mantle. *Earth and Planet. Sci. Letters*, **5**, 401–412.

Ringwood, A.E. (1972a) Mineralogy of the deep mantle: Current status and future developments. In E.C. Robertson, editor, *The Nature of the Solid Earth*. New York: McGraw-Hill, pp. 67–92.

Ringwood, A.E. (1972b) Phase transformations and mantle dynamics. *Earth and Planet. Sci. Letters*, **14**, 233–241.

Ringwood, A.E. (1972c) Some comparative aspects of lunar origin. *Phys. Earth Planet. Interiors*, **6**, 366–376.

Ringwood, A.E. (1975) *Composition and Petrology of the Earth's Mantle*. New York: McGraw-Hill. 618 pp.

Ringwood, A.E. (1979) Composition and origin of the Earth. In M.W. McElhinny, editor, *The Earth: Its Origin, Structure, and Evolution*. London: Academic Press, pp. 1–58.

Ringwood, A.E., and Major, A. (1970) The system Mg_2SiO_4-Fe_2SiO_4 at high pressures and temperatures. *Phys. Earth and Planet. Interiors*, **3**, 89–108.

Rittman, A. (1962) *Volcanoes and Their Activity*. New York: Wiley, 305 pp.

Robertson, W.A., and Fahrig, W.E. (1971) The great Logan Paleomagnetic loop—the polar wandering path from Canadian Shield rocks during the Neohelikian Era. *Can. Jour. Earth Sci.*, **9**, 123–140.

Robinson, P.L. (1973) Paleoclimatology and continental drift. In D.H. Tarling and S.K. Runcorn, editors, *Implications of Continental Drift to the Earth Sciences*, Vol. 1. London: Academic Press, pp. 451–476.

Rona, P.A. (1973) Relations between rates of sediment accumulation on continental shelves, sea-floor spreading, and eustacy inferred from the central North Atlantic. *Geol. Soc. America Bull.*, **84**, 2851–2872.

Rona, P.A. (1977) Plate tectonics, energy and mineral resources: Basic research leading to payoff. *EOS*, **58**, 629–639.

Rona, P.A. (1978) Criteria for recognition of hydrothermal mineral deposits in oceanic crust. *Econ. Geol.*, **73**, 135–160.

Rona, P.A. (1980) TAG hydrothermal field: Mid-Atlantic ridge crest at latitude 26°N. *Jour. Geol. Soc. Lond.*, **137**, 385–402.

Ronov, A.B., and Migdisov, A.A. (1971) Geochemical history of the crystalline basement and the sedimentary cover of the Russian and North America platforms. *Sedimentology*, **16**, 137–185.

Ronov, A.B., and Yaroshevsky, A.A. (1969) Chemical composition of the earth's crust. *Am. Geophys. Union Mon.* No. 13, 37–57.

Rubey, W.W. (1951) Geologic history of sea water. *Geol. Soc. America Bull.*, **62**, 1111–1148.

Runcorn, S.K. (1965) Changes in the convection pattern in the earth's mantle and continental drift: Evidence for a cold origin of the earth. *Phil. Trans. Roy. Soc. Long. A.*, **258**, 228–251.

Russell, R.D., and Farquhar, R.M. (1960) *Lead Isotopes in Geology*. New York: Wiley-Interscience. 243 pp.

Rutten, M.G. (1971) *The Origin of Life by Natural Causes*. Amsterdam: Elsevier. 420 pp.

Sacks, I.S., and Okada, H. (1973) A comparison of the anelasticity structure between western South America and Japan. *Carnegie Inst. Wash. Yearbook* **72**, 226–233.

Safronov, V.S. (1972) *Evolution of the Protoplanetary Cloud and Formation of the Earth and Planets*. Jerusalem: Israel Program for Sci. Translations.

Salisbury, J.W., and Ronca, L.B. (1966) The origin of continents. *Nature*, **210**, 669–690.

Salop, L.I., and Travin, L.V. (1972) Archean stratigraphy in the central part of the Aldan Shield. *Intern. Geol. Rev.*, **14**, 29–40.

Sawkins, F.J. (1976) Widespread continental rifting: Some considerations of timing and mechanism. *Geology*, **4**, 427–430.

Schidlowski, M., Eichmann, R., and Junge, C.E. (1975) Precambrian sedimentary carbonates: carbon and oxygen isotope geochemistry and implications for the terrestrial oxygen budget. *Precamb. Research*, **2**: 1–69.

Schilling, J. (1973) Iceland mantle plume, geochemical study of Reykjanes Ridge. *Nature*, **242**, 565–571.

Schilling, J., and Winchester, J.W. (1969) Rare earth contribution to the origin of Hawaiian lavas. *Contr. Mineral Petrol.*, **23**, 23–37.

Schmidt, P.W., and Clark, D.A. (1980) The response of paleomagnetic data to Earth expansion. *Geophys. Jour.*, **61**, 95–100.

Scholl, D.W., von Huene, R., Vallier, T.L., and Howell, D.G. (1980) Sedimentary masses and concepts about tectonic processes at underthrust ocean margins. *Geology*, **8**, 564–568.

Scholz, D.H., Barazangi, M., and Sbar, M.L. (1971) Late Cenozoic evolution of the Great Basin, western United States, as an ensialic interarc basin. *Geol. Soc. America Bull.*, **82**, 2979–2990.

Schopf, J.W. (1976) Evidence of Archean life: A brief appraisal. In B.F. Windley, editor, *The Early History of the Earth*. New York: Wiley, pp. 589–593.

Schubert, G. (1979) Subsolidus convection in the mantles of terrestrial planets. *Ann. Revs. Earth Planet. Sci.*, **7**, 289–342.

Schwab, F.L. (1978) Secular trends in the composition of sedimentary rock assemblages—Archean through Phanerozoic time. *Geology*, **6**, 532–536.

Sclater, J.G. (1972) New perspectives in terrestrial heat flow. *Tectonophys.*, **13**, 257–291.

Sclater, J.G., Jaupart, C., and Galson, D. (1980) The heat flow through oceanic and continental crust and the heat loss of the Earth. *Revs. Geophys. Space Phys.*, **18**, 269–311.

Scotese, C.R., Bambach, R.K., Barton, C., VanderVoo, R., and Ziegler, A.M. (1979) Paleozoic base maps. *Jour. Geology*, **87**, 217–277.

Sengör, A.M.C. (1976) Collision of irregular continental margins: Implications for foreland deformation of Alpine-type orogens. *Geology*, **4**, 779–782.

Sengör, A.M.C., and Burke, K. (1978) Relative timing of rifting and volcanism on Earth and its tectonic implications. *Geophys. Res. Lett.*, **5**, 419–421.

Sengör, A.M.C., Burke, K., and Dewey, J.F. (1978) Rifts at high angles to orogenic belts: Tests for their origin and the upper Rhine Graben as an example. *Am. Jour. Sci.*, **278**, 24–40.

Shackleton, R.M. (1976) Pan-African structures. *Phil. Trans. Roy. Soc. Long. A.*, **280**, 491–497.

Shanti, M., and Roobol, M.J. (1979) A late Proterozoic ophiolite complex at Jabal Ess in northern Saudi Arabia. *Nature*, **279**, 488–491.

Shaw, D.M. (1972) Development of the early continental crust, Part 1. *Can. Jour. Earth Sci.*, **9**, 1577–1595.

Shaw, D.M. (1976) Development of the early continental crust. Part 2. In B.F. Windley, editor, *The Early History of the Earth*. New York: Wiley, pp. 33–54.

Shaw, H.R., Kistler, R.W., and Evernden, J.F. (1971) Sierra Nevada plutonic cycle, Part II: Tidal energy and a hypothesis for

orogenic-epeirogenic periodicities. *Geol. Soc. America Bull.*, **82**, 869–896.

Sheinmann, Y.M. (1971) *Tectonics and the Formation of Magmas*. New York: Consultants Bureau. 173 pp.

Shor Jr., G.G., Kirk, H.K., and Menard, H.W. (1971) Crustal structure of the Melanesian area. *Jour. Geophys. Res.*, **76**, 2562–2585.

Siever, R. (1978) Plate-tectonic controls on diagenesis. *Jour. Geol.*, **87**, 127–155.

Sillitoe, R.H. (1972) Relation of metal provinces in western America to subduction of oceanic lithosphere. *Geol. Soc. America Bull.*, **83**, 813–818.

Sillitoe, R.H. (1976) Andean mineralization: A model for the metallogeny of convergent plate margins. *Geol. Assoc. Canada Spec. Paper 14*, 59–100.

Silver, E.A., and Beutner, E.C. (1980) Melanges. *Geology*, **8**, 32–34.

Sleep, N.H. (1971) Thermal effects of the formation of Atlantic continental margins by continental break-up. *Geophys. Jour.*, **24**, 325–350.

Sleep, N., and Toksoz, N.M. (1971) Evolution of marginal basins. *Nature*, **233**, 548–550.

Smith, A.G., and Hallam, A. (1970) The fit of the southern continents. *Nature*, **225**, 139–144.

Smith, J.V., and Mason, B. (1970) Pyroxenegarnet transformation in Coorara meteorite. *Science*, **168**, 832–833.

Smith, R.B., and Christensen, R.L. (1980) Yellowstone Park as a window on the Earth's interior. *Scient. Amer.*, **242**, 104–117.

Smithson, S.B., Brewer, J.A., Kaufman, S., Oliver, J.E., and Zawislak, R.L. (1980) Complex Archean lower crustal structure revealed by COCORP crustal reflection profiling in the Wind River Range, Wyoming. *Earth and Planet. Sci. Lett.*, **46**, 295–305.

Smithson, S.B., and Brown, S.K. (1977) A model for lower continental crust. *Earth and Planet. Sci. Lett.*, **35**, 134–144.

Spudich, P., and Orcutt, J. (1980) A new look at the seismic velocity structure of the oceanic crust. *Revs. Geophys. Space Phys.*, **18**, 627–645.

Steiger, R.H., and Jäger, E. (1977) Subcommission on geochronology: Convention on the use of decay constants in geo- and cosmochemistry. *Earth and Planet. Sci. Lett.*, **36**, 359–362.

Stern, T.W., Phair, G., and Newell, M.F. (1971) Boulder Creek Batholith, Colorado, Part II: Isotopic age of emplacement and morphology of zircon. *Geol. Soc. America Bull.*, **82**, 1615–1634.

Stewart, J.H. (1972) Initial deposits in the Cordilleran Geosyncline evidence of a late Precambrian (< 850 m.y.) continental separation. *Geol. Soc. America Bull.*, **83**, 1345–1360.

Stewart, J.H. (1978) Basin-Range structure in western North America: A review. *Geol. Soc. America Mem.* **152**, 1–32.

Stockwell, C.H. (1965) Structural trends in Canadian Shield. *Bull. Amer. Assoc. Petroleum Geol.*, **49**, 887–893.

Stommel, H.E., and Graul, J.M. (1978) Current trends in geophysics. *Arab. Jour. Sci. Eng.*, Spec. Issue, pp. 41–63.

Strong, D.F. (1979) Proterozoic tectonics of northwestern Gondwanaland: New evidence from eastern Newfoundland. *Tectonophys.*, **54**, 81–101.

Sugimura, A., Matsuda, T., Chinzei, T., and Nakamura, K. (1963) Quantitative distribution of late Cenozoic volcanic materials in Japan. *Bull. Volcanolog.*, **26**, 125–140.

Sun, S.S., and Nesbitt, R.W. (1977) Chemical heterogeneity of the Archean mantle, composition of the earth and mantle evolution. *Earth Planet. Sci. Lett.*, **35**, 429–448.

Sun, S.S., Tatsumoto, M., and Schilling, J.G. (1975) Mantle plume mixing along the Reykjaues ridge axis: Lead isotopic evidence. *Science*, **190**, 143–147.

Sutton, G.H., Maynard, G.L., and Hussong, D.M. (1971) Widespread occurrence of a high-velocity basal layer in the Pacific crust found with repetitive sources and sonobuoys. *Am. Geophys. Union Mon.* No. 14, 193–209.

Sutton, J. (1963) Long-term cycles in the evolution of continents. *Nature*, **198**, 731–735.

Swanberg, C.A., and Blackwell, D.D. (1973) Areal distribution and geophysical significance of heat generation in the Idaho Batholith and adjacent intrusions in eastern Oregon and western Montana. *Geol. Soc. America Bull.*, **84**, 1261–1282.

Talwani, J., LePichon, X., and Ewing, M. (1965) Crustal structure of the mid-ocean ridges, part 2. *Jour. Geophys. Res.*, **70**, 341–352.

Talwani, J., Sutton, G.H., and Worzel, J.L. (1959) A crustal section across the Puerto Rico Trench. *Jour. Geophys. Res.*, **64**, 1545–1555.

Tarling, D.H. (1971) *Principles and Applications of Paleomagnetism.* London: Chapman and Hall. 164 pp.

Tarney, J., Dalziel, I.W.D., and deWitt, M.J. (1976) Margin basin "rocas verdes" complex from southern Chile: a model for Archean greenstone belt formation. In B.F. Windley, editor, *The Early History of the Earth.* New York: Wiley, pp. 131–146.

Tatsumoto, M. (1978) Isotopic composition of lead in oceanic basalt and its implication to mantle evolution. *Earth and Planet. Sci. Lett.*, **38**, 63–87.

Taylor, S.R. (1964a) Trace element abundances and the chrondritic earth model. *Geochim. Cosmochim. Acta*, **28**, 1989–1998.

Taylor, S.R. (1964b) Abundance of chemical elements in the continental crust: A new table. *Geochim. Cosmochim. Acta*, **28**, 1273–1285.

Taylor, S.R. (1967) The origin and growth of continents. *Tectonophys.*, **4**, 17–34.

Taylor, S.R. (1969) Trace element chemistry of andesites and associated calc-alkaline rocks. *Oregon Dept. Geol. and Min. Indust., Bull.* **65**, 43–63.

Taylor, S.R. (1975) *Lunar Science: A Post-Apollo View.* New York: Pergamon Press. 372 pp.

Taylor, S.R. (1979a) Chemical composition and evolution of the continental crust: The rare earth element evidence. In M.W. McElhinny, editor, *The Earth: Its Origin, Structure, and Evolution.* London: Academic Press, pp. 353–376.

Taylor, S.R. (1979b) Structure and evolution of the moon. *Nature*, **281**, 105–110.

Taylor, S.R., et al. (1969) Genetic significance of Co, Cr, Ni, Sc, and V content of andesites. *Geochim. Cosmochim. Acta*, **33**, 275–286.

Thiessen, R., Burke, K., and Kidd, W.S.F. (1979) African hotspots and their relation to the underlying mantle. *Geology*, **7**, 263–266.

Toksoz, N.M., Minear, J.W., and Julian, B.R. (1971) Temperature field and geophysical effects of a downgoing slab. *Jour. Geophys. Res.*, **76**, 1113–1137.

Towe, K.M. (1978) Early Precambrian oxygen: A case against photosynthesis. *Nature*, **274**, 657–661.

Tozer, D.C. (1959) The electrical properties of the earth's interior. *Phys. and Chem. Earth*, **3**, 414–436.

Turcotte, D.L. (1980) On the thermal evolution of the Earth. *Earth and Planet. Sci. Lett.*, **48**, 53–58.

Turcotte, D.L., and Burke, K. (1978) Global sea-level changes and the thermal structure of the continents. *Earth and Planet. Sci. Lett.*, **41**, 341–346.

Turcotte, D.L., and Oxburgh, E.R. (1972a) Statistical thermodynamic model for the distribution of crustal heat sources. *Science*, **176**, 1022–1023.

Turcotte, D.L., and Oxburgh, E.R. (1972b) Mantle convection and the new global tectonics. *Ann. Rev. Fluid Mechanics*, **4**, 33–68.

Ullrich, L., and VanderVoo, R. (1981) Minimum continental velocities with respect to the pole since the Archean. *Tectonophys.*, **74**, 17–27.

Uyeda, S. (1977) Some basic problems in the trench-arc–back-arc system. In M. Talwani and W.C. Pittman III, editors, *Island Arcs, Deep-sea Trenches, and Back-arc Basins.* Washington, D.C.: Am. Geophys. Union, Maurice Ewing Series 1, pp. 1–14.

Uyeda, S., and Miyashiro, A. (1974) Plate tectonics and the Japanese Islands: A synthesis. *Geol. Soc. America Bull.*, **85**, 1159–1170.

Valentine, J.W. (1973) *Evolutionary Paleoecology of the Marine Biosphere*. Englewood Cliffs, N.J.: Prentice-Hall. 510 pp.

Valentine, J.W., and Moores, E.M. (1970) Platetectonic regulation of faunal diversity and sea level: A model. *Nature*, **228**, 657–659.

Valentine, J.W., and Moores, E.M. (1972) Global tectonics and the fossil record. *Jour. Geol.*, **80**, 167–184.

Veizer, J. (1973) Sedimentation in geologic history: recycling vs. evolution or recycling within evolution. *Contr. Mineral. Petrol.*, **38**, 261–278.

Veizer, J. (1976) $^{87}Sr/^{86}Sr$ evolution of seawater during geologic history and its significance as an index of crustal evolution. In B.F. Windley, editor, *The Early History of the Earth*. New York: Wiley, pp. 569–578.

Veizer, J., and Compston, W. (1976) $^{87}Sr/^{86}Sr$ in Precambrian carbonates as an index of crustal evolution. *Geochim. Cosmochim. Acta*, **40**, 905–914.

Vine, F.J. (1966) Spreading of the ocean floor: Evidence. *Science*, **154**, 1405–1415.

Vine, F.J., and Matthews, D.H. (1963) Magnetic anomalies over oceanic ridges. *Nature*, **199**, 947–949.

Vitorello, I., and Pollack, H.N. (1980) On the variation of continental heat flow with age and the thermal evolution of continents. *Jour. Geophys. Res.*, **85**, 983–995.

VonHuene, R. (1979) Structure of the outer convergent margin off Kodiak Island, Alaska, from multichannel seismic records. *Am. Assoc. Petrol. Geol.*, *Mem. 29*, 261–272.

Wagner, G.A., and Reimer, G.M. (1972) Fission-track tectonics: A tectonic interpretation of fission track apatite ages. *Earth and Planet. Sci. Lett.*, **14**, 263–268.

Walter, M.R. (1977) Interpreting Stromatolites. *American Sci.*, **65**, 563–571.

Walter, M.R., Buick, R., and Dunlop, J.S.R. (1980) Stromatolites 3400–3500 myr old from the North Pole area, Western Australia. *Nature*, **284**, 443–445.

Watanabe, T., Langseth, M.G., and Anderson, R.N. (1977) Heat flow in back-arc basins of the western Pacific. In M. Talwani and W.C. Pittman III, editors, *Island Arcs, Deep-sea Trenches, and Back-arc Basins*. Washington, D.C.: Am. Geophys. Union, Maurice Ewing Series 1, pp. 137–162.

Waters, A.C. (1961) Stratigraphic and lithologic variations in the Columbia River basalt. *Am. Jour. Sci.*, **259**, 583–611.

Watson, J.V. (1976) Vertical movements in Proterozoic structural provinces. *Phil. Trans. Roy. Soc. Long. A.*, **280**, 629–640.

Wegener, A. (1912) Die Entstehung der Kontinente. *Geol. Rund.*, **3**, 276–292.

Wetherill, G.W. (1977) Accumulation of the terrestrial planets. *Carnegie Inst. Washington Year Book* **76**, 761–767.

Wetherill, G.W., and Bickford, M.E. (1965) Primary and metamorphic Rb-Sr chronology in central Colorado. *Jour. Geophys. Res.*, **70**, 4669–4686.

Whitcomb, J.H., and Anderson, D.L. (1970) Reflection of p'dp' seismic waves from discontinuities in the mantle. *Jour. Geophys. Res.*, **75**, 5713–5728.

White, I.G. (1967) Ultrabasic rocks and the composition of the upper mantle. *Earth and Planet. Sci. Letters*, **3**, 11–18.

Whitford, D.J., Compston, W., Nicholls, I.A., and Abott, M.J. (1977) Geochemistry of late Cenozoic lavas from eastern Indonesia: Role of subducted sediments in petrogenesis. *Geology*, **5**, 571–575.

Wilcox, R.E. (1954) Petrology of Paricutin volcano, Mexico. *U.S. Geol. Survey Bull.* 965-C.

Williams, G.E. (1973) Geotectonic cycles, lunar evolution, and the dynamics of the earth-moon system. *Modern Geol.*, **4**, 159–183.

Wilson, J.T. (1965) Transform faults, oceanic ridges and magnetic anomalies southwest of Vancouver Island. *Science*, **150**, 482–485.

Wilson, J.T. (1968) A revolution in the earth sciences. *Geotimes*, **13**(10), 10–16.

Wilson, J.T. (1973) Mantle plumes and plate motions. *Tectonophys.*, **19**, 149–164.

Windley, B.F. (1970) Anorthosites in the early crust of the earth and on the moon. *Nature*, **226**, 333–335.

Windley, B.F. (1973) Crustal development in the Precambrian. *Phil. Trans. Roy. Soc. Long. A.*, **273**, 321–341.

Windley, B.F. (1977) *The Evolving Continents*. New York: Wiley, 385 pp.

Windley, B.F., and Bridgewater, D. (1971) The evolution of Archean low- and high-grade terrains. *Geol. Soc. Aust. Spec. Publ.* 3, 33–46.

Wise, D.U. (1963) An origin of the moon by rotational fission during formation of the Earth's core. *J. Geophys. Res.*, **68**, 1547–1554.

Wise, D.U. (1973) Freeboard of continents through time. *Geol. Soc. America Mem.* **132**, 87–100.

Wood, D.A., Jorou, J., and Treul, M. (1979) A reappraisal of the use of trace elements to classify and discriminate between magma series erupted in different tectonic settings. *Earth and Planet. Sci. Lett.*, **45**, 326–336.

Wood, J.A., and Mitler, H.E. (1974) Origin of the moon by a modified capture mechanism, or half a loaf is better than a whole one. *Lunar Science*, **5**: 851–853.

Woollard, G.P. (1968) The interrelationship of the crust, the upper mantle, and isostatic gravity anomalies in the United States. *Am. Geophys. Union Mon.* No 12, 312–341.

Woollard, G.P. (1972) Regional variation in gravity. In E.C. Robertson, editor, *The Nature of the Solid Earth*. New York: McGraw-Hill, pp. 463–505.

Wyllie, P.J. (1970) Ultramafic rocks and the upper mantle. *Mineral. Soc. America Spec. Paper 3*, 3–32.

Wyllie, P.J. (1971a) Experimental limits for conditions for melting in the earth's crust and upper mantle. *Am. Geophys. Union Mon.* No. 14, 279–302.

Wyllie, P.J. (1971b) The role of water in magma generation and initiation of diapiric uprise in the mantle. *Jour. Geophys. Res.*, **76**, 1328–1338.

Wyllie, P.J. (1973) Experimental petrology and global tectonics—preview. *Tectonophys.*, **17**, 189–209.

Wynne-Edwards, H.R. (1972) The Grenville Province. *Geol. Assoc. Canada, Spec. Paper, 11*, 264–334.

Wynne-Edwards, H.R. (1976) Proterozoic ensialic orogenesis: the millipede model of ductile plate tectonics. *Am. J. Sci.*, **276**, 927–953.

Yungul, S.H. (1971) Magnetic anomalies and the possibilities of continental drifting in the Gulf of Mexico. *Jour. Geophys. Res.*, **76**, 2639–2642.

Ziegler, A.M., Scotese, C.R., McKerrow, W.S., Johnson, M.E., and Bambach, R.K. (1979) Paleozoic paleogeography. *Ann. Rev. Earth Planet. Sci.*, **7**, 473–502.

Zietz, I., et al. (1969) Aeromagnetic investigation of crustal structure for a strip across the western United States. *Geol. Soc. America Bull.*, **80**, 1703–1714.

Zietz, I., et al. (1971) Interpretation of an aeromagnetic strip across the northwestern United States. *Geol. Soc. America Bull.*, **82**, 3347–3372.

Zoback, M.L., and Zoback, M. (1980) State of stress in the coterminous United States. *Jour. Geophys. Res.*, **85**, 6113–6156.

Index